T5-DIJ-722

FRONTIERS OF FREE RADICAL CHEMISTRY

ACADEMIC PRESS RAPID MANUSCRIPT REPRODUCTION

This book is based on papers prepared by speakers in a symposium entitled "Frontiers of Free Radical Chemistry," held at Louisiana State University in Baton Rouge, Louisiana, in April 10–11, 1979, and sponsored by Exxon Education Foundation.

FRONTIERS OF FREE RADICAL CHEMISTRY

edited by

WILLIAM A. PRYOR

Boyd Professor of Chemistry
Louisiana State University
Baton Rouge, Louisiana

ACADEMIC PRESS 1980
A Subsidiary of Harcourt Brace Jovanovich, Publishers
NEW YORK LONDON TORONTO SYDNEY SAN FRANCISCO

ACADEMIC PRESS, INC.
111 Fifth Avenue, New York, New York 10003

United Kingdom Edition published by
ACADEMIC PRESS, INC. (LONDON) LTD.
24/28 Oval Road, London NW1 7DX

Library of Congress Cataloging in Publication Data

Main entry under title:

Frontiers of free radical chemistry.

Papers presented at a symposium held at Louisiana
State University, Apr. 10–11, 1979 in the 50th anniversary of the Exxon Research and Development Laboratory.

Includes index.
1. Radicals (Chemistry)—Congresses. I. Pryor,
William A. II. Louisiana State University, Baton Rouge.
III. Exxon Research and Development Laboratory.
QD471.F76 541.2′24 80-19007
ISBN 0-12-566550-4

PRINTED IN THE UNITED STATES OF AMERICA

80 81 82 83 9 8 7 6 5 4 3 2 1

To the Exxon Research and Development Laboratories of Baton Rouge on its 50th anniversary, and to all the personnel that make that Laboratory the eminently successful industrial research laboratory that it is.

CONTENTS

FUEL CHEMISTRY, OXIDATION, PEROXIDES, AND INHIBITION

ORGANIC PROCESSES

CONTRIBUTORS

The numbers in parentheses indicate the pages on which the authors' contributions begin.

T. N. Bell (139), Department of Chemistry, Simon Fraser University, Burnaby, British Columbia, Canada

Sidney W. Benson (1), Hydrocarbon Research Institute, Chemistry Department, University of Southern California, Los Angeles, California 90007

John N. Bradley (73), Department of Chemistry, University of Essex, Colchester, Essex, England

John I. Brauman (23), Department of Chemistry, Stanford University, Stanford, California 94305

H. K. J. Choi (139), Department of Chemistry, University of Alberta, Edmonton, Alberta, Canada

Daniel F. Church (355), Department of Chemistry, Louisiana State University, Baton Rouge, Louisiana 70803

David M. Golden (31), Department of Chemical Kinetics, SRI International, Menlo Park, California

Robert N. Hazlett (195), Chemistry Division, Naval Research Laboratory, Washington, D. C.

Richard R. Hiatt (225), Department of Chemistry, Brock University, St. Catharines, Ontario, Canada

Kendall N. Houk (43), Department of Chemistry, Louisiana State University, Baton Rouge, Louisiana 70803

J. A. Howard (237), Division of Chemistry, National Research Council of Canada, Ottawa, Ontario, Canada K1A OR9

F. C. James (139), Department of Chemistry, University of Alberta, Edmonton, Alberta, Canada

J. Alistair Kerr (171), Department of Chemistry, University of Birmingham, Birmingham, B15 2TT, England

Jay K. Kochi (297), Department of Chemistry, Indiana University, Bloomington, Indiana 41401

William A. Pryor (355), Department of Chemistry, Louisiana State University, Baton Rouge, Louisiana 70803

J. Howard Purnell (93), Department of Chemistry, University College of Swansea, Swansea, Wales, U.K.

Charles Rebick (117), Corporate Research Science Laboratories, Exxon Research and Engineering Company, Linden, New Jersey 07036

B. Ruzsicska (139), Department of Chemistry, University of Alberta, Edmonton, Alberta, Canada

O. P. Strausz (139), Department of Chemistry, University of Alberta, Edmonton, Alberta, Canada

Felicia Tang (355), Department of Chemistry, Louisiana State University, Baton, Rouge, Louisiana 70803

Robert Tang (355), Department of Chemistry, Louisiana State University, Baton Rouge, Louisiana 70803

James G. Traynham (283), Department of Chemistry, Louisiana State University, Baton Rouge, Louisiana 70803

PREFACE

The year 1977 was the 50th anniversary of the Exxon Research and Development Laboratory in Baton Rouge. In order to celebrate this felicitous occasion, the Chemistry Department of Louisiana State University had the great honor and I had the very considerable personal pleasure of organizing a symposium on chemistry of interest both to LSU and to the industrial community. With generous support from the Exxon Education Foundation, this symposium was held in the LSU Chemistry Department in April 1979.

The papers in this volume represent contributions from the speakers in that LSU Exxon Symposium. As is indicated by the table of contents, a very prestigious and well-known group of international experts gathered to discuss topics that include the mechanisms of hydrocarbon cracking, oxidation processes, inhibition, the theory of organic radical reactions, halogenations, and organic peroxide chemistry.

The authors have outlined their research areas in a broad manner. Thus, the chapters in this volume do not represent the talks given at the symposium so much as surveys and reviews of areas of interest both to academia and to chemical industry. As such, these chapters should find wide and lasting utility.

I believe that free radical chemistry represents a particularly fortunate choice for a symposium honoring this outstanding industrial research laboratory. Radicals are involved in many basic industrial processes, including air oxidations, thermal cracking, vinyl polymerizations, halogenations, and many other important processes. Research in free radical chemistry is a more active and varied field than ever before, with current work covering organic reaction mechanisms, gas phase thermochemistry, cracking, catalysis, coal liquifaction and gasification, and free radicals in biological systems. This volume represents some of the current work underway in the free radical field.

I want to express my own thanks as well as those of LSU and of the LSU Chemistry Department to the Exxon Research and Development Laboratories in Baton Rouge for choosing LSU to host this symposium. I also want to express my thanks to the Exxon Education Foundation for the generous grant that made this meeting possible. In addition, I would like to thank some of the Exxon personnel who envisioned this event, chose me as the organizer, and then helped in every way to make this symposium the success that it was. In particular, I would like to mention Dr. Roby Bearden,Scientific Advisor at Exxon, Mr. Jerry Bernstein, Manager of Exxon Baton Rouge Laboratories, and Mr. Robert L. Payton of the Exxon Education Foundation. I also want to thank Professor Richard D. Gandour and Ms. Doreen Maxcy of LSU for their very considerable help in the planning and organizing of the symposium.

William A. Pryor
Baton Rouge, Louisiana
April 1980

Introduction

The Exxon Research and Development Laboratory was founded in Baton Rouge in 1927 and marked one of the first efforts by a petroleum refining company to establish a fundamental long-range research facility. The step was taken at a time when the connection between research and petroleum refining was not at all well recognized.

It is amusing that in the mid-1920s there was a national concern that a shortage of oil was imminent. In fact, one authority predicted that the total oil reserves would only last seven more years, and Senator LaFollette of Wisconsin predicted the price of gasoline would soon reach $1 a gallon!

In this setting, Exxon initiated its petroleum research laboratory in Baton Rouge with the hope that its research would discover ways to make American industry less dependent on foreign oil. The initial research effort was aimed at developing processes that would increase the yield of gasoline from heavy American crudes, and this challenge was met with the development of high-pressure catalytic hydrogenation technology. In the 1930s, research was broadened to include a range of types of catalytic processing of petroleum. Many discoveries in petroleum catalytic refining have come from the Exxon laboratories over the past 50 years: These developments include fluid catalytic cracking, the Butyl rubber process, fluid coking, Hycracking (a hydrocracking process that converts heavy gas oils to high octane gasoline), as well as discoveries in hydrodesulfurization.

The Baton Rouge Exxon Research and Development Laboratory has remained one of the most creative laboratories of its type, contributing substantially to the Exxon Corporation and to American petroleum chemistry. In no small way, this Laboratory also contributes to the strength of the Baton Rouge area, supplying a powerful focus of technical and scientific work. The Louisiana State University in general, and the Chemistry Department in particular, have continued to benefit very markedly from the support and cooperation of this Laboratory.

William A. Pryor
Baton Rouge, Louisiana
April 1980

FRONTIERS OF FREE RADICAL CHEMISTRY

THE THERMOCHEMISTRY AND KINETICS OF GAS PHASE REACTIONS

SIDNEY W. BENSON

Hydrocarbon Research Institute
Chemistry Department
University of Southern California
Los Angeles, California

The history of free radical chemistry leading to current modeling efforts for complex kinetic processes is examined briefly. Current understanding of molecular and free radical mechanisms is summarized as is the current state of the art in predicting thermochemical data for radicals and molecules. An application of these techniques to understanding the reaction of thermal electrons with SF_6 gas is given. Arrhenius parameters are deduced for the thermal reaction $SF_6^- \rightarrow SF_6 + e^-$, the first such data for an electron ejection.

Ion pathways in gas reactions are discussed and the intermediacy of intimate ion pairs in such reactions is demonstrated in a few cases. Applications are made to the pyrolysis of cyclopropyl halides, the low temperature (-80°C) reaction of O_3 with saturates and the initiation reaction in oxidation.

INTRODUCTION

In the short space of about 50 years the chemistry of free radicals has passed from the stage of a tentative and descriptive science of limited interest to chemists to that

ISBN 0-12-566550-4

of a quantitative science filling an important niche in the broad spectrum of chemistry and impinging on almost every area of current chemistry and chemical technology. This development of free radical chemistry has paralleled a slightly older development in the field of chemical kinetics, a science which is itself only about 100 years old. These two sciences have had important interactions; discoveries in one have frequently triggered discoveries in the other.

Free radical chemistry can be assigned a formal beginning with the classical research paper of Gomberg on the triphenyl methyl radical [1] in 1900. By a totally unrelated coincidence, free radical kinetics finds its origin in the equally classic study by Lind and Bodenstein [2] of the $H_2 + Br_2$ chain reaction done shortly after Gomberg's work. It was not however until the radical—mirror removal experiments of Paneth et al. [3,4] and others that free radical kinetics really opened up. In these experiments, a rapidly moving stream of organic molecules passed through a hot quartz furnace (900-1200°K; $\sim$ 1 second residence) and underwent pyrolysis, producing some active intermediates capable of reacting with metallic mirrors (Pb, Zn, Be, Sb, As, Se, etc.). The products of these mirror removal experiments were metal organic compounds such as PbR_4 or ZnR_2 ($R=CH_3$, Et, ϕ, etc.). The rates and products observed made possible the first measures of radical concentrations and of radical reactivity. The climax of this early pioneeering period was achieved in 1934 with the publication of the Rice-Herzfeld mechanisms [5] inaugurating for the first time a completely detailed scheme of elementary radical reactions to account quantitatively for the rates and products of complex pyrolysis reactions.

SOME COMPLEX RADICAL REACTIONS—KINETIC MODELING

From its very beginning, free radical chemistry has had significant technological application. Hydrocarbon cracking as practised since the early discovery of petroleum, leaned increasingly towards quantitative modeling. With the development of the Rice-Herzfeld mechanisms it became possible to predict theoretically, the multivaried product mix in petroleum refining. More important, for the first time a sufficient insight was given to the mechanism of pyrolysis that it became possible to make systematic chemical and physical variations in the reactor conditions in an attempt to change the product yields.

Early efforts in the modeling of hydrocarbon cracking were restricted by the necessity to obtain explicit solutions for product yields and rate constants. Chemical models consisted of from 10 to 30 elementary step reactions. For convenience, approximate solutions were empirically generated and provided with "lumped parameters" to simplify the involved algebra. The recent development of rapid inexpensive electronic computers has made this type of simplification unnecessary and a recent industrial advertisement [6] announces the availability of a computing program designed to handle up to 1,000 elementary step reactions, under adiabatic or isothermal conditions. This announcement claims considerable accuracy in the prediction of product yield from the pyrolysis of complex hydrocarbon mixtures.

While such detail may be excessive for the description of the kinetics of many complex chemical processes, there is no doubt but that the future trend in chemical engineering will be in the direction of such detailed modeling. Early examples of such a trend appeared soon after the end of World War II with the development of a kinetic model for the chemical reactions taking place in the high temperature (3000-6000°K) boundary layer of a re-entry vehicle [7]. This was followed by the considerable effort to model the detailed kinetics of smog (tropospheric air pollution) [8] and its logical extension to the stability of the ozone layer in the stratosphere [9]. The developing energy crisis has inspired comparable efforts to develop kinetic models for the oxidation and combustion [10,11] of simple hydrocarbons and 250 step models for $CH_4 + O_2$ oxidation have been prepared.

A long range program including many hundreds of elementary reactions of both ions and neutrals has been in existence for some time, devoted to the understanding of the chemistry of the upper atmosphere (E and F layers) and their electromagnetic properties [7]. This had its stimulus in the early observations of the effects of atom bomb blasts in the upper atmosphere. More recently, the spectacular developments of chemical lasers have inspired the detailed modeling (80-100 steps) of the $H_2 + F_2$ chain reaction in which each ro-vibronic state of HF and H_2 has been given individual and separate, kinetic representation [12].

With the growing importance of energy and the concomitant importance of transforming coal and other carbon sources into liquid and gaseous fuels, there has been an increased development of the relatively minor field of plasma chemistry [13] and the detailed modeling of electron-ion-molecule reactions both at low and at high temperatures.

The political importance which has been attached to some of these modeling efforts, particularly in areas such as air pollution (troposphere and stratosphere) suggest a word of caution. The success of a kinetic model in fitting known data does not guarantee the accuracy of the model. Kinetics history is replete with examples of "well-understood" model reactions which have not been understood at all. Perhaps most famous is the $H_2 + I_2$ reaction to give 2HI, first studied by Bodenstein in 1894 [14]. Represented for 60 years as the model, 4-center reaction it was shown first to be about 1/2 atom-chain reaction [15] and then finally not to happen at all [16]. Our knowledge of the detailed chemistry of a complex reaction is always in principle, incomplete. One can always disprove a given model (or theory); one can never irrevocably, prove it. In science and its applications such as air pollution, one must always be prepared for surprises, for new evidences of our ignorance.

Mechanism of Chemical Reactions

Chemical reactions seldom occur in one elmentary step. Instead, they generally proceed by way of a number of intermediate steps involving reactive species which behave as catalysts in the overall reaction. These intermediates may be ions or ion-pairs as is generally true for reactions at ambient temperatures in condensed phases. In the gas phase usually above 100°C these intermediates are free radicals. Sometimes, but relatively rarely, ions and radicals may participate in the same process.

It is the complexity of mechanism which distinguishes kinetics from its older sister science, thermodynamics. In thermodynamics or preferably (from the chemist's point of view) thermochemistry, there is only one stoichiometric equation which connects reactants and products. The rate of the reaction, the complexity of the pathway is of no consequence to the thermochemist. For each reaction there is only one signpost, K_{eq} or alternatively $\Delta G°$, the Gibbs Free Energy Change, marking the ultimate direction, uphill or downhill!

The rapid growth of chemical kinetics has given us the prospect that quite possibly we may be close to a complete understanding of mechanism. New mechanisms in free radical kinetics all relatively rare. They are qually rare in ionic reactions. We can today make complete categories of reaction processes for molecular and free radical reactions and we have developed both theoretical and empirical tools which permit us to predict many of these with almost current state-of-art experimental accuracy [17]. Let us consider some examples of this sophistication.

Prior to the advent of the Rice-Herzfeld mechanisms, it was common for kineticists to write the simplest possible mechanism to explain the basis of molecular structure the production of products from reactants. Even up to 1950, it was common to write for a pyrolysis pathway for butane, the single step:

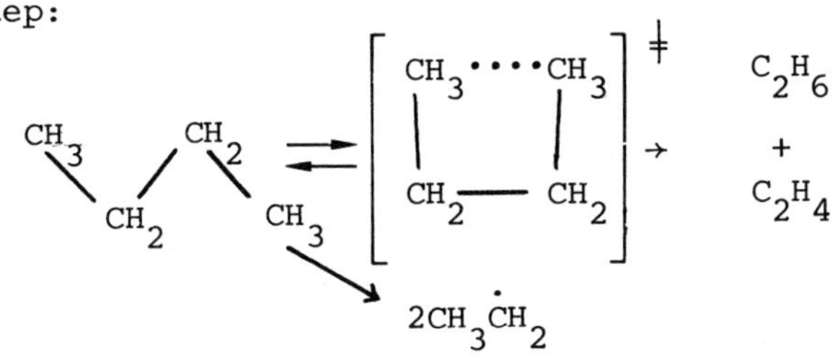

$$CH_3CH_2CH_2CH_3 \rightleftharpoons \left[\begin{array}{cc} CH_3\cdots & CH_3 \\ | & | \\ CH_2 - & CH_2 \end{array}\right]^{\ddagger} \rightarrow C_2H_6 + C_2H_4 \qquad (1)$$

$$CH_3CH_2CH_2CH_3 \rightarrow 2CH_3\dot{C}H_2 \qquad (2)$$

In the absence of any direct knowledge about the intermediate stages of such a reaction except that a high activation energy would be required, there was little experience to suggest how fast or slow process 1 might be relative to the competing process 2. Today, we know that the 4-center complex required for path 1 will impose a strain energy of the order of that found in cyclobutane (∿27 kcal/mole) and the promotion of carbon to a pentavalent state will add of the order of 15 kcal/mole for each of the two C-atoms so involved. This suggests an intrinsic activation energy of the order of 2x15 + 27 = 57 kcal for the exothermic direction of this reaction, namely, the reverse process of $C_2H_6 + C_2H_4 \rightarrow C_4H_{10}$. If to this we add the endothermicity of reaction 1, namely, 22.5 kcal/mole we arrive at an estimated activation energy of about 80 kcal/mole. Our knowledge of the thermochemistry of free radicals [17] allows us to assign an activation energy of 82 kcal/mole to path 2. However, we can also calculate that the A-factor of path 2 is about 10^{17} sec^{-1} while that for path 1 is about 10^{12} sec^{-1}, so that there are no conditions under which path 1 is ever expected to be observed relative to path 2.

Similar considerations lead us to reject each of the following reactions as significant pathways:

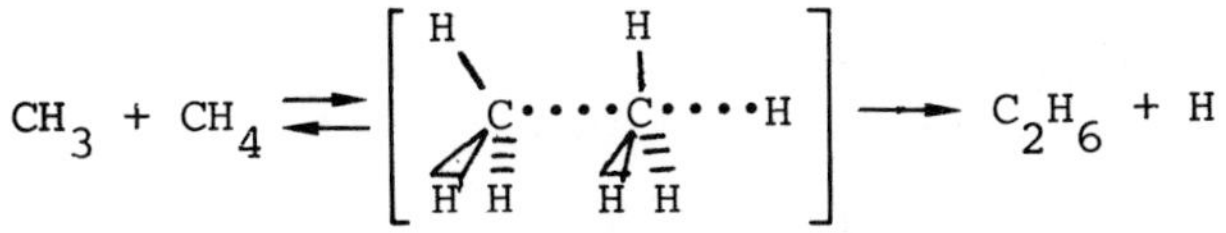

$$CH_3 + CH_4 \rightleftharpoons \left[H_3C\cdots CH_3 \cdots H \right] \longrightarrow C_2H_6 + H \qquad (3)$$

$$CH_3 + O_2 \rightleftharpoons \left[\begin{array}{c} H \\ H \diagup C - O \\ | \quad | \\ H \cdots O \end{array} \right] \rightarrow CH_2{=}O + OH \quad (4)$$

$$2RH + O_2 \rightleftharpoons [R-H\cdots O\text{-}O\cdots H-R] \rightarrow 2\dot{R} + H_2O_2 \quad (5)$$

Reaction 3 can be shown to have a low Arrhenius A-factor ($\sim 10^8$ l/mole-sec) and an activation energy of about 30 kcal/mole. Reaction 4 which had its origin in flame reactions some 60 years ago as a pedagogical convenience has been almost impossible to remove from the oxidation literature despite direct evidence for its non-occurrence [18]. Reaction 5 which has been suggested as a possible source of radicals in low temperature liquid phase oxidation can be shown to be much too slow to be significant, if only for its endothermicity of $\geq$ 46 kcal/mole.

Let us examine in a little more detail some of the modern techniques which have made some of these conclusions possible. They start with the extension of empirical thermochemistry into the domain of free radicals and large molecules.

Current Thermochemistry

Over the past 20 years the author and his colleagues have been able to develop a series of empirical laws which today permit the estimation of thermochemical properties of molecules and free radicals with great accuracy [17]. Thermodynamics gives us the following relationships between equilibrium constants and thermochemical quantities:

$$-RT \ln K_{eq\,(T)} = \Delta G^\circ_{(T)} \quad (6)$$

$$\Delta G^\circ_{(T)} = \Delta H^\circ_{(T)} - T\Delta S^\circ_{(T)} \quad (7)$$

$$\Delta H^\circ_{(T)} = \Delta H^\circ_{(T_o)} + \int_{T_o}^{T} C_{P(T)}\,dT \quad (8)$$

$$\Delta S^\circ_{(T)} = \Delta S^\circ_{(T_o)} + \int_{T_o}^{T} C_{P(T)}\,dT/T \quad (9)$$

If we knew for each chemical species that might take part in a stoichiometric equilibrium reaction, its heat of formation $\Delta H^\circ_{fT_o}$, its entropy $S^\circ_{T_o}$, both at some reference temperature T_o, and its heat capacity C_{pT} over a large temperature range, we could predict the equilibrium constant for that stoichiometric reaction at any temperature.

TABLE I

SENSITIVITY ANALYSIS—THERMOCHEMICAL ESTIMATES

Quantity	*Uncertainty*	*Uncertainty in k or K (or other)*
ΔS° or $\Delta S^\ddagger$	± 1 e.u.	factor of 1.6 (x 1.6)
ΔH° or $\Delta H^\ddagger$	± 1 kcal	x 5 (300°K) x 1.6 (1000°K)
	± 0.3 kcal	x 1.6 (300°K) x 1.16 (1000°K)
internal frequency	± 20%	max. of ± 0.4 e.u. in S°_T max. of ± 0.2 e.u. in C°_{pT}
bond length	± 5%	max. of ± 0.2 e.u. in S°_T [a] no direct effect on C_{pT}
bond angle	± 5%	max. of ± 0.2 e.u. in S°_T no direct effect on C_{pT}

[a] *In polyatomic molecules it is only the largest distances between the two or three heaviest atoms that are important in determining the moments of inertia.*

TABLE II

Standard Entropies and Heat Capacities at 300°K of Some Structurally Similar Molecules With No Hindered Internal Rotors

Molecule	$S^\circ_{300(int)}$ [b]	(σ,η) [c]	S°_{300}	C°_{p300}
CHF_3	64.2	(3,0)	62.0	12.2
BF_3	64.3	(6,0)	60.7	12.1
NF_3	64.5	(3,0)	62.3	12.8
COF_2	63.3	(2,0)	61.9	11.3
FNO_2	63.6	(2,0)	62.2	11.9

[a] *All values in cal/mole-°K.*
[b] $S^\circ_{300(int)} = S^\circ_{300} + R\ln(\sigma/\eta)$.
[c] η *is the number of optical isomers,* σ *is the total symmetry number.*

We can in fact, on the basis of our current knowledge of molecular structure use the methods of quantum statistical mechanics to calculate C_{pT} and $S^\circ_{T_o}$ to reasonable accuracy of about ±1 cal/mole-°K for each. This basically requires an assignment of bond lengths, bond angles, barriers to internal torsion and internal frequencies to any given molecular structure. This is possible even with rather large uncertainty in our assignments because C_{pT} and $S^\circ_{T_o}$ are relatively insensitive to the precise values of these quantities. The last three items in Table I indicate this insensitivity. Even the least accurate empirical rules permit us to fix bond lengths and angles to an uncertainity of ±5% and vibration frequencies to an uncertainty of ±20%. We can see from Table I that such crudeness leads to minor uncertainties in ΔS° and subsequently in K_{eq}.

The insensitivity of S°_{300} and C°_{pT} to the details of structure also permit us to deduce values of both of these quantities from either structural analogies or else simple additivity rules with speed and accuracy. Tables II and III

TABLE III

Standard Entropies and Heat Capacities[a] *at 300°K of Some Structurally Similar Molecules With Two Hindered Internal Rotors*

Molecule	$S^{\circ b}_{300(int)}$	$(\sigma,\eta)^c$	S°_{300}	$C^{\circ\ d}_{p300}$
$CH_3CH_2CH_3$	70.3	18	64.6	17.6
CH_3OCH_3	70.4	18	63.7	15.8
CH_3CH_2OH	69.7	3	67.5	15.7
CH_3NHCH_3	69.8	9	65.4	16.6
$CH_3CH_2NH_2$	69.9	3	67.7	17.4
CH_3NHNH_2	68.8	3	66.6	17.0

[a] *All values in cal/mole-°K.*
[b] $S^\circ_{300} = S^\circ_{300} + R\ln(\sigma/\eta)$.
[c] *η is number of optical isomers, σ is the total symmetry number.*
[d] *Note that a correction of about 0.8 for each H atom will narrow the spread in values.*

illustrate this for species containing zero and two hindered, internal rotations but otherwise the same number of heavy atoms of about the same mass. With such strong evidence we would have great assurance in predicting that for fluorine azide, S°_{300} (FN_3) would be 63.9±0.7 cal/mole-°K while C°_{p300} (FN_3) = 12.0±1.0 cal/mole-°K. Here σ = 1 and η = 1.

In similar vein, we would predict S°_{300} $(HOCH_2NH_2)$ = 70.1 ±0.4 cal/mole-°K with C°_{p300} = 15.0±0.8 while $S^\circ_{300}(HOCH_2OH)$ = 68.7±0.4 and C°_{p300} = 14.9±0.8.

Analogies such as these give us great confidence in estimating S°_{300} and C°_{pT} for free radicals and ionic species in the gas phase.

The most important barrier to our ability to predict K_{eq} lies in our knowledge of ΔH°_f for molecules and for radicals. These lie in the region of ±1 kcal/mole for small molecules and between ±1 and ±2 kcal/mole for free radicals. From this data base, simple additivity laws permit us to predict ΔH_f° for more complex molecules and free radicals [17]. The most important conclusion one draws from these additivity laws is that with a few carefully made measurements of one or two compounds/or radicals in a homologous series one can deduce reliable values for the sample property for the entire series.

CHEMICAL KINETICS - TRANSITION STATE THEORY

Elementary processes can be divided into two distinct categories. The most common are "chemical" processes while the less common are "energy transfer" processes. In the latter, the observed rate corresponds to the rate of transfer of energy, either intramolecularly between different degrees of freedom of a single species or else intermolecularly between two different molecules. The quenching by collision of vibrationally excited HF in a HF laser is an example of an energy transfer process (intermolecular). The intersystem crossing of stilbene from singlet to triplet states is an example of unimolecular energy transfer. The rates of thermal decomposition of small molecules such as $I_2 \rightarrow 2I$; $NOCl \rightarrow NO + Cl$; $HONO \rightarrow HO + NO$, under usual gas phase pressures of 1 atmosphere or less are dominated by the rate of transfer of energy in a bimolecular collision involving highly excited vibrational states of these molecules. At sufficiently high tempratures all unimolecular, gas phase reactions tend to become energy-transfer processes [20].

The more common, chemical processes can be described as the reaction of a Boltzmann energy distribution of energized species. Transition State Theory, originally developed by Eyring and his collaborators [21] can be shown to give an accurate and quantitative account of these reactions, either in gas phase or in solution [17,22]. For an elementary reaction of any molecularity:

$$A + B + C + \rightleftarrows [A\cdots B\cdots C]^{\ddagger} \rightarrow \text{products} \tag{10}$$

Transition State Theory yields a rate constant of the form:

$$k_{obs} = \left(\frac{kT}{h}\right) K^{\ddagger} = \frac{kT}{h} \exp\left\{\frac{\Delta S^{\ddagger}}{R} - \frac{H^{\ddagger}}{RT}\right\} \tag{11}$$

where k is the Boltzmann Constant, h is Planck's Constant and T is the absolute temperature. $K^{\ddagger}$ is the modified equilibrium constant between reactants and the species having the structure corresponding to the maximum in the potential energy diagram for the system. The modification in $K^{\ddagger}$ is that the internal degree of freedom corresponding to passage across the energy maximum, the so-called "reaction coordinate" has been factored out of the statistical mechanical formulation of $K^{\ddagger}$.

If we can generate, a priori, the potential energy surface for the reaction, then we could in principle calculate $K^{\ddagger}$ and hence k_{obs} for any elementary reaction. This approach has proven to be somewhat elusive. However, if we can consider the transition state structure to correspond to a new molecular entity then we can turn rate data around and from measured k_{obs} and its Arrhenius parameters A and E, calculate $\Delta H^{\ddagger}$ and $\Delta S^{\ddagger}$. Knowing ΔH°_{f} and S° for the reactant species we can then compute $\Delta H^{\circ\ddagger}_{f}$ $S^{\circ\ddagger}$ for the transition state species.

To do this we must use the equations 12 and 13 derivable from 11:

$$A = \left(\frac{ekT_m}{h}\right) e^{S^{\ddagger}/R} \quad (12)$$

$$E = \Delta H^{\ddagger} + RT \quad (13)$$

where T_m is the mean temperature (°K) of the experiments, A is the Arrhenius A-factor and E, the Arrhenius Activation Energy.

Such a development makes the transition state amenable to the same estimation methods we have used to predict the thermochemistry of ordinary chemical species, in particular statistical methods for $S^{\ddagger}_{300}$ (and $C^{\ddagger}_{pT}$) as well as simple additivity methods for $H^{\ddagger}_{f300}$, $S^{\circ\ddagger}_{300}$ and $C^{\ddagger}_{pT}$ [17]. Where enough reliable, kinetic data are available this makes it possible to estimate Arrhenius parameters for elementary reactions with experimental state-of-the-art accuracy. In fact, one can argue that because such rules average over many diverse kinetic techniques, the estimation methods are probably better in any given instance than the experimental accuracy. The latter corresponds in most cases to about a factor of 2 to 3 in A and about 5 to 10% in E.

To proceed in such a direction, we must first elaborate the mechanisms for elementary processes, identify the reaction coordinate so as to properly correct $K^{\ddagger}$, $H^{\ddagger}$ and $S^{\ddagger}$ and deduce a set of empirical rules for each mechanistic process, to describe these quantitatively. Table IV shows the categories of mechanism (with examples) which have been identified, for elementary unimolecular processes.

Table V illustrates the various categories for bimolecular reactions. Termolecular and higher order reactions can be dealt with in similar fashion although most of these belong to the energy transfer type of process.

TABLE IV

REACTION MECHANISMS: UNIMOLECULAR

Simple Fission: (1 coordinate)	$R\text{-}X \rightarrow R + X$
Complex Fission: (polycoordinate)	2, 3, 4, 5, 6 center
	$CF_3H \rightarrow F_2C\text{—}F$ (bridged by H) $\rightarrow \ddot{C}F_2 + HF$
Simple Isomerization: (1 coordinate)	cis- trans isomerization, bond rotation
Complex Isomerization: (polycoordinate)	cyclo-$(CH_2)_3 \rightarrow CH_3\text{-}CH\text{=}CH_2$
	1,5 H-atom shift

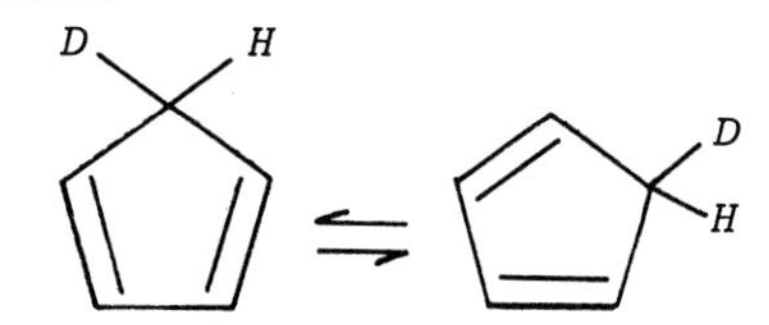

(stable intermediates)

The major distinction which has been discovered in these categories is between the "loose" and "tight" transition states. In simple bond fission where only a single internal bond is broken, the separating fragments pass through a transition state in which the internuclear distance is about 2 to 3 times the ground state bond length and in which the interaction energy has diminished to about RT. This is a "loose" transition state and is described by almost free

TABLE V

BIMOLECULAR REACTION MECHANISMS

A. *Metathesis*

B. *Addition to Pi Bonds (Electron Metathesis)*

C. *Combination (Acid-base)*

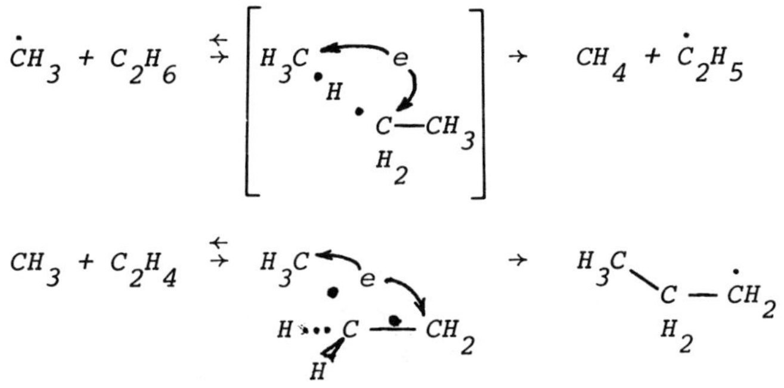

$$\dot{C}H_3 + C_2H_6 \rightleftarrows \left[H_3C \cdot H \cdot \overset{}{C}H_2\text{—}CH_3 \right] \rightarrow CH_4 + \dot{C}_2H_5$$

$$CH_3 + C_2H_4 \rightleftarrows H_3C \cdot \; H\cdots \underset{H}{C} \cdot CH_2 \rightarrow H_3C\text{—}\underset{H_2}{C}\text{—}\dot{C}H_2$$

$$\dot{C}H_3 + \dot{C}H_3 \rightarrow C_2H_6 \qquad \left\{ \begin{array}{l} r_{coll} \sim 4.5\text{Å} \\ r_{C\text{-}C} = 1.54\text{Å} \end{array} \right.$$

rotations [17] for the separating species. The inverse process, recombination of radicals or atoms has the same "loose" transition state. All other processes have "tight" transition states involving only small changes in the frequencies close to the reacting centers.

Of the categories of bimolecular reactions we can see that only metathesis is truly unique and distinct from unimolecular pathways. Addition to multiple bonds, association of radicals, n-center ($n > 3...6$) molecular-concerted reactions are all related to their tight transition states for the inverse, unimolecular, complex fission reactions. We may also note that addition to a multiple bond is a type of electron-metathesis reaction involving metathesis of a pi-electron in a double bond. Both atom metathesis and electron metathesis involve similar tight, transition states characterized by 3-atom centers and three electrons, two bonding and one non-bonding. The non-bonding electron is shared between the terminal atoms and the activation energies for these metathesis reactions seem to be very sensitive to the electron affinities of the entering and leaving groups [23,17].

It is interesting to give an example of the application of those methods to a different kind of reaction than one normally encountered, an electron-molecule reaction. Fehsenfeld [24] has studied the reaction of thermal electrons with SF_6 in the range 300-500°K. He found that the reaction went predominantly by a 2nd order path (down to 0.2 torr) to give SF_6^- with from 0.01% (300°K) to 4% (500°K) production of SF_5^-. The chemistry is:

$$e^- + SF_6 \underset{-1}{\overset{1}{\rightleftharpoons}} SF_6^-$$

$$\underset{-2}{\overset{2}{\rightleftharpoons}} SF_5^- + F$$

SF_6^- and SF_5^- are both little known species with no spectroscopic data available for either. We may reasonably assume that SF_6^- has a distorted octahedral structure not too different from SF_6 and an "effective" symmetry number the same as SF_6. We can further assume that SF_5^- has a pyramidal structure with $\sigma=4$ and is otherwise very similar in bonding to SF_6. On this basis we assigned $S^\circ_{300}(SF_6^-) = 71.1$ e.u. and $S^\circ_{300}(SF_5^-) = 69.7$ e.u.

From the observation that reaction 1 is 2nd order down to 0.5 torr pressure, we can deduce using RRK Theory [17] that the electron affinity of SF_6 is in excess of 1.4 eV. Other data fix it at 1.4 eV [24]. From the observed value of $k_1 = 10^{13.5-32/\theta} sec^{-1}$. Where $\theta = 2.303RT$ (kcal/mole). This A-factor is very close to that which we would have estimated for the thermal fission of an electron from an ion if there is very little change in entropy of the transition state from that of the ion. To my knowledge, this would represent the first thermal rate constant (k_{-1}) ever measured for unimolecular electron emission.

Fehsenfeld did not calculate k_2 but from the ratio of SF_5^-/SF_6^- he computed an activation energy difference of 10 kcal. Using his data we can estimate:

$$k_2 = 10^{17.1-10/\theta} \text{ l/mole-sec}$$

and hence from our estimated $\Delta S^\circ_2 = 31.5$ e.u.; $\Delta H_2^\circ = 10$ kcal/mole:

$$k_{-2} = 10^{10.4} \text{ l/mole-sec.}$$

This latter is very close to the rate constant we would have calculated for the exothermic back reaction (-2) if the F^- collision with pyramidal SF_5 must take place on the empty orbital (base of the pyramid) and there is no activation energy for the reaction. It is interesting to observe that the Arrhenius A-factor for reaction 2 is the largest bimolecular A-factor ever reported. While surprising it is in fact expected for a light election and is deducible from A_{-2} and the loose transition state for this process. The relatively high A-factor for reaction 1 gives a collision diameter to the electron SF_6 pair of about 6.5Å which is just what one might calculate for the Langevin cross-section.

Ionic Pathways in Free Radical Reactions

It has been a truism in chemical kinetics that gas phase reactions usually involve radicals while condensed phase reactions generally involve ions. If we think of free radicals as the catalytic intermediates which provide low energy pathways in gas phase reactions, then ions or ion-pairs may be considered as the catalytic intermediates which provide the low energy pathways for condensed phase reactions.

Since the days of Arrhenius, chemists have been accustomed to thinking of the solvent as providing the driving force (solvation energy) making possible the low energy formation of ion-pairs and free ions in condensed phases. However, recent studies of ionic reactions and thermochemistry in the gas phase [25,26] have provided data which suggest that ionic and free radical pathways may not be very far apart in energies. Some time ago, the author noted that the covalent interactions of free radicals which is of the order of RT at 2 to 3 times their bond lengths may have significant contributions from ionic states such as $R^\circ....R^\circ \longleftrightarrow R^+....R^-$ [27].

The first direct observations of such interaction in gas phase reactions comes from molecular beam studies of alkali metal-halogen reactions. The mechanism has been termed "harpooning" [28]. When K atoms come to a critical distance r_i of an electron attracting species, an electron can jump from K, generating an ion-pair. For K + Cl_2 we can write:

$$K + Cl_2 \rightleftharpoons [K^+ \quad {}^-Cl_2]_{rc}$$

$$\downarrow\uparrow$$

$$[K^+ \ldots Cl^- \ldots Cl]^{\ddagger} \rightarrow K^+Cl^- + Cl$$

The distance r_i at which this can happen is given by solving for r_i at $V(r) = o$ the equation:

$$V(r) = I.P.\ (K) - E.A.(Cl_2) - \frac{e^2}{r} + E_{pol} \tag{14}$$

Where $V(r)$ is the interaction of the K/Cl_2 pair, I.P. is the ionization potential of K, E.A. the electron affinity of Cl_2 and e^2/r the Coulombic energy of the ion pair. The polarization energy, E_{pol}, is negligible for $r > 5Å$. At $r = r_i$, $V(r_i) = 0$ and the electron jump becomes possible, forming the ion pair. The resultant Coulombic attraction brings the pair into proximity. From known values of I.P. and E.A. we can calculate $V(r) = o$ for $r = 7.6Å$, an unusually large "collision" diameter for neutral particles in the gas phase. Note that this is a genuine tunneling process in that the idealized two-center system, $[K^+ \ldots Cl_2]$ has a potential well for the single electron with a double minimum, one near the K^+ and the other near the Cl_2. Between these two, nearly equally deep minima lies a potential barrier for the migrating electron.

Despite the tunneling restriction on the electron, at distances where tunneling does occur, it is sufficiently rapid that it is the much slower relative motion of the various nuclei which determines the overall rate of the process.

If we now consider a free radical system in which two different free radicals $D^\bullet$ and $A^\bullet$ have been formed, a situation comparable to the $K \ldots Cl_2$ interaction can occur. For the pair $[tBu^\bullet \ldots Cl^\bullet]$ one can estimate from known data that at about 4.6Å separation, the ion pair $[tBu^+ \ldots Cl^-]$ will have the same binding energy. In a gas phase encounter between the two radicals, one thus expects ion pair formation with a collision diameter of 4.6Å and a probability of 1/4 arising from the necessity for a singlet, electron configuration on collision.

$$tBu + Cl \rightleftharpoons [tBu \ldots\ldots Cl] \nearrow i\text{-}C_4H_8 + HCl$$

$$r_i \sim 4.6Å \qquad \searrow tBuCl^* \xrightarrow[M]{} tBuCl \quad (tBuCl^* \uparrow i\text{-}C_4H_8 + HCl) \tag{15}$$

One consequence of such a property is that it can be shown that no thermal excitation can ever lead to the simple homolysis of the tertiary C-X bond (x = Cl, Br, I) but must always take the lower energy path yielding olefin + HX.

Over the past decade [29], investigations of the gas phase pyrolytic isomerizations of halogen substituted cyclopropanes have yielded evidence for an ionic pathway for what are considered to be biradical processes. In the particular case shown in equation 16, the pyrolysis of gem-dichlorocyclopropane proceeds via breaking of the strongest C-C bond not the weakest involving the CCl_2 group:

$$\mathrm{Cl_2C(CH_2)_2} \underset{}{\overset{1}{\rightleftharpoons}} \left[\mathrm{Cl_2C(\dot{C}H_2)(\dot{C}H_2)} \overset{2}{\rightleftharpoons} \mathrm{ClC(H_2\overset{+}{C})(CH_2)\;\overset{-}{Cl}} \right] \longrightarrow \mathrm{H_2C{=}C(Cl)CH_2Cl} \qquad (16)$$

$$\Delta H_2^\circ \simeq 6 + \mathrm{I.P.\ (allyl)} - \mathrm{EA\ (Cl)} - E_c\mathrm{(ion\ pair)} \simeq -17\ \mathrm{kcal/mole}$$

Using known values one can estimate that ΔH_2° is exothermic by about 17 kcal/mole for $r_i = 2.6$Å. If step 2 occurred concertedly with the endothermic step 1, one could anticipate a lowering of the overall activation energy by anywhere up to the entire 17 kcal. The observed isomerization has been shown to be concerted [29,30] and to have an activation energy lower by about 8 kcal than that anticipated. Cyclopropyl bromide and iodide would be expected to behave similarly and in fact they do decompose at much lower temperatures [31].

The first serious considerations of ion-pair pathways in gas phase reactions were presented by Maccoll [32] in trying to rationalize the data on 4-center, HX elimination reactions from alkyl halides. This was later shown to be tenable only for the tertiary alkyl halides and a modified model, termed the semi-ion pair model was proposed by Benson and Bose [33]. In this model the transition state is considered to be compared of two semi-ion pairs $\overset{+}{H}\cdot\overset{-}{X}\cdot$ and $R\overset{+}{C}H\cdot\overset{-}{C}H_2\cdot$ which interact as an acid base pair with no further activation energy.

$$H - X + RCH = CH_2 \rightleftharpoons \left[\begin{array}{cc} (+\frac{1}{2}) & (-\frac{1}{2}) \\ \multicolumn{2}{c}{H \cdot X} \\ \multicolumn{2}{c}{\cdot \quad \cdot} \\ \multicolumn{2}{c}{H_2\dot{C} \cdots CHR} \\ (-\frac{1}{2}) & (+\frac{1}{2}) \end{array}\right]^{\ddagger} \rightleftharpoons H_3C - CHRX \quad (17)$$

With this model one can estimate the activation energy to ± 1 kcal. It has been extremely successful in accounting for substitutional effects, such as Markownikoff Addition [34,35] and has been extended to the cases where X=F, OH, OMe, SH, NH_2 and PH_2.

In recent work on the mechanism of the kinetics of O_2 and O_3 reactions in the author's laboratory another ionic mechanism has emerged for a reaction up to now considered to go by a radical pathway. O_3 has been shown to react with primary and secondary alcohols at -80°C, to form compounds of the type $HOCR_1R_2OOOH$ which decompose above -40°C. It also reacts with saturated hydrocarbons at low temperatures, although very slowly, to form hydrotrioxides [36,37].

It can be shown that no radical pathway can account for such behavior. The radical metathesis reaction:

$$RH + O_3 \longrightarrow R^{\cdot} + HO_3^{\cdot} \quad (18)$$

is endothermic by from 22 to 32 kcal for RH = alkane or alcohol and could not proceed at a measurable rate even at room temperature. Nor could it give the observed products. A concerted reaction to yield the insertion product RO_3H has no precedent. Thus the only remaining alternative is an intimate ion-pair path [38,39] proceeding by way of a hydride ion transfer. We can show that electron transfer to form O_3^- is highly endothermic. In the very rapid reaction of iPrOH + O_2 at -80°C we can write [35]:

$$(CH_3)_2\,CHOH + O_3 \underset{}{\overset{1}{\rightleftharpoons}} \left[\begin{array}{c} (CH_3)_2\,\overset{+}{C}\text{-}OH \\ \\ HO_3^- \end{array}\right]_{cage} \overset{2}{\longrightarrow} (CH_3)_2\,\underset{OOOH}{\underset{|}{C}}\text{-}OH \quad (19)$$

$$H_1^{\circ} \simeq 6\ \text{kcal} - |\Delta H^{\circ}_{solv}| \quad (20)$$

where the net solvation enthalpy of the caged, ion-pair, ΔH°_{solv} can be estimated to be about 15 kcal/mole in the alcohol solution [38]. The net uncertainties in these estimates is about 5 kcal, but even the least favorable number gives a plausible rate at -80°C.

A similar elimination in oxidation mechanisms leads to the conclusion that the induction period observed in spontaneous oxidation reactions in the range 30-300°C can be attributed to the production of hydroperoxides via an ion pair route [39]:

$$RH + O_2 \rightleftharpoons \left[R^+ \; HO_2^-\right]_{cage} \longrightarrow RO_2H \qquad (21)$$

For hydrocarbons with tertiary C-H, aldehydic RCO-H or allylic C-H bonds, even the hydrocarbon solvent can provide sufficient solvation energy (±10 kcal/mole) to permit the reaction to proceed in the range 30-130°C [38]. For stronger C-H bonds or R radicals with higher ionization potentials, polar impurities or polar solvents may compensate by providing higher solvation energies. In the gas phase reactions in the cool flame region 250-300°C the surface is the seat of the ion-pair reaction. Polar and ionic surfaces can solvate the ion-pair formation. There are many indications that oxidation in biological systems may also proceed via a similar hydride ion transfer. In this case NADH (nicotinamide adenine dinucleotide) provides the easily abstractable hydride ion.

REFERENCES

1. Gomberg, M., *J. Am. Chem. Soc. 22,* 757 (1900); *Ber. 33,* 3150 (1900).

2. Bodenstein, M., and Lind, S.C. *Z. Physik. Chem. 57,* 168 (1906).

3. Paneth, F., and Hofeditz, W. *Ber. 62B,* 1335 (1929).

4. Rice, F.O., and Rice, K.K. The Aliphatic Free Radicals, The Wilkins and Wilkins Co., Balt., (1935).

5. Rice, F.O., and Herzfeld, K.F. *J. Am. Chem. Soc. 56,* 284 (1934).

6. Protec Co., Processes and Technology, Milan, Italy (1977).

7. Reaction Rate Handbook, 2nd Ed. Defense Nuclear Agency, Santa Barbara, Calif. (1972-present).

8. Heicklen, J., *Atmospheric Chemistry*, Academic Press, Inc., N.Y. (1976).

9. Proc. Fourth Conf. on Oimatic Impact Assessment Program, U.S. Dept. of Trans., NTIS Document (1975), Springfield, VA.

10. Baulch, D.L., Drysdale, D.D., Horne, D.G., and Lloyd, A.C., *Evaluation Kinetic Data for High Temperature Reactions*, Vol. I, II, III. Butterworth, London (1972-76).

11. Engelman, V.S., Survey and Evaluations of Kinetic Data on Reactions in CH_4/Air Combustion, U.S. EPA (Jan. 1976), NTIS Documents, Springfield, VA.

12. Cohen, N., and Bott, J.F., *Kinetics of Hydrogen-Halide Chemical Lasers*, Chapter 2 in *Handbook of Chemical Lasers*, Ed. by R. Gross and J.F. Bott, J. Wiley & Sons, N.Y. (1976).

13. Koyano, I., *Ion-Molecule Reactions*, Chapter 6 in *Comp. Chem. Kinetics*, Vol. 18, Ed. by C.H. Bamford and C.F.H. Tipper, Elsevier Publ. Co., N.Y. (1976).

14. Bodenstein, M., *Z. Physik. Chem. 13*, 56 (1894).

15. Benson, S.W., and Srinivasan, R., *J. Chem. Phys. 23*, 300 (1955).

16. Sullivan, J.H., *J. Chem. Phys. 30*, 1292, 1577 (1959).

17. Benson, S.W., *Thermochemical Kinetics*, 2nd Ed., John Wiley & Sons, Inc., N.Y. (1976).

18. Golden, D.M., Spokes, B.N., and Benson, S.W., *Angew. Chem. (Int. Ed.*, in English) *12*, 534 (1973).

19. Benson, S.W., *Angew. Chem.* (*Int. Ed.*, in English), *17*, 812 (1978).

20. Ref. 18, p. 539.

21. Glasstone, S., Laidler, K., and Eyring, H., *The Theory of Rate Processes*, McGraw-Hill Book Co., N.Y. (1940).

22. *Physical Chemistry, An Advanced Treatise,* Volume VII, Chap. 2, S.W. Benson and D.M. Golden in *Reactions in Condensed Phases,* Ed. by H. Eyring, D. Henderson and W. Jost, Academic Pres, Inc., N.Y. (1975).

23. Alfassi, Z.B., and Benson, S.W., *Int. J. Chem. Kin. 5,* 879 (1973). See also ref. 17, pg. 190-197.

24. Fehsenfeld, F., *J. Chem. Phys. 53,* 2000 (1970).

25. Kebarle, P., *Ann. Rev. of Phys. Chem. 28,* 445 (1977). *Ion Thermochemistry and Solvation from Gas Phase Ion Equilibria.*

26. Arnett, E.M., *Accts. Chem. Rev. 6,* 404 (1973); Beauchamp, J.L., *Ann. Rev. Phys. Chem. 22,* 527 (1971); Brauman, J.I., and Blair, L.K., *J. Am. Chem. Soc. 92,* 5986 (1970).

27. Benson, S.W., *Adv. in Photochem. 2,* 1 (1964).

28. Evans, M.G., and Polanyi, M., *Trans. Far. Soc. 31,* 875 (1935); Magee, J.L., *J. Chem. Phys. 8,* 687 (1940).

29. Robinson, P.J. and Waller, M.J., *Int. J. Chem. Kin. 11,* (1979).

30. Robinson, P.J. and Holbrook, K.A., *Unimolecular Reactions,*

31. Benson, S.W., and O'Neal, H.E., *Kinetic Data on Gas Phase Unimolecular Reactions,*

32. Maccoll, A., *Gas Phase Heterolysis,* Chapter 2 in *Adv. in Phys. Org. Chem. 3* (1965), Acad. Press, N.Y.

33. Benson, S.W., and Bose, A.N., *J. Chem. Phys. 39,* 3463 (1963).

34. Haugen, G., and Benson, S.W., *J. Phys. Chem. 74,* 1607 (1970); *Idem. Int. J. Chem. Kin. 2,* 235 (1970).

35. Benson, S.W., *Reaction Transition States,* Symposium: Gordon and Breach Science Publ. Co., N.Y. (1972).

36. Whiting, M.C., Bolt, A.J.N., and Parish, J.H., *Oxid. of Organic Comp'ds. III, Adv. in Chem. Series 77,* 4 (1968).

37. Kovac, F., and Plesnicar, B., *J. Am. Chem. Soc., 101*, 2677 (1979).

39. Benson, S.W., and Nangia, P., paper in preparation.

40. Nangia, P. and Benson, S.W., *Accts. of Chem. Research, 12* (1979).

41. Dr. Lee Mahoney (Ford Motor Co. Laboratories) has recently shown that ROOH are the primary products of the liquid phase oxidation of n-hexadecane in the range 130-180°C. He has also shown that in this range (0.1% reaction) of temperatures the total RO_2H yield is autocatalytically related to the concentration of RO_2H produced, suggesting a radical supported secondary process. (Private communication)

NEUTRAL THERMOCHEMISTRY FROM IONIC REACTIONS

JOHN I. BRAUMAN

Department of Chemistry
Stanford University
Stanford, California 94305

Summary

Thermochemistry of reactive neutrals can be obtained by combining thermodynamic data for ionic equilibria with electron affinities of the reactive neutrals. Recent advances in gas phase ion methodology make this method an attractive alternative way to obtain neutral thermochemistry.

Knowledge of the thermochemistry of highly reactive neutral species is essential in understanding the kinetics and mechanisms of many chemical reactions. Having such information allows us to predict the position of equilibria involving free radicals, to rule out possible intermediates in proposed mechanisms, and to predict the rates of individual steps of reactions. Free radicals are among the most interesting and ubiquitous of the reactive intermediates. Nevertheless, there are relatively few ways of determining radical thermochemistry, there are very few checks available to confirm accepted values, and there are a number of radicals whose thermochemistry is not known and cannot easily be determined by conventional methods.

Briefly, there have been three major methods in use for determining radical thermochemistry. The first of these is direct spectroscopic determination of bond strengths. This is undoubtedly the most accurate and unambiguous method available, but it is limited, in effect, to diatomic molecules and consequently is of no value in determination of thermochemistry for polyatomic systems.

In situations in which some ionic heats of formation are known, photoionization and mass spectrometry are often used to provide neutral thermochemistry through determination of the threshold for reactions, Eq. [1].

ISBN 0-12-566550-4

$$AB \longrightarrow A^{+} + B \qquad [1]$$

Photoionization results, while limited in number appear to provide accurate values (although there is no obvious reason to have predicted this). Mass spectrometric determinations, on the other hand, are less reliable. While correct values can be obtained with this technique, there are many examples in which these experiments have provided incorrect bond strengths. In addition, the numbers obtained often suffer in precision owing to the general difficulty in determining accurate appearance potentials.

The most commonly used methods are thermokinetic ones. These include both equilibria (1), as used in Eqs. [2] or [3]

$$Cl\cdot + CH_4 \rightleftharpoons CH_3\cdot + HCl \qquad [2]$$

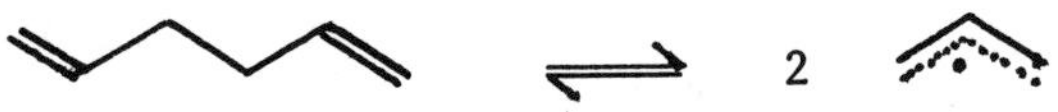

$$\rightleftharpoons 2 \qquad [3]$$

as well as kinetics as used in the iodination technique (2,3).

$$RH + I_2 \rightleftharpoons RI + HI \qquad [4]$$

In Eq. [4], one measures the activation energy for hydrogen abstraction by I• and determines the thermochemistry with the assumption of a small activation energy for the back reaction. These methods have provided a significant number of the reliable bond energies currently available (4).

In view of its importance, it would clearly be valuable to have some additional methods for the determination of radical thermochemistry. With the recent development of techniques for measurement of gas phase ionic equilibria (5) and for accurate determination of electron affinities (6), it is now possible to determine neutral thermochemistry by another completely independent method. Basically, one determines relative heats of formation of ions _via_ an equilibrium constant. Then one connects the ion thermochemistry to the neutral thermochemistry by measuring the energy for electron removal or attachment (ionization potential or electron affinity).

If we measure the equilibrium constant, K, in the gas phase,

$$AH + B^{-} \overset{K}{\rightleftharpoons} A^{-} + BH \qquad [5]$$

we can either determine ΔH^{o} from the temperature dependence of

K or obtain it from ΔG^o via an estimate of ΔS^o using standard statistical methods. Having done so we can obtain bond dissociation energies (D^o) from the relationship

$$\Delta H^o = D^o(A\text{-}H) - D^o(B\text{-}H) + EA(B) - EA(A). \quad [6]$$

(One can carry out a similar exercise for positive ions using bond energies and ionization potentials, but this method has less applicability since a threshold for reaction rather than an equilibrium constant must be measured.)

The attractiveness of utilizing ionic gas phase reactions arises for a number of reasons. First, the neutral radicals are not directly involved in the chemical steps, so that properties of highly reactive species can be determined. Second, properties of less stable neutral radicals can be determined if they give rise to more stable ions. For example, in acetaldehyde the $H\text{-}CH_2\overset{O}{\overset{\|}{C}}H$ bond strength can be determined even though the $CH_3\overset{O}{\overset{\|}{C}}\text{-}H$ bond is weaker, because the most stable anion is $^-CH_2\overset{O}{\overset{\|}{C}}H$. Finally, bond strengths which are very large can be determined in cases where the acidity of the molecule is substantial. Thus, $HC\equiv C\text{-}H$ is a strong enough acid to allow study although the C-H bond is very strong.

Methods

There are three methods by which most of the ionic gas phase equilibrium acidities have been measured. These are high pressure mass spectrometry, the flowing afterglow technique, and ion cyclotron resonance spectrometry. A general review of gas phase acidities makes extensive discussion unnecessary (5). It should be noted that while each of these methods has both its strengths and weaknesses, there has generally been very good agreement (± 2 kcal/mole) for most compounds which have been studied by more than one method.

The other key to using ionic methods is to connect the ions to the reactive neutral radical of interest. This has become possible through measurement of the threshold for electron photodetachment, Eq. [7].

$$A^- + h\nu \longrightarrow A + e^- \quad [7]$$

In favorable cases, the onset for photodetachment can be related to the electron affinity (EA), since the EA corresponds to the minimum energy of a photon required to remove an electron from the negative ion. This area, too, has been reviewed recently (6). The field has progressed greatly in recent years owing to advances in instrumentation and in light sources, especially tunable lasers. The method used in our laboratory is to generate, trap, and detect the ions with an ion cyclotron resonance spectrometer. The spectrometer is fitted with an optical window at one end, and the number of ions present is measured with and without light incident on the cell. The relative photodetachment cross section is extracted from the raw data by dividing the change in signal by the number of photons (measured with a thermopile). A theoretical treatment of the photodetachment process (7) provides a helpful guide for extrapolating to threshold, and enables us to deal with systems of very high symmetry which would otherwise be difficult to evaluate (8).

Results

Some results obtained by ionic equilibrium measurements, combined with electron affinities are considered below. The gas phase acidities, referenced to known compounds such as HF or HCl are expressed as D^o-EA where D^o is the bond dissociation energy. (See Eq. [6].) Values of gas phase acidities have been taken from the recent review by Bartmess and McIver (5). Unless otherwise indicated, EA's are taken from the review by Janousek and Brauman (6). The "accepted" values of D^o's are those chosen by Kerr (4), unless otherwise indicated. All values are in kcal/mole.

TABLE 1

Hydrocarbons

Compound	EA	D^o-EA	D^o
$C_6H_5CH_2$-H	20.4	65.4	86
CH_2=CH-CH_2-H	13	77	90
H H	41.2	42.5	84

These are in quite good agreement with the values of 87, 87 and 81 kcal/mole for toluene (2), propylene (1b,2) and cyclopentadiene, respectively.

TABLE 2

Ketones

Compound	EA	D^{o}-EA	D^{o}
CH_3COCH_2-H	41.3	~ 56	~ 97
$HCOCH_2$-H	41.7	~ 53	~ 95

The acetone number agrees well with the accepted value of 98; the acetaldehyde bond strength can be compared with 86 for the carbonyl C-H.

TABLE 3

Alcohols

R-O-H	EA	D^{o}-EA	D^{o}
Methyl	36.7	65.5	102
tert-Butyl	43.4	59.3	103
Neopentyl	44.5	58.0	103

These values (5,9) can be compared with the accepted values of ~ 104. The flowing afterglow acidity gives even better agreement for methanol.

TABLE 4

Thiols

R-S-H	EA	D^o-EA	D^o
Methyl	42.9	45.4	88
Ethyl	45.0	43.8	89
n-Propyl	46.1	42.8	89
i-Propyl	46.6	42.0	89
tert-Butyl	47.7	41.1	89

These (10) are in very good agreement with the accepted value for CH_3S-H. The recent result of Benson (11), when corrected for the recently revised value of Rossi and Golden (2) for the heat of formation of benzyl radical also agrees.

Acetylene. EA = 68, D^o-EA = 64 (flowing afterglow) or 61.8 (ICR); D^o = 132 or 130. This very large bond energy had not previously been determined by thermokinetic methods. See ref. (12). It is somewhat larger than the photon impact and mass spectrometrically derived values (4,12).

Ammonia. EA = 17.1, D^o-EA = 89.8, D^o = 107. This value for the NH_2-H bond strength can be compared with the reported value (4) of 110. Again, this number relies on the heat of formation of the benzyl radical and when corrected using Rossi and Golden's new value (2) gives better agreement at 107-108.

Silane. EA = 33.2, D^o-EA = 57.9, D^o = 91. The value is in excellent agreement with the recent determination (13) of 90 for SiH_3-H by Doncaster and Walsh.

Conclusion

The general method for determining bond energies (or radical heats of formation) by combining ionic equilibrium acidities with electron affinities has been shown to give values in good agreement with those determined by "conventional" thermochemical methods. The technique is especially useful for compounds which are strong acids in spite of having large bond strengths. Consequently, it is now possible to determine values which were previously inaccessible.

Acknowledgments

I am greatly indebted to my co-workers who have carried out work in this area, especially K. C. Smyth, J. H. Richardson, K. J. Reed, A. H. Zimmerman, H. L. McPeters, R. Gygax, and B. K. Janousek. M. J. Pellerite provided a number of helpful discussions about acidities and thermochemistry. Hans Andersen has made many important contributions to our understanding of the theory of these experiments, and David Golden has provided continuing cousel and guidance on a variety of relevant and not so relevant topics. Our research has been supported by the National Science Foundation. This was written during the tenure of a fellowship from the John Simon Guggenheim Foundation, for which I express my thanks.

References

1. (a) Baghal-Vayjooee, M. H., Colussi, A. J., and Benson, S. W. Int. J. Chem. Kinet., 11, 147 (1979); J. Am. Chem. Soc., 101, 2838 (1979). (b) Rossi, M., King, K. D., and Golden, D. M. J. Am. Chem. Soc., 101, 1223 (1979).

2. Rossi, M., and Golden, D. M. J. Am. Chem. Soc., 101, 1230 (1979).

3. Golden, D. M., and Benson, S. W. Chem. Rev., 69, 125 (1969).

4. Kerr, A., "Bond Strengths in Polyatomic Molecules" in R. C. Weast, Ed., "Handbook of Chemistry and Physics," Chemical Rubber Co., Cleveland, 59th edition, 1978-79.

5. Bartmess, J. E., and McIver, Jr., R. T. in M. T. Bowers, Ed., "Gas Phase Ion Chemistry, Vol. 2," Academic Press, New York, 1979, Chap. 11. Also, see Bartmess, J. E., Scott, J. A., and McIver, Jr., R. T. J. Am. Chem. Soc., 101, 6046 (1979).

6. Janousek, B. K., and Brauman, J. I., in M. T. Bowers, Ed., "Gas Phase Ion Chemistry, Vol. 2," Academic Press, New York, 1979, Chap. 10.

7. Reed, K. J., Zimmerman, A. H., Andersen, H. C., and Brauman, J. I. J. Chem. Phys., 64, 1368 (1976).

8. Gygax, R., McPeters, H. L., and Brauman, J. I. J. Am. Chem. Soc., 101, 2567 (1979).

9. Janousek, B. K., Zimmerman, A. H., Reed, K. J., and Brauman, J. I. J. Am. Chem. Soc., 100, 6142 (1978).

10. Janousek, B. K., Reed, K. J., and Brauman, J. I. J. Am. Chem. Soc. (in press).

11. Colussi, A. J., and Benson, S. W. Int. J. Chem. Kinet., 9, 295 (1977).

12. Janousek, B. K., Brauman, J. I., and Simons, J. J. Chem. Phys., 71, 2057 (1979).

13. Doncaster, A. M., and Walsh, R. J. C. S. Chem. Commun., 904 (1979).

THERMOCHEMISTRY AND KINETICS OF AROMATIC RADICALS

DAVID M. GOLDEN

Department of Chemical Kinetics
SRI International
Menlo Park, California

Present knowledge of the thermochemistry of aromatic radicals is rather sparse. A combination of conventional and exciting new techniques is being applied to alleviate this lack of information. The goal is the extension of codification and extrapolation techniques to aromatic systems.

INTRODUCTION

Kineticists have compiled quantitative understanding of complex gas-phase systems involving aliphatic free radicals over a good number of years. The driving force has often been related to petroleum processing and thus the appropriateness of this symposium. Recently, interest has grown in understanding aromatic systems due to the probability of using fuels (and feedstocks?) derived from coal.

The lack of fundamental thermochemical and kinetic parameters for aromatic and alkyl-aromatic systems is a barrier to the development of a basic understanding of coal liquefaction and gasification, the combustion of alternate fuels having substantial aromatic content, and any other processes involving the thermal reactions of these compounds. For instance, the complex relationship between the strength of individual bonds in coal molecules and the overall rate at which coals undergo degradation or hydrogenation under various process conditions is of great technological importance. A better understanding of this relationship could lead to the development of processes in which maximum degradation to lower molecular weight species under relatively mild conditions can be achieved. This possibility is supported by a number of facts:

- Postulated coal structures contain not only very strong bonds (phenyl-phenyl), but also carbon-carbon bonds weak enough to spontaneously rupture at 400 °C in a matter of seconds (benzyl-benzyl types or bonds convertible to benzyl-benzyl bonds).

ISBN 0-12-566550-4

- Conversion of strong bonds to very much weaker bonds is often only a matter of the appropriate hydrogen transfer within the coal structure.
- Hydrogen transfer from donor solvent to coal and transfer within the coal structures themselves have been shown to be quite facile.

A prime datum for mechanistic purposes is the heat of formation of the radicals, which allows knowledge of the value of ΔH for elementary processes.

In this article we will discuss some values for $\Delta H_f(R^\cdot)$ that we have measured, some values that are predicted, and some new experimental techniques that will aide in determining new values.

In some cases, I will also take the liberty of discussing nonaromatic systems.

EXPERIMENTAL METHODS AND REPRESENTATIVE RESULTS

Iodination Kinetics

In any discussion of aromatic radicals, phenyl radical seems a good starting point. The heat of formation of phenyl ($C_6H_5\cdot$) is well established (1) as 78.5 ± 1 kcal $mole^{-1}$, corresponding to a bond dissociation energy in benzene of DH°(-H) = 111 ± 1 kcal $mole^{-1}$. This value has been determined using a technique developed by Benson and coworkers (2) and largely exploited in these laboratories.

In general, for the reaction of organic iodides with HI, the mechanism which applies is (2):

$$I_2 \rightleftarrows 2I$$

$$RI + I \underset{2}{\overset{1}{\rightleftarrows}} R + I_2$$

$$R + HI \underset{4}{\overset{3}{\rightleftarrows}} RH + I$$

In the particular case where R = C_6H_5, this reaction was studied with UV-Vis spectrophotometry over the range 375-500°C and over a wide range of pressures. The rate constant for reaction (1) may be expressed,

$$\log k_1/M^{-1}\ s^{-1} = (11.36 \pm 0.06) - (28.4 \pm 0.2)/\theta$$

$$\theta = 2.303\ RT \text{ in kcal mole}^{-1}$$

The value of the activation energy E_1, combined with that of the reverse reaction E_2, would yield a value of $\Delta H_{1,2}$, and thus of $\Delta H^\circ_f(C_6H_5^\cdot)$ if values of ΔH_f are known for C_6H_5I and HI.

In the literature (2) on this technique, the seemingly reasonably well-founded value E_2 = 1 ± 1 kcal $mole^{-1}$ has been assumed. This assumption may now be subjected to experimental verification (3) (Vida Infra).

Another prototype is the benzyl radical, which typifies the class of stabilized aromatic systems.

The value for the heat of formation of benzyl to be found in the literature (2) has led to a BDE ($\emptyset CH_2$-H) = 85 ± 1 kcal $mole^{-1}$ which corresponds to a stabilization energy of 13 kcal $mole^{-1}$ when compared to the BDE (CH_3CH_2-H) = 98 kcal $mole^{-1}$. Recent evidence (4) (discussed below) suggests however that BDE ($\emptyset CH_2$-H) = 88 ± 2 kcal $mole^{-1}$ may be a better value.

Very Low-Pressure Pyrolysis (VLPP)

Another technique that can be used to advantage in determining the heats of formation of free radicals is the Very Low-Pressure Pyrolysis (VLPP) technique developed at SRI International (5). VLPP is ideally suited for the measurement of the rate of initial bond-breaking reactions in the pyrolysis of organic molecules where secondary reactions often interfere with the characterization of the initial step. The technique has been described in detail previously (5,6).

The procedure consists of allowing the reactant to flow through the inlet system at a controlled rate and switching valves in the inlet lines so that the reactant alternately flows through a Knudsen cell reactor or through the bypass directly into the mass spectrometer. Since the flow rate is held constant, the difference between the mass spectrometer signals for the reactant in the bypass position and the reactor position corresponds to the amount of reactant that is decomposed as it flows through the reactor.

The data are interpreted with the aid of various steady-state expressions derived as shown below. At low flow rates, the treatment for a simple irreversible unimolecular decomposition is appropriate, and the extraction of rate parameters is straightforward. At higher flow rates (higher reaction pressures) and smaller escape apertures, rapid bimolecular reactions can compete with unimolecular decomposition and with escape from the reactor. These interactions must be included in the analysis. The reason for using VLPP at pressure high enough for secondary reactions to occur is shown by the description given below of our recent study for the pyrolysis of 1-ethyl naphthalene (7). Briefly, observation of competition between unimolecular bond scission and radical recombination amounts to measurement of an equilibrium constant:

$$\text{1-ethylnaphthalene} \underset{k_r}{\overset{k_d}{\rightleftharpoons}} \text{1-naphthylmethyl radical } (CH_2^{\cdot}) + CH_3 \quad (1)$$

Reliable measurement of an equilibrium constant can provide very good third-law values for $\Delta H°$, since these values are not subject to the systematic errors that can markedly affect the slope of Arrhenius plots. Thus, by providing for measurement of an equilibrium constant that is otherwise not readily measured at higher pressures or in static systems, the VLPP system allows two largely independent measurements of bond strength and, therefore, a valuable internal consistency check.

For irreversible, unimolecular decomposition of a substance A, only three rate processes are considered:

$$\begin{aligned} \text{inlet} &\xrightarrow{R_A} A \\ A &\xrightarrow{k_d} \text{products} \\ A &\xrightarrow{k_{eA}} \text{mass spectrometer} \end{aligned} \quad (2)$$

where R_A is the rate at which the reactant is allowed to flow into the reactor and k_{eA} is the first-order rate constant describing escape from the reactor.

Steady-state analysis provides the expression:

$$\frac{1}{F} \equiv \frac{A_{escaped}}{A_{reacted}} = \frac{(A)_{ss}}{(A)_{o,ss} - (A)_{ss}} = \frac{k_{eA}}{k_d} \quad (3)$$

where $(A)_{o,ss}$ is the steady-state concentration in the reactor when there is no decomposition.

When recombination of the radical fragments produced by unimolecular decomposition competes with escape from the reactor the following reactions must be considered:

$$\begin{aligned} \text{inlet} &\xrightarrow{R_A} A \\ A &\underset{k_r}{\overset{k_d}{\rightleftharpoons}} B^{\cdot} + C^{\cdot} \\ A &\xrightarrow{k_{eA}} \\ B^{\cdot} &\xrightarrow{k_{eB}} \\ C^{\cdot} &\xrightarrow{k_{eC}} \end{aligned} \left.\vphantom{\begin{aligned} A \\ B \\ C \end{aligned}}\right\} \text{escape to mass spectrometer} \quad (4)$$

The controlled flow rate into the reactor is given by R_A, and the first-order escape rate constants for the various fragments are related by square roots of their masses so that the escape constant for the 1-naphthylmethyl radical (B·) is related to that for 1-ethylnaphthalene by

$$k_{eB} = k_{eA}\left(\frac{156}{141}\right)^{\frac{1}{2}} \cong k_{eA}(1.05) \qquad (5)$$

Steady-state analysis of this sequence for 1-ethyl naphthalene provides an expression similar to equation (3):

$$\frac{1}{F} \equiv \frac{A_{escaped}}{A_{reacted}} = \frac{(A)_{ss}}{(A)_{o,ss} - (A)_{ss}} = \frac{k_{eA}}{k_d} + \frac{k_r F(A)}{k_d(3.39)} \qquad (6)$$

Equation (6) indicates that decomposition and competitive recombination will provide a straight line of slope k_r/k_d and intercept k_{eA}/k_d. In the limiting case where recombination is unimportant, equation (6) reduces to equation (3).

The VLPP technique can also be applied to the study of rapid radical-molecule bimolecular reactions (2,4) This allows the verification of assumptions concerning the activation energy of R + HI reactions if the radicals can be produced from a convenient precursor.

Thermal Radical Production

We have recently applied (4,3) this technique to the cases of R = allyl, benzyl, and t-butyl radicals. These experiments do not quite "close the loop" because they are performed at different temperatures than the iodinations. However, the temperature correction is not difficult and the value used above for benzyl heat of formation comes from these recent experiments.

Photolytic Radical Production

We have recently (8) combined the phenomenon of infrared multiphoton dissociation of organic molecules with the VLPP technique to produce a method for determining the rate constant for radical-molecule reactions at temperatures determined by reactor wall temperatures and completely independent of any need to heat the radical precursor. (Current powerful dye lasers will allow this technique to be useful for UV-fis photochemical radical production as well.) The current application has been to reactions of CF_3 radicals from CF_3I, but extension to aliphatic and aromatic systems requires only time and funds.

In this experiment, as previously, the effusive molecular beam was mechanically chopped in the second (differentially pumped) chamber before it reached the ionizer of the quadrupole mass filter (Finnigan 400). The signal was demodulated by a lock-in amplifier (PAR 128A) whose output was now stored in a signal averager (PAR 4202), which also served to trigger the laser. The two-aperture Knudsen cell was fitted with KCl windows and had an optical pathlength of 20.5 cm and a volume of approximately 105 cm^3. The cell was coated with Teflon by rinsing with a finely dispersed Teflon slurry in a water/aromatic solvent mixture (Fenton Fluorocarbon, Inc.) and curing at 360°C.

The Lumonics TEA-laser (Model K-103) was operated at the R(16) line of the 9.6 μ transition at a pulse repetition frequency of .25 Hz, slow enough to permit > 99% of the reaction products to escape from the cell before the next laser shot. The multimode output of the laser consisted of a pulse of approximately 5.0 J, directed through the photolysis cell after being weakly focussed by a concave mirror (fl = 10 m), which gave a beam cross section of 2.67 cm^2 at the KCl-entrance window of the cell.

A typical experiment consisted of averaging the time-dependent mass spectroscopic signal intensity of the products for a number of laser shots (10-100) as a function of the flow rate of the reactant gases at constant CF_3I flow rate and constant energy per pulse. Although experiments could be performed on a single-shot basis, the signal/noise ratio was improved by averaging the results of a number of laser shots.

The total yield of product formed in the reaction of interest was then determined by integration of the accumulated time-dependent signal of the signal averager on a strip-chart recorder fitted with an electronic integrator (Linear Instruments, Inc.).

Using the apparatus described above, reaction products could be observed as well as the transient depletion of CF_3I. An advantage of the low-pressure technique is that the effects of secondary reactions are minimized, although they must be considered in the data analysis.

The following chemical reaction mechanism is appropriate for studying the reaction $CF_3 + Br_2 \rightarrow CF_3Br + Br$:

$$CF_3I + nh\nu \rightarrow CF_3 + I$$

$$CF_3 + Br_2 \xrightarrow{1} CF_3Br + Br$$

$$CF_3 \xrightarrow{k_w} \text{(wall loss \underline{not} yielding } CF_3Br\text{)}$$

No heterogeneous first-order reaction of CF_3 to produce CF_3Br is included, since the data interpretation does not suggest it. As in the usual data treatment for VLPP studies, each molecular and radical species escapes from the reactor with a characteristic first-order escape rate constant; for CF_3I, Br_2, CF_3, and CF_3Br, the escape rate constants are k_2, k_3, k_4, and k_5.

Analysis of the reaction mechanism and solution of the appropriate differential equations give an expression for the time-dependent mass spectrometer signal due to CF_3Br. The total yield of CF_3Br (Y) is related to the rate constants as follows:

$$Y^{-1} = \left(\alpha\beta\left[CF_3I\right]_\circ\right)^{-1}\left[1 + \frac{k_4 + k_w}{k_1\left[Br_2\right]}\right]$$

where pseudo-first-order conditions are assumed to hold. In this expression, α is a mass spectrometric sensitivity factor, β is the fraction of the initial CF_3I that is dissociated by the laser pulse, and V is the volume of the cell. The initial $[CF_3I]_\circ = F_{CF_3I}/(V\bullet k_2)$, where F_{CF_3I} is the flow rate of CF_3I into the reactor, and V is the reactor volume; similarly, $[Br_2] = F_{Br_2}/(V\bullet k_3)$. For each F_{Br_2}, two escape apertures can be used, giving two different values for the escape rate constants, corresponding to the "big" and "small" apertures. Plots of Y^{-1} versus $F_{Br_2}^{-1}$ give two straight lines with intercepts c_s and c_b (small and big apertures) given by (i = s or b)

$$c_i = \frac{k_{2i}}{\alpha\beta V F_{CF_3I}}$$

The slopes of the straight lines are given by

$$m_i = \frac{c_i V(k_{4i} + k_w)k_{3i}}{k_1}$$

Laser-Powered Homogeneous Pyrolysis (LPHP)

In VLPP experiments aimed at obtaining bond scission information, we rely on the walls of the VLPP reactor to be the source of heat through gas-wall collisions, while at the same time being non-catalytic for the destruction of the substrate. Sometimes this latter condition is not met. Therefore, we may, in addition to VLPP, use the technique of laser-powered homogeneous pyrolysis (9). This technique is a valuable complement to the VLPP procedure, since it essentially provides a "wall-less" reactor. The total pressure in the reactor is on the order of 100 torr, consisting mostly of bath gas and SF_6. An IR laser is used to heat the strongly absorbing sensitizer (SF_6), using a wavelength at which the substrate does not absorb. The SF_6 transfers its thermal energy by collision to the substrate molecules, and decomposition takes place. As described by Shaub and Bauer (9), the technique worked well for compounds of relatively high vapor pressure, using a static reactor system and a CW laser; we are currently adapting the technique for application to poly-nitro aromatics and other low-vapor-pressure substrates, using a pulsed CO_2 laser and a flow system with GC detection. The reasons for these modifications are: (1) When operation is with a pulsed laser, reaction times are short because of rapid cooling by contact with the off-axis cell contents, and secondary reactions are either unimportant or can be minimized by suitable choice of a scavenger, and (2) when low-vapor-pressure substrates are being studied, quantitative recovery and measurement of products and unreacted starting material is simpler with a flow system.

We have preliminary evidence from this technique which indicates that nitrotoluenes decompose by NO_2-aromatic bond scission. This seems to be true of those with ortho-methyl substitution as well, in contrast to earlier reports.

The presence of ortho-substituted methyl groups is well known to increase the decomposition rates of nitrobenzenes. Consistent with this trend, the decomposition of nitrobenzene in a nitrogen-carrier stream results in biphenyl and other products traceable to initial homolytic bond scission to form phenyl radicals and NO_2, but o-nitrotoluene reacts under these same conditions by a different pathway. In the latter case, the principal product has been shown to be aniline, resulting from the initial formation and decarboxylation of anthranilic acid (10):

(1) o-nitrotoluene (NO_2, CH_3) $\xrightarrow[\text{packed vessel, } P_{tot} = 1 \text{ atm}]{600°C,\ N_2}$ [anthranilic acid (NH_2, CO_2H)] ⟶ aniline (NH_2)

[anthranilic acid] $\xrightarrow{CH_3OH}$ methyl anthranilate (NH_2, CO_2CH_3)

Similarly, reported Arrhenius parameters reflect an apparent change of mechanism in gas-phase pyrolysis: nitrobenzene and meta- and para-substituted nitrobenzenes exhibit A-factors and activation energies consistent with initial homolytic bond scission of the weakest bond in the molecule, but orthonitrotoluene exhibits a much lower A-factor and activation energy, as would be consistent with an initial step involving a complex intramolecular rearrangement (11):

(2) nitrobenzene (NO_2) $\xrightarrow[.05 \text{ torr};\ \sim 400^{o}C]{k_d}$

$$\log k_d\ (sec^{-1}) = 17 - 69/2.303\ RT$$

(3) [o-nitrotoluene, six-membered ring transition state] $\xrightarrow[.05\ \text{torr};\ \sim 400^{\circ}C]{k_d}$ [o-nitrosobenzyl radical $CH_2^{\bullet}$ + $\dot{O}H$]

$$\log k_d\ (sec^{-1}) = 12.4 - 49.5/2.303\ RT$$

Although the six-membered ring transition state suggested in reaction (3) is, on steric grounds, quite plausible, it does not lead directly to sufficiently stable products to be associated with the observed activation energy: The first step, as written, is estimated to be ca. 80 kcal/mole endothermic, and therefore could not give rise to an activation energy in the range of 40 kcal/mole. On the other hand, the stable products observed in reaction (1) require substantial movement of at least four atoms, a kind of rearrangement which is difficult to imagine as taking place in a single elementary step in the gas phase. For these reasons, it seemed worthwhile to study the decomposition of nitroaromatics under conditions where the rates and products of the initial decomposition steps can be measured, and where a clear distinction can be made between homogeneous gas-phase decomposition processes and those processes aided by association with solid durfaces or condensed phases. In order to understand the heterogeneous decomposition processes, it is necessary to first understand the behavior that is intrinsic to an isolated molecule.

Preliminary experiments with the decomposition of nitrobenzene and o-nitrotoluene using the laser-powered homogeneous pyrolysis technique support our speculation that the low Arrhenius A-factors and activation energies and "abnormal" products exhibited by ortho-substituted nitrobenzenes cannot arise from a homogeneous gas-phase decomposition process: The pyrolysis of o-nitrotoluene in the presence of excess benzene gives only products expected from homolytic C-N scission and no aniline or other products resulting from an internal oxidation-reduction process. More precise experiments in which decomposition is carried out under flow conditions will provide a confirmation of the C-N bond strength in o-nitrotoluene and a measure of the C-N bond strength in polynitro aromatics, where the effect of the additional strongly electron-withdrawing groups is now unknown. This information will provide a basis for understanding the high-temperature decomposition of nitro compounds where homolytic bond scission of suitably weakened phenyl-NO_2 bonds could become the principal decomposition pathway.

Theoretical Considerations

We have extended Herndon's (13) simple methods for computing heats of formation of stable species to those of radicals (14). The essence of our predictions of the stabilization energy is

$$E\pi(SE) \equiv E\pi(R^{\cdot}) - E\pi(RH)$$
$$E\pi(RH) = 27.33\ \ell n[CSC(RH)] \quad (13)$$
$$E\pi(SE) = 22.68\ \ell n[CSC(R^{\cdot})] - 27.33\ \ell n[CSC(R\text{-}1)] \quad (14)$$
$$E_{SE} = E\pi(SE) + E_{\sigma} \text{ scaled to } \phi CH_2 = 10 \text{ kcal mole}^{-1}.$$

These considerations predict that in 1-methylnaphthalene the methyl C-H bond strength ($D\phi\phi CH_2$-H) would be 4 kcal $mole^{-1}$ more stable than in toluene compared to the value measured (7) of 3 kcal $mole^{-1}$ using the VLPP technique as described. A good test will be the value in 9-methylanthracene where the prediction is for an extra 10 kcal $mole^{-1}$ of stability. We are planning such a test using both VLPP and LPHP in the near future.

CONCLUSION

We have indicated the rather sparse knowledge of thermochemistry for aromatic systems. We have pointed out how the combination of conventional and exciting new techniques will, and are, making these values more amenable to measurements. It is to be hoped that with a sufficient data bank, we shall be able to extend the codification and extrapolation embodied in Thermochemical Kinetics (15) to aromatic systems.

ACKNOWLEDGMENTS

The work discussed herein has been supported by several agencies over several years. Primary support from AFOSR, NASA, DOE, NSF, and ARO is appreciated. The bulk of the credits should go to my colleagues, past and present, in the Department of Chemical Kinetics at SRI International. Special thanks goes to D. F. McMillen, M. J. Rossi, J. R. Barker, P. K. Trevor, and K. E. Lewis.

REFERENCES

1. Rodgers, A. S., Golden, D. M., and Benson, S. W., J. Amer. Chem. Soc., 89, 4578 (1967). [Corrected for correct ΔH_f of C_6H_5I (38.9, not 40.5 kcal $mole^{-1}$)].

2. Golden, D. M., and Benson, S. W., Chem. Rev., 69, 125 (1969).

3. Rossi, M. J., and Golden, D. M., Int. J. Chem. Kinetics, 11, 969 (1979).

4. Rossi, M. J., and Golden, D. M., J. Amer. Chem. Soc., 101, 1230 (1979).

5. Golden, D. M., Spokes, G. N., and Benson, S. W., Angew. Chem., 85, 602 (1973).

6. Rossi, M., and Golden, D. M., J. Amer. Chem. Soc., 101, 1223 (1979).

7. McMillen, D. F., Trevor, P. K., and Golden, D. M., to be published.

8. Rossi, M. J., Barker, J. R., and Golden, D. M., J. Chem. Phys., 71, 0000 (1979).

9.. Shaub, W. M., and Bauer, S. H., Int. J. Chem. Kinetics, 7, 509 (1975); Shaub, W. M., Ph.D. Thesis, Cornell University, 1975.

10. Fields, E. K., and Meyerson, S., Tetrahedron Letters, 1201 (1968).

11. Matveev et al., Inst. Khim. Fiz., Cheronogolovka, USSR (4) 7836 (1978).

12. Matveev, B. G., et al., Akad. Nauk., Ser. Khim., 2, 474 (1978).

13. Herndon, W. C., J. Amer. Chem. Soc., 98, 887 (1976), and references therein.

14. Stein, S. E., and Golden, D. M., J. Org. Chem., 42, 839 (1977).

15. Benson, S. W., Thermochemical Kinetics, John Wiley and Sons, Inc. New York, 1968.

Frontiers of Free Radical Chemistry

MOLECULAR DISTORTIONS AND ORGANIC REACTIVITY: ADDITIONS, CYCLOADDITIONS, AND FREE RADICAL REACTIONS

K. N. HOUK

Department of Chemistry
Louisiana State University
Baton Rouge, Louisiana

It is a pleasure for me to participate in this symposium honoring the Exxon Research Laboratories in Baton Rouge. It may seem odd for me to be here speaking in the context of "Frontiers of Free Radical Chemistry", because until recently, all of the efforts of my research group were devoted to the study of concerted reactions. Our closest approach to free radicals came in various efforts to prove that diradicals were not intermediates in Diels-Alder and 1,3-dipolar cycloaddition reactions. The Chairman of this symposium, Bill Pryor, invited me to contribute soon after my group had begun a theoretical study of an intriguing area of free radical chemistry, the competition between additions to alkenes and hydrogen abstractions by radicals. This work will be described later herein, but I would first like to put our investigation of free radical reactivity into a more general context, and to describe the relationship between distortions of organic molecules and the reactions--pericyclic, ionic, and radical, that organic molecules undergo.

The relationship between molecular distortions and reactivity is a subject about which I am sure that every chemist knows a great deal. The distortions that molecules undergo in the course of a reaction are of pivotal importance in determining the facility of that particular reaction. It is frequently stated that the activation energy of a reaction arises from the energy required to distort reactants away from equilibrium geometries, or to cause rehybridization of the atoms undergoing bonding changes. What I will discuss here is more specific, in that we have investigated theoretically the energies required for specific types of distortions, and have determined the influence that these distortions have on the energies and shapes of particular

[1]*This work was supported financially by the National Science Foundation and the National Institutes of Health.*

ISBN 0-12-566550-4

molecular orbitals. Since the frontier orbitals, the HOMO and LUMO of a molecule, are well known to have a critical role in reactivity (1,2), we have paid particular attention to the influence of distortions on these orbitals. Here, I will use a model pericyclic reaction, the trimerization of acetylene, to demonstrate general concepts, and then describe the role of molecular distortions upon nucleophilic additions to acetylenes and olefins, carbene cycloadditions, and, finally, the competition between radical additions and hydrogen abstractions.

The importance of molecular distortions upon reactivity is not a new concept. It was discussed by Evans and Polanyi in the '30's (3); they showed that there is a relationship between the potential energy surfaces for bond dissociation of reactants and products and the activation energies for atom transfer reactions. Evans and Polanyi described a reaction path as consisting of two phases: early on the reaction path, the reacting molecules resemble distorted reactants, while after the transition state, this ensemble resembles distorted products. The transition state is that point where the distorted reactants evolve into distorted products. The applications of such ideas to free radical reactions are described in this Symposium volume by Pryor and Church.

Orbital interaction models of organic reactivity

Before describing our studies of the influence of molecular distortions on reactivity, I will give a brief background of the type of thinking that we have applied to organic reactivity; once the model is laid out, it is much more obvious, I think, why we can isolate molecular distortions as something different and special from orbital interactions.

The frontier molecular orbital theory of organic reactivity has been quite useful in correlating reactivities with electronic structures, and in predicting products of reactions (1,2,4). The Woodward-Hoffmann rules, which can couched in frontier orbital language, are perhaps the finest examples of this. Fukui has stated succinctly the principle tenet of frontier molecular orbital theory: organic reactions occur in such a way as to maximize overlap of the highest occupied molecular orbital of one system with the lowest unoccupied molecular orbital of another (4).

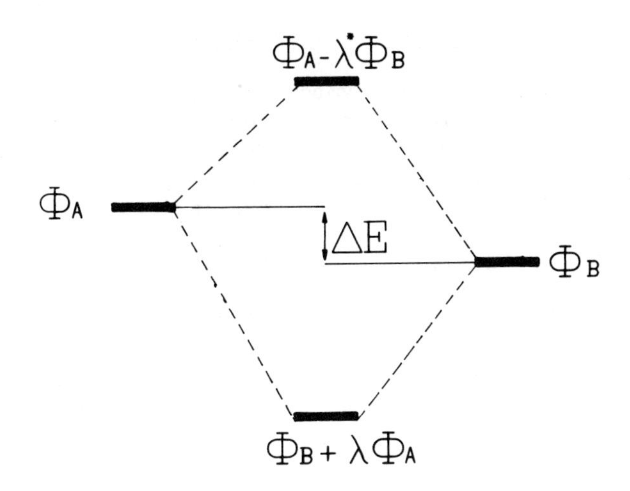

FIG. 1. Schematic representation of orbital interactions.

Figure 1 summarizes the general principles upon which frontier molecular orbital ideas are based. When two molecules interact, each orbital ϕ_A on one molecule can interact with the various orbitals, ϕ_B, on the second molecule. The mixing of ϕ_A and ϕ_B generates a bonding combination and an antibonding combination. The extent of mixing of ϕ_A and ϕ_B depends upon the overlap between these orbitals and upon their energetic separation: those orbitals closest in energy and overlapping best, will interact with each other to the greatest extent. If both of these orbitals are filled, then the interaction between them is repulsive. This closed-shell repulsion depends approximately on the square of the overlap integral involving these orbitals. If only one orbital is doubly occupied, these two electrons go into the stabilized orbital ($\phi_A + \phi_B$), and the interaction is stabilizing. Similarly, for a total of one or three electrons in these two orbitals, stabilization results, since more electrons end up in stabilized orbitals than into destabilized orbitals.

The smaller the energy gap (ΔE) between ϕ_A and ϕ_B, the more these orbitals interact. This is the origin of the apparent dominance of frontier orbital interactions upon organic reactivity (4). If the HOMO of one molecule overlaps well with the LUMO of a second molecule, then this interaction will lead to more stabilization than the interaction of any other pair of filled and vacant orbitals.

Perturbation treatments of radical reactivity

In the context of this symposium, it is relevant to describe several applications of these ideas to free radical reactivity. Some time ago, Fukui defined quantities called delocalizability and superdelocalizability to serve as reactivity indices, and applied these indices to radical reactions.

According to Huckel theory, a radical, R•, has a singly occupied molecular orbital (SOMO) of energy, α. Upon interaction of this SOMO with a filled orbital (ϕ_o) of energy ε_o, residing on another molecule, the stabilization which occurs is directly proportional to the square of the overlap between these orbitals, and is inversely proportional to the energy separation between these orbitals,

$$E(\text{SOMO},\ \phi_o) \cong \frac{S^2}{\varepsilon_o - \alpha}\ .$$

To consider the interaction between the SOMO of a given radical with ϕ_o of a series of molecules, the coefficient, C_o, at the site of interaction of the radical with the molecule can be substituted for S. That is,

$$E(\text{SOMO},\ \phi_o) = \frac{C_o^{\ 2}}{\varepsilon_o - \alpha}\ ,$$

where C_o is the orbital coefficient at the site of interaction.

There is a similar term for the interaction of the SOMO of R• with each vacant orbital, ϕ_v, of the closed-shell

$$E(\text{SOMO},\ \phi_v) = \frac{C_v^{\ 2}}{\alpha - \varepsilon_v}\ ,$$

molecule. Combining these two terms, Fukui described radical delocalizability, D_R as follows:

$$D_R = \sum_o \frac{C_o^{\ 2}}{\varepsilon_o - \alpha} + \sum_v \frac{C_v^{\ 2}}{\alpha - \varepsilon_v}\ ,$$

where the sums are over all occupied and vacant orbitals of the closed-shell molecule.

Fukui found that delocalizability correlated with the relative rates of hydrogen abstraction from a series of alkanes (5). That is, the hydrogen coefficients of the MOs of alkanes are good predictors of the preferred site of hydrogen abstraction.

Recently, Bartels _et al._ found a similar correlation for hydrogen abstractions from alkanes by the methyl radical (6). Using MINDO/3 orbitals for eight alkanes, and $\alpha = -4.23eV$, the following correlation with activation energy was observed:

$$E_a = 66.0 + 466D_R$$

For hydrogen abstraction by the trifluoromethyl radical ($\alpha = -6.25eV$) the following relationship was found:

$$E_a = 63.0 + 458D_R$$

These workers also found correlations between delocalizabilities and rates of addition of the dicyanomethyl and trifluoromethyl radicals to alkenes (7). The use of a single number for α, the energy of the SOMO, is of some concern, as described later.

Of particular relevance to our discussion is the finding by Shinohara _et al._ (8) that the energies required to stretch CH bonds in eight alkanes are linearly related to the delocalizabilities calculated for the interactions of each hydrogen with the methyl radical. This observation suggests that a correlation of reactivity with delocalizability is not necessarily a proof of the dominance of charge-transfer interaction, since several molecular properties of importance to reactivity may be linearly correlated. Nevertheless, these examples indicate that perturbation theory is of some value in correlating radical reactivity.

Barriers to thermally allowed reactions: the acetylene trimerization

I will return to radical reactions later, but I will begin by describing a theoretical study of a pericyclic reaction. The prototype 6π electron pericyclic reaction is the trimerization of acetylene to form benzene. Even though this reaction is exothermic by 144 kcal/mol, the reaction takes place only to a small extent at temperatures above 200°C, and many other hydrocarbons are formed in large amounts. The termolecular reaction would be disfavored entropically, but intramolecular cases are also reluctant, as described below.

Although our primary concern here is with understanding why the concerted reaction does not occur, it is relevant in this symposium to consider the actual mechanism occuring in the acetylene pyrolysis. The reaction was studied first by Berthelot (9) and there are hundreds of investigations since then, including studies by several participants in this symposium. The first steps in a radical chain mechanism proposed by Back (10) is shown below:

$$2HC \equiv CH \rightarrow HC \equiv C\cdot + H_2C = \dot{C}H$$

$$HC \equiv C\cdot + HC \equiv CH \rightarrow HC \equiv C - CH = \dot{C}H$$

The disappearance of acetylene gives an extraordinarily linear Arrhenius plot over a range of 600-2400°K, leading to the activation parameters, E_{act} = 60 kcal/mol and log A = 9. The mechanism shown involves a molecule-assisted homolysis as the first step. If a concerted trimerization were to occur, the A factor should be about 100 times less than that for this bimolecular reaction. In other words, the activation energy of the non-observed termolecular reaction must be greater than 30 kcal/mol at a temperature of 600°K. This is not inordinate, but is perhaps surprisingly high for such a highly exothermic, thermally allowed, reaction.

Even cyclic triacetylenes do not readily cyclize. Both 1 and 2 have been synthesized (11,12), and group equivalent calculations suggest that the conversion of 1 to 2 is still exothermic by 44 kcal/ mol. Nevertheless, neither molecule

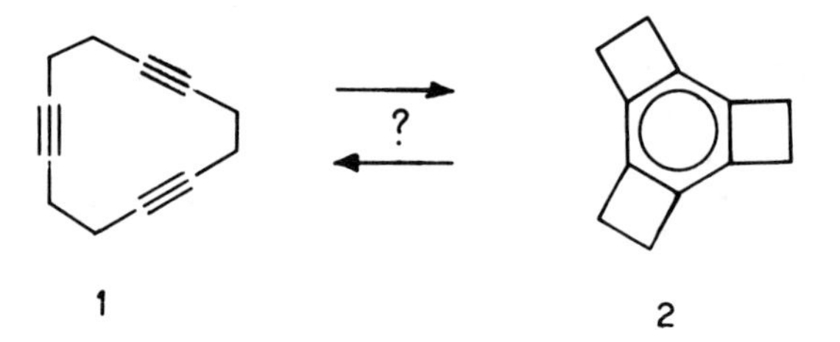

readily converts to the other upon pyrolysis. No significant entropic problem should be present here, so there must be a substantial activation energy barrier preventing interconversion of these molecules.

Why should a highly exothermic reaction which is doubly allowed due to the presence of both in-plane and out-of-plane 6π electron aromatic sextets have such a high activation energy? We have investigated both the trimerization of acetylene and the interconversion of 1 and 2 by both MINDO/3 and *ab initio* STO-3G methods (13-15). For both methods, we only studied the pathways with D_{3h} symmetry, and accurate activation energy predictions cannot be made on the basis of these calculations. The D_{3h} "transition state", 3, calculated by MINDO/3 for the trimerization of acetylene, shows substantial CCH bending, but only slight CC stretching

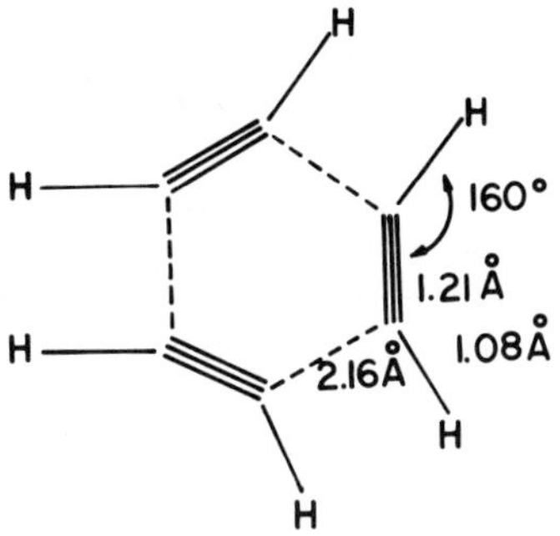

H
H
H
H
H
H
160°
1.21 Å
1.08 Å
2.16 Å
3

(MINDO/3 acetylene has a CC bond length of 1.20Å). The D_{3h} "transition state" for the interconversion of 1 and 2 is similar in geometrical parameters.

Before the "transition state", 3, is reached, the C_6H_6 ensemble can be considered to be three distorted acetylenes, whereas after the transition state, the system is clearly a distorted benzene. The major CC bonding changes occur rather suddenly at the transition state. We have analyzed this reaction in some detail, and have also considered the simpler six electron pericyclic "bond-switching" reaction of the $3H_2$ ensemble (14). In both cases, electron density initially residing in more or less localized regions of the reactants, the in-plane orbitals acetylenes, or the HH bonds of hydrogen, must be transferred to different regions in the products. Formally, this requires mixing of occupied orbitals of reactants with vacant orbitals of reactants, a process which creates new filled orbitals localized in different regions in space from those in reactants.

For the acetylene trimerization, the calculations show that as the molecules approach one another, there is a smooth evolution of the three out-of-plane occupied and three out-of-plane vacant π MOs of three acetylenes into the six of benzene. However, the orbitals which are initially π in the isolated actylenes, but lie in the plane of the three acetylenes for this reaction path, exhibit behavior like that first observed for the Diels-Alder reaction by Jorgenson (16), and more recently by Townshend et al. (17). That is, the occupied in-plane MOs increase in energy as the transition state is approached. At this point, they are 2eV higher in energy than in acetylene. After the transition state, these orbitals are suddenly stabilized and evolve into relatively low-lying orbitals of benzene. The LUMOs exhibit the opposite behavior, dropping in energy until the transition state, and then rising to form the σ^*_{CC} orbitals of benzene.

What occurs in this case, as well as in the Diels-Alder reaction and the $3H_2$ bond-switching, is that the initial nuclear motions which occur in the reaction begin to disrupt bonds, destabilizing the vacant orbitals. In language akin to that used in the Evans-Polanyi discussion of atom transfer (3), the beginning of the reaction can be considered to be essentially the pure distortion of reactants. Reactant bonds are being broken, and these distortions require energy. Just past the transition state, the atoms are on a distorted product surface. At the transition state, there is a transition from the distorted reactant surface to the distorted product surface. In terms of orbitals, the initial motion involves an intended correlation of one or more bonding orbitals with antibonding orbitals in the product. This is avoided at the transition state, where substantial mixing of the filled and vacant orbitals of reactants occur.

In the acetylene trimerization, at early stages of the reaction, there is no appreciable stabilizing interaction of the HOMOs and LUMOs of acetylenes because of the very large HOMO-LUMO gap. The acetylenes must distort appreciably before the HOMO-LUMO gap decreases enough for stabilizing interactions between these orbitals to overcome the destabilizing distortions. Once the charge-transfer, or delocalization (HOMO-LUMO), interaction has become sufficient to transfer appreciable electron density into product bonding regions, further distortions are stabilizing, leading to the equilibrium geometry of benzene. Calculations on the distorted acetylenes with the minimal STO-3G basis indicate that 60 kcal/mol of the 80 kcal/mol barrier is due to the energy required to distort the three acetylenes to the "transition state" geometries. The remaining 20 kcal/mol must arise mainly from the excess of exchange repulsion over charge-transfer stabilization until the "avoided HOMO-LUMO crossing" occurs.

This example demonstrates some general principles of the relationship between molecular distortions and organic molecules toward electrophiles. The activation energy of a polymolecular reaction is the point where stabilizing interactions, arising from mixing of filled and vacant orbitals, overcome distortion intermolecular repulsive energies. The energy of distortion is critical, but the influence of distortion upon orbital energies and the facilitation of stabiliizing interactions is important as well.

I should mention here tha Bader (18), Salem (19), and Pearson (20) have said similar things before in discusssions of reactivity in the configuration interaction formalism (21).

Alkyne vs. alkene reactivity

Let us turn to a consideration of the influence of distortion energies and orbital behavior upon the relative reactivities of different types of molecules. It is well-known to organic chemists that acetylenes are much more reactive toward nucleophiles than are alkenes, whereas the opposite is true of reactivities of these two types of molecules toward electrophiles. As examples, Br_2 in CCl_2 is not decolorized by simple alkynes, but can be used to titrate alkenes. Phenylacetylene adds 0.1N methoxide in methanol at 124°C with a half-life of 6 hours, while only 2% addition to styrene occurs after 10-20 days in 0.6N methoxide (22). Of further interest, in reactions of nucleophiles with alkylacetylenes and arylacetylenes, anti addition stereochemistry is invariably followed (23).

Simple frontier orbital arguments are able to account for the greater reactivity of electrophiles such as Br_2 with alkenes: as shown in Figure 2, the HOMO of ethylene is higher in energy than that of acetylene, as measured by ionization potentials of these two molecules (24). Since the HOMO of ethylene is closer in energy to the LUMO of an electrophile, more charge transfer stabilization is conferred upon the transition state than with acetylene.

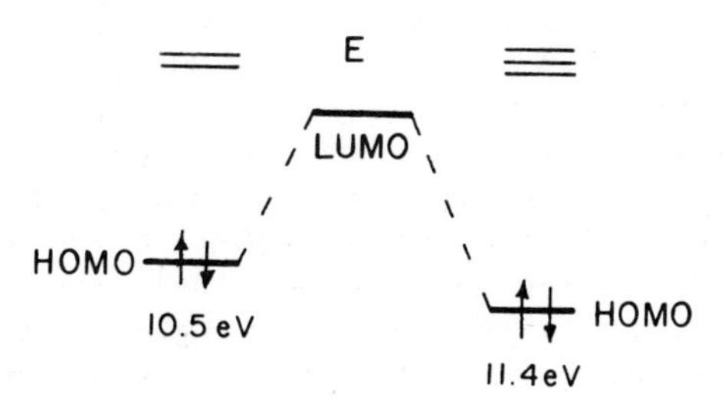

FIG. 2. HOMOs of acetylene and ethylene, and LUMO of an electrophile.

Frontier molecular orbital theory cannot, however, successfully explain the greater reactivity of nucleophiles toward acetylenes than toward ethylenes. As shown in Figure 3, the acetylene LUMO is higher in energy than that of ethylene. The numbers in Figure 3 are negatives of electron affinities, measured by electron transmission spectroscopy by Burrow and Jordan (25). Acetylene has a higher LUMO energy than ethylene, in accord with the shorter CC bond length in the former, and concomitant greater antibonding in the π^* orbital. The HOMO of a nucleophile should interact more strongly with the LUMO of ethylene than with that of acetylene contrary to the greater reactivity of the latter.

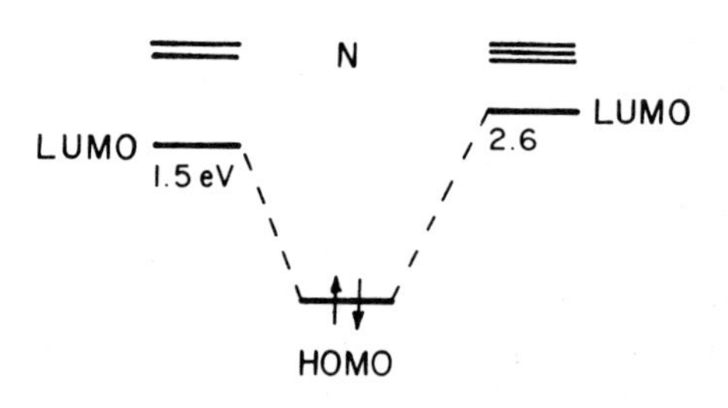

FIG. 3. LUMOs of acetylene and ethylene, and HOMO of a nucleophile.

Indeed, a 4-31G calculation on the interaction of hydride, a model nucleophile, with undistorted ethylene or acetylene at 2Å separation predicts that the interaction of hydride with acetylene is 5 kcal/mol less repulsive (+30 kcal/mol) than the interaction with ethylene (+25 kcal/mol), due to smaller charge-transfer stabilization with acetylene than with ethylene (24).

However, the situation changes dramatically in the transition states. Figure 4 shows the transitions states obtained from 4-31G calculations carried out at L.S.U. by Robert Strozier (24). Independent studies on hydride plus acetylene by Dykstra, Arduengo, and Fukunaga give a similar transition state (26). Using the 4-31G basis set, activation energies of only 3 kcal/mol are obtained for the two reactions. Thus, the bending apparent in the acetylene transition states shown here causes the difference in interaction energy of hydride with ethylene and acetylene to disappear. Note especially the much greater bending of acetylene than ethylene, and the fact that both bend trans in the transition state. This accounts for the preferred stereochemistry of addition.

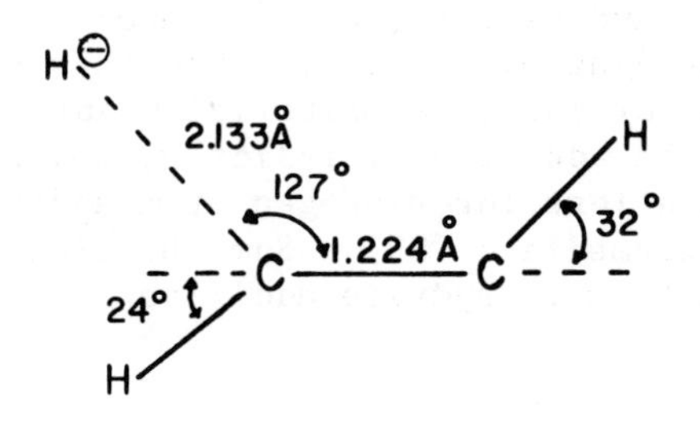

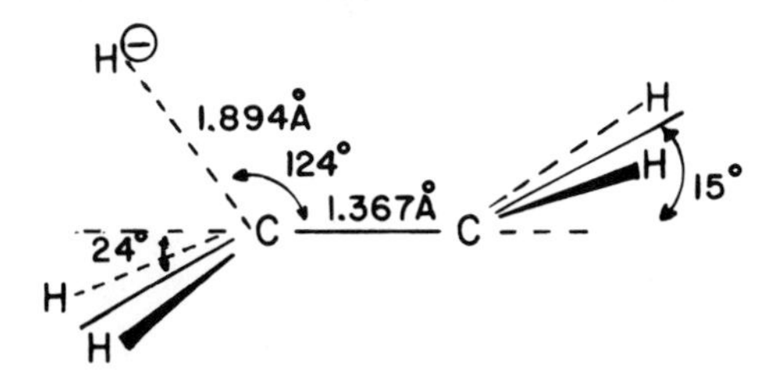

FIG. 4. Transition states for hydride attack on acetylene and ethylene.

It is known that out-of-linearity bending of acetylene is easier than out-of-planarity bending of ethylene, which accounts, in part, for the greater bending of acetylene than ethylene in these transition states. However, there is also a big difference in the behavior of the LUMOs of these species upon bending.

Although the LUMO of acetylene is higher in energy than that of ethylene in the isolated molecules, bending of these molecules to the geometries they have in the transition states drops both LUMOs to the same energy. The increase in charge transfer upon bending acetylene nearly exactly compensates for the distortion energy of 11 kcal/mol for acetylene.

It is also notable that the HOMOs of ethylene and acetylene do not change in energy significantly upon bending, so that attack of electrophiles is accompanied by no driving force for bending. Thus, bending distortions are important upon attack by nucleophiles, but not by electrophiles. Our hydride addition transition state shows drastic bending of acetylene, because the LUMO drops and the consequent increase in charge transfer more than compensates for the energy of distortion. For protonation of acetylene, calculated by

Pople et al. (27) and others (28), (Figure 5) there is essentially no bending distortion, since this requires energy, but does not increase stabilizing interactions with electrophiles. The addition of radicals, modelled by Nagase and Kern's calculations for hydrogen atom addition to acetylene, is intermediate (29). Some bending occurs, but not nearly as much as in hydride addition.

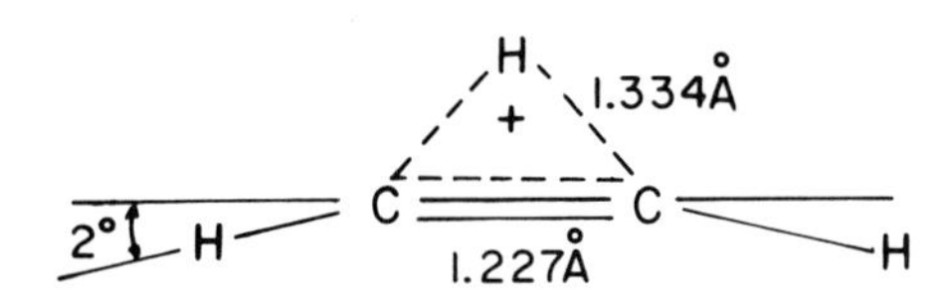

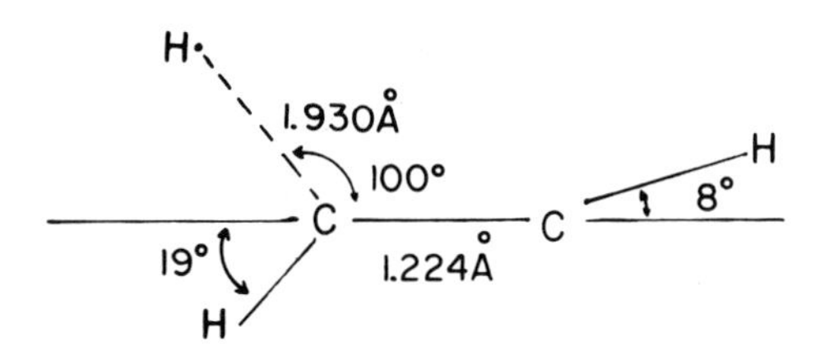

FIG. 5. Geometries of protonated acetylene (27) and the transition state for attack of the hydrogen atom on acetylene (29).

A comparison of these transition states reveals a general principle: in reactions of hydrocarbons with nucleophiles, significant distortions must occur in order to drop the hydrocarbon LUMO sufficiently for stabilizing interactions with the nucleophile HOMO. These distortions simultaneously relieve closed-shell repulsion between the filled orbitals of the two molecules. Little or no distortion is required, however, during the reaction of a hydrocarbon with electrophiles. The distortion occuring upon reactions of radicals with hydrocarbons will depend upon the nature of the radical: hydrocarbons will be substantially distorted in transition states of reactions with nucleophilic radicals, while hydrocarbons will be essentially undistorted in transition states for reactions with electrophilic radicals. This will be discussed in more detail later.

Transition states of singlet carbene cycloadditions

The difference in distortion of an alkene or acetylene upon attack by nucleophiles or electrophiles show up in a novel way in carbene cycloadditions. Nelson Rondan has recently investigated singlet carbene cycloadditions. These are representative of reactions in which one atom is simultaneously nucleophilic and electrophilic in perpendicular directions, so that some molecular gymnastics have to be present in the transition state in order that both of these interaction can occur. Other cheletropic reactions, such as reactions of allenes, ketenes, and other cumulenes, or of singlet oxygen may also be examples of this type.

Figure 6 shows the frontier MOs of a singlet carbene, which qualitatively indicate the reactivity characteristics of such species. The HOMO is an in-plane hybrid MO, and any nucleophilic reactions of carbenes should be dominated by this orbital. The LUMO is a vacant p orbital, which should dominate electrophilic reactivity of carbenes. Methylene and

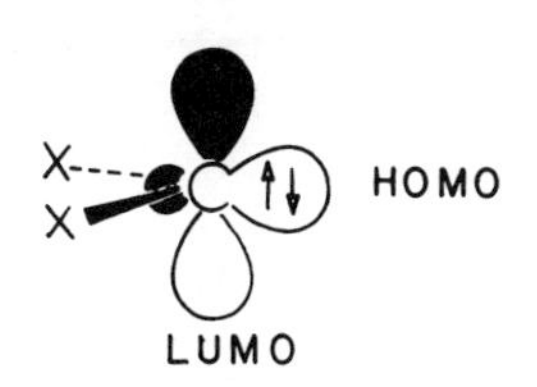

FIG. 6. The frontier orbitals of a carbene.

many substituted carbenes are electrophilic, while some highly donor-substituted carbenes can be nucleophiles. A large number of theoretical studies, both qualitative (30) and numerical (31) have been carried out on carbene reactions. For hydrogen abstractions by singlet and triplet methylene, rather accurate potential surfaces are available as a result of work by Schaefer and others (32). Hoffmann gave the first detailed theoretical description of the non-least-motion cycloaddition of singlet methylene to alkenes (33). His conclusions were essentially identical to those deduced empirically earlier by Skell and Doering (34), and indicate that the initial approach of methylene to alkenes is electrophilic, maximizing overlap of the carbene LUMO with the alkene HOMO, and only later does the carbene rotate to form the second cyclopropane bond. Last year, Kutzelnigg and Zurawski reported an *ab initio* surface for the addition of singlet methylene to ethylene (35). As in some earlier empirical treatments, no activation energy was found, and the initial approach of the LUMO of methylene toward the ethylene orbital is followed by methylene rotation to form cyclopropane. We decided that it would be valuable to study the cycloadditions of substituted carbenes which are known to have non-zero activation energies. Although a surface for a reaction without an activation energy gives some qualitative idea of the preferred pathway of approach, first π-complex-like, and later product-like, no knowledge about the position of the transition state, early or late, is obtained. However, in a reaction with an activation energy, the transition state can be determined, and from this the electrophilic and nucleophilic characteristics of the transition state, and the substituent effects on this transition state, can be explained.

Ab initio calculations using the STO-3G minimal basis set (36) were used to locate the transition state, and then computations on various points on the surface were recalculated using the 4-31G extended basis set (37). Four carbenes were chosen for initial studies: dichlorocarbene, a very reactive electrophile, difluorocarbene, a more selective but still electrophilic species, fluorohydroxycarbene, which provides a model for an ambiphilic species, and dihydroxycarbene, an unknown carbene, but a reasonable model for nucleophilic carbenes such as dimethoxycarbene. Table I summarizes energetic results of these calculations.

The energies of these reactions and positions of the transition states are directly related to the stabilities of the carbenes. That is, the more the substituent stabilizes the carbene, the greater the activation energy for reaction. Figure 7 shows the transition states for the electrophilic and nucleophilic extremes. Table II summarizes the geometries of the four transition states. For dichlorocarbene, the transition state is remarkably similar to that proposed by Skell and Doering over 25 years ago (34). The distance, *d*, from the carbene carbon to the double bond decreases from 2.12Å for CCl_2 to 1.83Å for $C(OH)_2$, so that there is a very slight shift to a later transition state along this series as the carbene becomes more stable. However, the very small shift in transition state lateness as measured by the reaction coordinate, *d*, represents rather

Table I

Summary of ab initio calculations on singlet carbene cycloadditions

	STO-3G		4-31G	
Carbene	ΔE_{rxn} [a]	ΔE^* [b]	ΔE_{rxn} [a]	ΔE^* [b]
CH_2	-138	0	-96	0
CCl_2	-97	5	-70	8
CF_2	-64	24	-46	27
CFOH	-54	29	-31	37
$C(OH)_2$	-47	33	-18	45

[a] ΔE_{rxn} = E(products) - E(reactants).

[b] ΔE^* = E(transition state) - E(reactants).

Table II

Geometries of transition states for singlet carbene cycloadditions[a]

	Carbene			
Geometrical Parameter	CCl_2	CF_2	CFOH	$C(OH)_2$
d	2.01	1.90	1.90	1.85
$r(C_1C_2)$	2.31	2.22	2.35	2.23
$r(C_2C_3)$	1.96	1.80	1.78	1.71
$r(C_1C_3)$	1.36	1.38	1.38	1.43
α	13.8	22.2	25.0	29.7
β	1.9	1.3	2.8	4.2
γ	109.7	114.2	117.6	125.4
δ	105.8	108.8	110.6	112.8
ζ	38	43	48	58

[a] Distances in Å, angles in degrees.

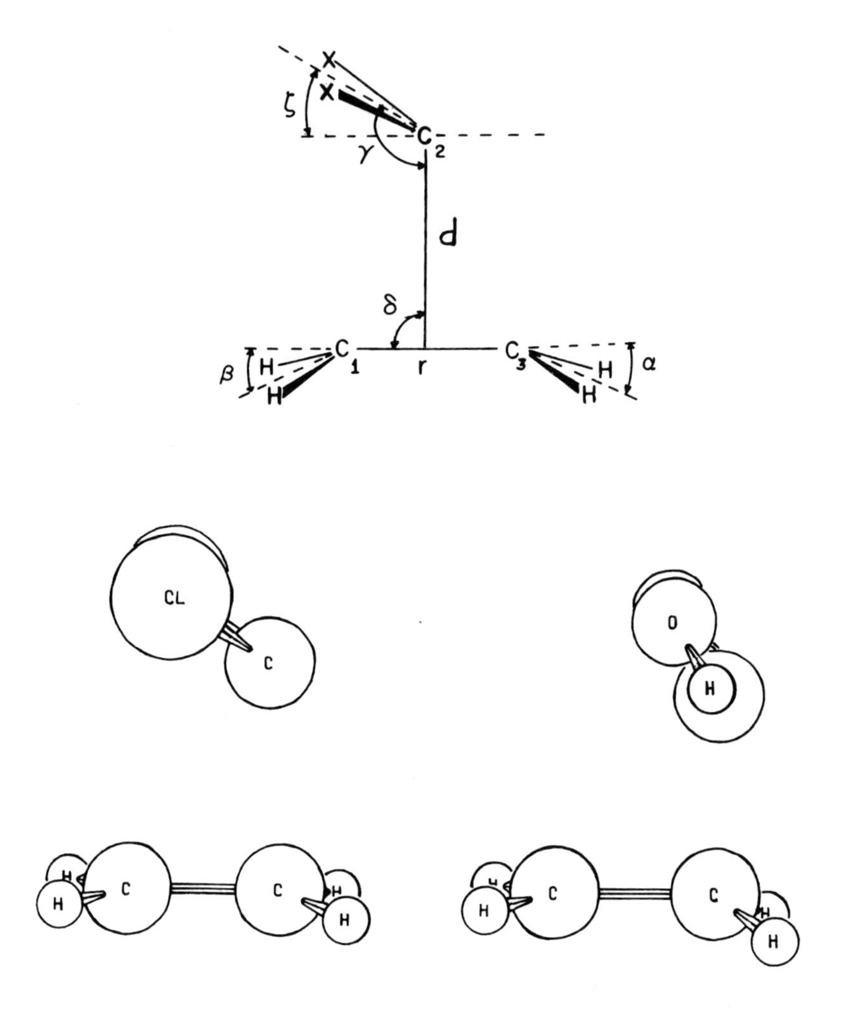

FIG. 7. Top: geometrical variables optimized as a function of d, the reaction coordinate. Bottom: transition states for cycloadditions of CCl_2 and $C(OH)_2$ to ethylene.

profound changes in the electrophilic or nucleophilic character of the transition state. Even for CCl_2, however, the transition state apparently involves only slightly more electrophilic carbene character than nucleophilic. That is, the extreme of electrophilic interactions would involve the carbene and ethylene in parallel planes, and the parameter, ζ, would be 0°. Dichlorocarbene is tilted away from parallel planes by 38° in this transition state. Qualitatively, a value of 45° would imply equal electrophilic and nucleophilic character. Bond formation is very unsymmetrical in all of these transition states, in contrast to the conclusions made by Hoffmann from EHT calculations on CF_2, and Dewar by MINDO/2 on CH_2.

The very small out-of-plane distortion of the methylene group reflects the fact that the weaker bond in the transition state is the result of a primarily electrophilic interaction between the carbene and ethylene terminus. That is, there is little out-of-plane bending (β) of the hydrogens at this terminus. However, the opposite terminus is distorted appreciably. The angles, α, are 20-30° in these transition states, and become only 30° in the fully formed cyclopropane product. Clearly the interaction is electrophilic at one terminus (C-1), but nucleophilic at the other (C-3). There is net charge transfer of electrons from the ethylene moiety to the dichlorocarbene or difluorocarbene moieties, as reflected by charges in various atoms, but with dihydroxycarbene, charge transfer is in the opposite direction. All of these things reflect the electrophilic character of CCl_2 and CF_2, but I would like to emphasize that both electrophilic and nucleophilic interactions are important in all of these transition states. If a hydroxy group replaces one of the fluorines, the carbene is stabilized, and the LUMO is increased in energy. This species becomes less electrophilic and more nucleophilic. This is reflected in the later transition state, and the increased rotation of the carbene away from the parallel planes geometry. The angle, ζ, is 48°, and the out-of-plane distortion at C-2 is correspondingly increased. One bond is stretched 3.4 times more than the other, and there is little charge transfer (0.1 electrons) from carbene to ethylene. Attachment of a methyl in place of one of the hydrogens in the ethylene causes essentially no change in the activation energy of the reaction. Finally, when both fluorines are replaced by hydroxyl groups, a very unsymmetrical transition states occurs. In this transition state, one bond is stretched 4.1 times more than the other, and one methylene is distorted essentially to the geometry it has in the product. Now, the charge transfer of 0.06 electrons in the direction of ethylene, and attachment of a methyl in place of one of the hydrogens destabilizes this transition state. This carbene is clearly a nucleophile in the transition state.

Once again, the pattern described earlier for nucleophilic and electrophilic attack on alkenes and acetylenes is manifested here. The interaction of nucleophiles with closed-shell molecules such as ethylene or acetylene is accompanied by large distortions, whereas the interactions of electrophiles are accompanied by essentially no distortions.

Competitions between oxyradical additions to alkenes and hydrogen abstraction

Finally, we turn to a subject more in keeping with the degree of radical character of this symposium. As described in our initial discussion of nucleophilic attack on acetylenes, frontier orbital theory is generally not capable of correctly predicting the relative reactivities of two different types of functional groups. I described earlier how molecular distortions, and accompanying orbital energy changes, must be considered explicitly in order to understand the differences in reactivities of alkenes and acetylenes toward nucleophiles, and here I would like to describe similar considerations in comparing reactivities of CH bonds and π bonds towards various types of reagents.

A particularly interesting case of such a competition arises in reactions of radicals with alkylethylenes and alkylaromatics. Figure 8 gives an interesting comparison between heats of hydrogen abstraction and addition by two different oxyradicals. With propene or other alkenes having allylic hydrogens, either hydrogen abstraction or radical addition may occur. The heats of reaction shown in Figure 8 indicate little preference for either type of reaction. In spite of this, the *tert*-butoxy radical is a notorious hydrogen abstractor. Walling and Thaler reported that abstraction is 30 times faster than addition, even though the latter is more exothermic (38). With the hydroxy radical, Hammond and coworkers reported that only addition occurs, in spite of the fact that both reactions are equally favorable energetically (39).

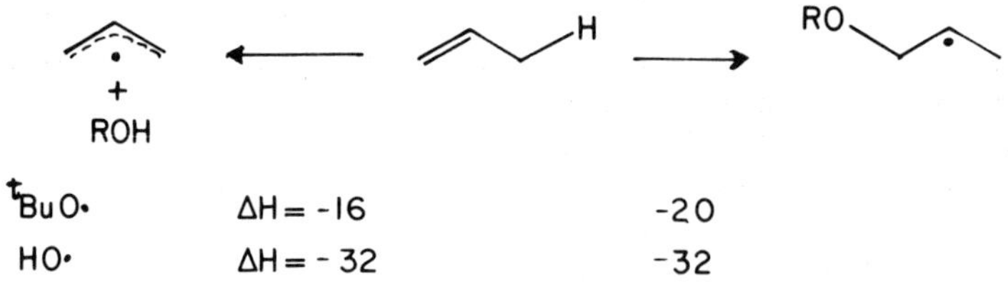

H

RO

\+
ROH

^{t}BuO· $\Delta H = -16$ -20
HO· $\Delta H = -32$ -32

FIG. 8. Energetics of radical addition and abstraction reactions.

Bertrand and Surzur recently showed that these two cases are a part of a more general pattern (40). Upon progressing from the very electron-rich oxygen radical anion to the relatively electron-deficient perfluoroalkyl and hydroxy

Table III

Relative reactivities of oxyradicals toward abstraction and addition, and ionization potentials and electron affinities of the radicals

Radical	k_{abstn}/k_{addn}	*IP(eV)*	*EA(eV)*
$O^{\overline{\cdot}}$	*v. large*	1.47	-9.1
t-BuO·	~30 @ 40°	~9.2	1.87
s-BuO·	1.2 @ -50°	~9.2	~1.9
i-PrO·	15 @ 60°	9.20	~1.73
n-BuO·	0.6 @ -50°	9.22	1.90
n-PrO·	12 @ 60°	9.2	1.87
MeO·	>1	9.3	1.59
ROO·	1	11.5	3.04
$PhCO_2\cdot$	0.2	>10	~3.6
$F_3CO\cdot$	*small*	>10	~2.5
$(CF_3)_3CO\cdot$	*v. small*	>10	~3.1
HO·	<0.1	13.2	2.1
O·	0	13.62	1.48
$H_3C\cdot$	0.1-6.1	9.48	0.1
Cl·	0.1-1.1	13.0	3.6
$Cl_3C\cdot$	0.01-0.1	8.78	1.4-2.1
$F_3C\cdot$	0.01	10.1	2.0
RS·(R=alkyl)	0	11.1	2.32
$NH_2\cdot$	0	11.4	1.2

radicals, there is an experimental trend toward a decrease in hydrogen abstraction relative to addition, as shown in Table III (14).

Bertrand and Surzur suggested that when the radical SOMO is higher in energy than the propene HOMO, as it is for radicals listed above the peroxy radical in Table III, then mixing of these two orbitals occurs, generating two new orbitals as shown in Figure 9. This stabilizing three electron interaction leads predominantly to abstraction, although this picture does not indicate why. For radicals with IPs greater than that of propene, such as those below the peroxy radical in Table III, significant or exclusive addition occurs. Bertrand and Surzur suggested that this situation, which corresponds to electron-transfer from propene to the radical, leads to addition. Again it is not clear from the diagram on the right of Figure 9 why this should be the case.

These authors erred in assuming that the IP of a radical measures the electron-accepting ability of the radical. In

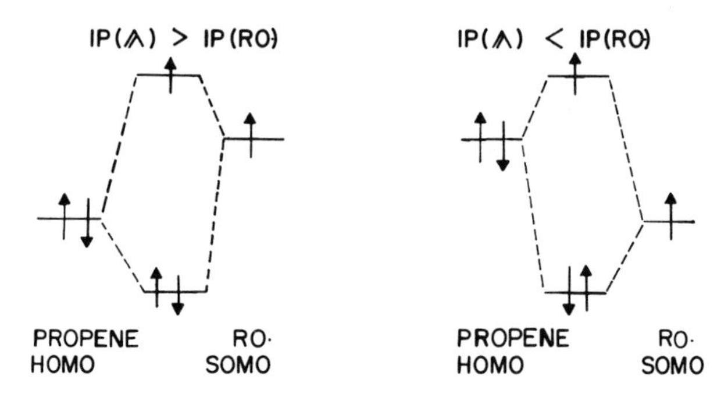

FIG. 9. Orbital interaction scheme suggested by Bertrand and Surzur (40) for nucleophilic (left) and electrophilic (right) radicals.

fact, electron-transfer is not expected in the cases summarized on the right of Figure 9. The electron affinities of radicals are the appropriate measures of electron-accepting abilities, and Table III shows that these are sufficiently low to preclude electron transfer with propene (IP = 9.9eV). In spite of this theoretical flaw in the Bertrand and Surzur argument, the empirical generalization about the trend in abstraction as compared to addition made by these authors is extremely valuable, and was the point of divergence for our studies.

The data given in Table III for abstraction and addition are quite approximate due to the great difficulties in measurement of such data. Nevertheless, there is a regular increase in addition relative to abstraction upon proceeding down the table, and this is the same general order as that of IP and EA increase. From the viewpoint of frontier molecular orbital theory, the radicals progress from nucleophilic character near the top of the table to electrophilic near the bottom. Also of note is the fact that there is a qualitative parallel of the last two columns in this table, but the difference between the IP and EA, which corresponds to the electron repulsion between two electrons in the SOMO varies from 12eV for localized radicals such as $O^{\overline{\cdot}}$ to about 6eV for delocalized radicals such as peroxy radicals.

Before attempting to explain the reactivity data in Table III, I must caution that the trend shown here is not generalizable to radicals other than oxyradicals. Our discussion today is confined, therefore, to trends for

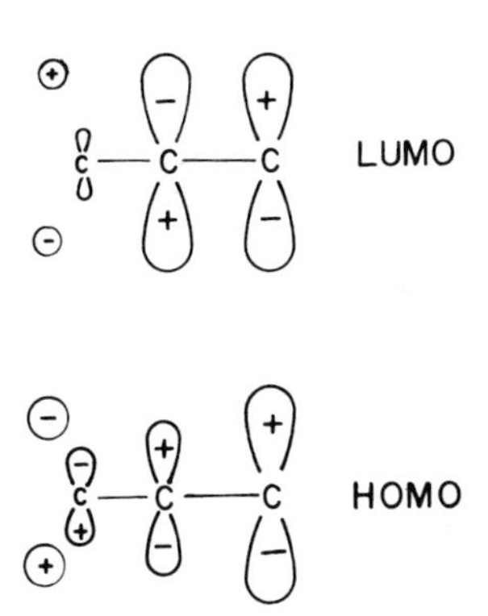

FIG. 10. The frontier molecular orbitals of propene.

oxyradicals. The HOMO and LUMO of propene are shown in Figure 10. Both are essentially pure π orbitals, although there is some hyperconjugative mixing of the π orbitals with the methyl σ_{CH_3} group orbitals. Based on the shapes of these frontier molecular orbitals, both nucleophiles and electrophiles, as well as nucleophilic and electrophilic radicals, should attack the bond. That is, for radicals, maximum overlap of the radical SOMO with either the HOMO or LUMO of propene should be greatest at C-1, as long as propene is undistorted. Similarly, nucleophiles and electrophiles can attain maximum stabilizing interactions upon overlap with the π orbital centers, in spite of the fact that nucleophiles (bases) generally remove allylic protons, rather than add to unactivated double bonds.

However, the distortions which occur along the reaction path for addition or abstraction change this picture considerably. I will first describe the changes that occur in these orbitals in a purely empirical way, and then describe calculations which support these empirical deductions.

Figure 11 shows qualitatively the behavior of the propene IPs and EAs upon CH stretching. Upon complete CH dissociation, the allyl radical and hydrogen atom are formed. There are four numbers on the right, because the two singly occupied MOs of the allyl radical of propene each have different IPs and EAs. As stretching of the CH bond occurs, the LUMO of propene can be considered to evolve gradually into the SOMO of the hydrogen atom, which has a rather high EA, while the HOMO can be considered to evolve into the allyl radical SOMO. Stretching the CH bond should thus facilitate the interaction of the propene LUMO with the SOMO of an attacking radical due to the dramatic LUMO drop and the concentration of this orbital on H; that is, there can be substantial charge-transfer stabilization involving the

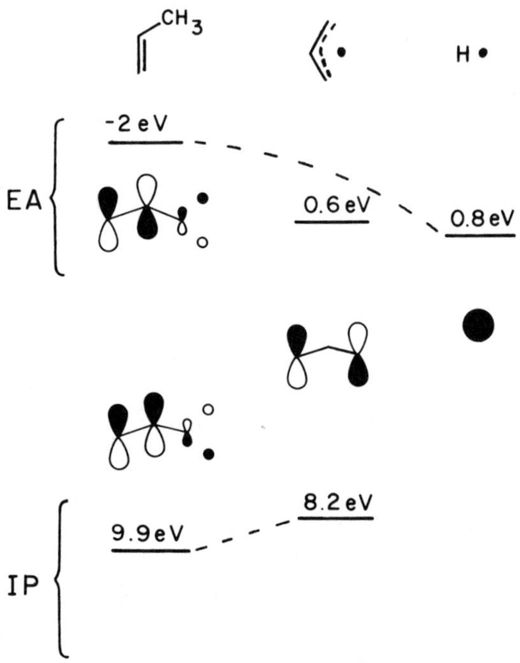

FIG. 11. *Ionization potentials and electron affinities of propene, allyl radical, and hydrogen atom.*

attacking radical SOMO and the propene LUMO during abstraction if the radical has a sufficiently high-lying LUMO. Stretching the CH bond does not facilitate interaction with an electrophilic hydrogen-abstractor. That is, although stretching raises the HOMO, little HOMO density drifts onto H during this distortion. An electrophilic radical SOMO will interact with the propene HOMO at the π-bonded carbons without the need for distortion.

Calculations summarized in Figure 12 reinforce this interpretation. These were carried out by Lynn Garcia using the 4-31G basis set. We have compared pyramidalization of one alkene terminus, which occurs upon addition, to CH stretching, which occurs upon hydrogen abstraction. Starting from equilibrium propene in the center of the figure, pyramidalization of one terminus has essentially no effect on the HOMO, and drops the LUMO by less than leV. On the other hand, CH stretching causes a precipitous LUMO drop. Even for

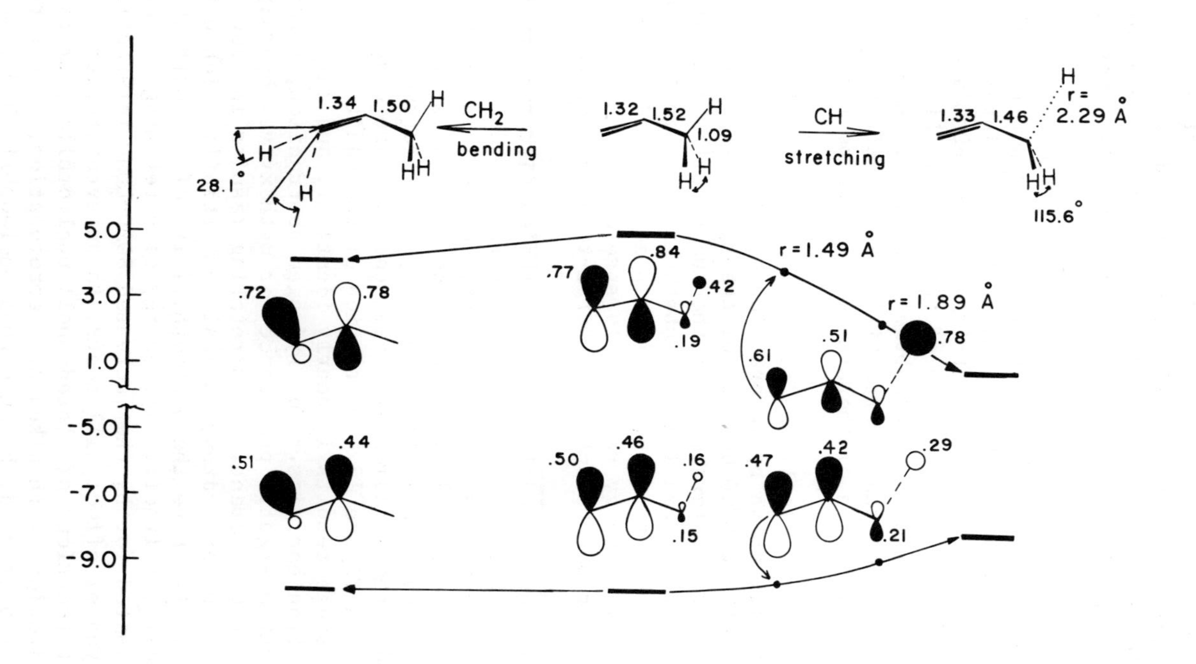

FIG. 12. The behavior of the propene frontier molecular orbitals upon pyramidalization (left) and CH stretching (right). Calculations were carried out with the extended 4-31G basis set.

rather minor stretching, the LUMO has already become concentrated on hydrogen. The origin of this effect is that as the CH bond stretches, the CH orbital is raised and the CH σ^* orbital is lowered in energy as both bonding and antibonding decrease. Even with minor CH stretchings, the π^* orbital rapidly evolves to a σ^*_{CH} orbital, and then to a hydrogen atom 1s orbital, because the π^* and σ^* orbitals are rather close in energy, and the σ^*_{CH} rapidly drops below the π^*.

Based on these arguments, we feel that electrophilic radicals, those with high EAs, preferentially add to alkenes because there can be effective charge-transfer interaction at the double bond between the high-lying alkene HOMO and the radical SOMO without distortion of the alkene. Like protonation which we described briefly earlier, the transition state of the electrophilic radical attack occurs with little distortion of the alkene, and at the site of highest HOMO density.

Nucleophilic radicals can obtain very little charge-transfer stabilization by interaction of the SOMO with the undistorted alkene LUMO because of the high energy of the latter. However, interaction of the nucleophilic radical at an allylic hydrogen can lead to substantial charge transfer stabilization when this interaction is accompanied by CH stretching. The energy required to stretch the CH bond is compensated for by the enhanced charge-transfer interaction that results.

Computations on model transition states give support to this view. We selected the transition state models shown in Figure 13 to test these ideas. First, computations with the 4-31G basis set on radicals interacting with undistorted propene at C and H indicate essentially no difference between hydroxy and methoxy radicals. These energies are given in parentheses in Figure 13. Distortion of the addition transition state results in nearly identical, and large, destabilization energies. For hydrogen abstraction, there is a very slight energetic preference for methoxy vs. hydroxy with undistorted propene. CH stretching results in some destabilization for hydroxy radical, but stabilization with methoxy radical. Here the nucleophilicity of the methoxy radical manifests itself. The energy of stretching is more than compensated for by the increased SOMO-LUMO interaction. The same type of effect will occur, to an even greater extent, for reactions of closed-shell nucleophiles, or bases, such as hydroxide, with alkenes. Deprotonation, rather than addition, is preferred, because good nucleophile HOMO, propene LUMO interaction can occur at hydrogen, when CH stretching occurs. Electrophilic radicals, or electrophiles, interact directly at the double bond without distortion.

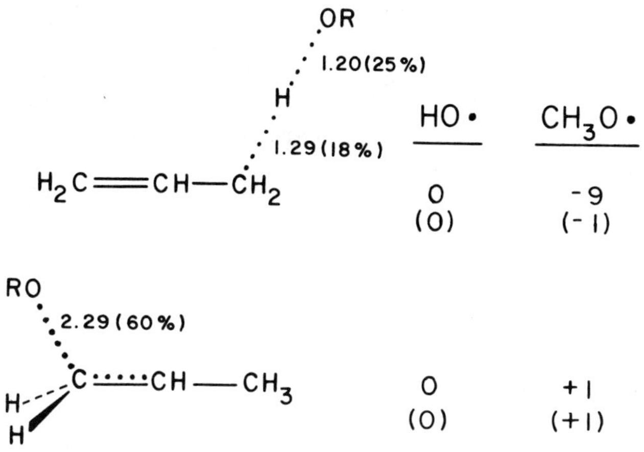

	HO•	CH_3O•
$H_2C{=}CH{-}CH_2 \cdots H \cdots OR$ (1.29 (18%); 1.20 (25%))	0 (0)	-9 (-1)
$RO \cdots CH_2{=}CH{-}CH_3$ (2.29 (60%))	0 (0)	+1 (+1)

FIG. 13. Transition state models for hydrogen abstraction and addition by HO• and MeO•. The numbers are all relative to the energy of interaction of HO•. Numbers in parentheses refer to interaction energies in the absence of distortion, while those not in parentheses are relative interaction energies after CH stretching or CH_2 pyramidalization.

We expect that similar trends should be observable for other series of radicals, as long as the odd electron is more or less localized on the same type of atom. Unfortunately, this generalization does not hold for comparison of different types of radicals, such as Cl•, HO•, and •CH_3. Nevertheless, we hypothesize that one factor which is important in determining the relative rate of hydrogen abstraction and addition is the radical nucleophilicity or electrophilicity.

Conclusion

I would like to conclude my part of this symposium with a risky generalization, but one which begins to emerge from several of the results that I have reported today. Neutral hydrocarbons, and most likely closed-shell molecules in general, react with electrophiles with relatively minor distortions in the transition state. Frontier orbital theory rather accurately predicts the facility and site of reactions of organic molecules with electrophiles. On the other hand, hydrocarbons and neutral closed-shell molecules, in general, must substantially distort in order to achieve stabilizing charge-transfer interactions with nucleophiles; unadorned frontier orbital is insufficient to predict the rates or sites of reaction in such cases. Instead, reactions of nucleophiles occur at sites which develop the largest vacancy when relatively easy distortions causing substantial LUMO energy drops occur. Although this generalization is somewhat vague at the moment, we hope that its refinement will lead to a deeper and more quantitative understanding of organic reactivity.

References

1. Houk, K.N. *Accounts Chem. Research 8, 361 (1975).*

2. Fleming, I., Frontier Orbitals and Organic Chemical Reactions, John Wiley and Sons, Inc., 1976.

3. Evans, M.G., and Polanyi, M. *Trans. Farad. Soc. 32, 1333 (1936); 34, 11 (1938).*

4. Fukui, K., Theory of Orientation and Selection, Springer-Verlag, Berlin, 1975.

5. Fukui, K., Kata, H., and Yonezawa, T. *Bull. Chem. Soc. Japan 34, 442, 1111 (1961).*

6. Bartels, H., Eichel, W., Riemenschneider, K., and Boldt, P. *J. Am. Chem. Soc. 100, 7740 (1978).*

7. Riemenschneider, K., Bartels, H., Eichel, W., and Boldt, P. *Tetrahedron Lett., 189 (1979).*

8. Shinohara, H., Imamura, A., Masuda, T., and Kinda, M. *Bull. Chem. Soc. Japan 51, 98 (1978).*

9. Berthelot, M. *Ann. Chim. Phys. 9, 445, 469, 481 (1866).*

10. Back, H.M. *Can. J. Chem. 49, 2199 (1971).*

11. Barkovich, A.J., and Vollhardt, K.P.C. *J. Am. Chem. Soc. 98, 2667 (1976);* Vollhardt, K.P.C. *Accounts Chem. Res. 10, 1 (1977);* Barkovich, A.J., Strauss, E.S., and Vollhardt, K.P.C. *J. Am. Chem. Soc. 99, 8321 (1977).*

12. Nutakul, W., Thummel, R.P., and Taggert, A.D. *J. Am. Chem. Soc. 101, 770 (1979).*

13. Houk, K.N., Strozier, R.W., Santiago, C., Gandour, R.W., and Vollhardt, K.P.C. *J. Am. Chem. Soc. 101, 5183 (1979).*

14. Houk, K.N., Gandour, R.W., Strozier, R.W., Rondan, N.G., and Paquette, L.A. *J. Am. Chem. Soc. 101, 6797 (1979).*

15. Houk, K.N., Gandour, R.W., Strozier, R.W., Dewar, M.J.S., and Ritchie, J.W., in preparation.

16. Jorgenson, W.L., unpublished results referred to in reference 17.

17. Townshend, R.E., Ramunni, G., Segal, G., Hehre, W.J., and Salem, L. *J. Am. Chem. Soc. 98, 2190 (1976).*

18. Bader, R.F.W. *Mol. Phys. 3, 137 (1960); Can. J. Chem. 40, 1164 (1962).*

19. Salem, L. *Chem. Phys. Lett. 3, 99 (1969).*

20. Pearson, R.G., Symmetry Rules for Chemical Reactions, Wiley-Interscience, New York, 1976.

21. Houk, K.N. in Marchand, A.P., and Lehr, R.E., Pericyclic Reactions, Academic Press, New York, 1978, Vol. 2.

22. Miller, S.I. *J. Org. Chem. 21, 247 (1956).*

23. Dickstein, J.L., and Miller, S.I. in Patai, S., Ed., The Chemistry of the Carbon-Carbon Triple Bond, Wiley, New York, 1978, Pt. 2.

24. Strozier, R.W., Caramella, P., and Houk, K.N. *J. Am. Chem. Soc. 101, 1340 (1979).*

25. Jordan, K.D., and Burrow, P.D. *Accounts Chem. Res. 11, 341 (1978);* Burrow, P.D., and Jordan, K.D. *Chem. Phys. Lett. 36, 594 (1975).*

26. Dykstra, C.E., Arduengo, A.J., and Fukunaga, T. *J. Am. Chem. Soc. 100, 6007 (1978).*

27. Lathan, W.A., Curtiss, L.A., Hehre, W.J., Lisle, J.B., and Pople, J.A. *Prog. Phys. Org. Chem. 11, 175 (1974).*

28. Weber, J., and McLean, A.D. *J. Am. Chem. Soc. 98, 875 (1976).*

29. Nagase, S., and Kern, C.W. *J. Am. Chem. Soc. 101, 2544 (1979).*

30. (a) Woodward, R.B., and Hoffmann, R. *Angew. Chemie Int. Ed. Engl. 8, 781 (1969);* (b) Fueno, T., Nagase, S., Tatsumi, K., and Yamaguchi, K. *Theoret. Chim. Acta, 26, 43 (1972);* Fujimoto, H., Yamabe, S., and Fukui, K. *Bull. Chem. Soc. Japan 45, 2424 (1972);* (c) Goddard, W.A. III *J. Am. Chem. Soc. 94, 793 (1972);* (d) Klopman, G. in Chemical Reactivity and Reaction Paths, Klopman, G., Ed., Wiley-Interscience, New York, 1974, pp. 135-138; (e) Zimmerman, H.E. *Accounts Chem. Res. 5, 393 (1972);*

Pericyclic Reactions, Marchand, A.P., and Lehr, R.E., Ed., Academic Press, New York, 1977, Vol. I, pp. 93-97; (f) Jones, W.M., and Brinker, U.H. in Pericyclic Reactions, Marchand, A.P., and Lehr, R.E., Ed., Academic Press, New York, 1977, Vol. I, pp. 110-118.

31. Bodor, N., Dewar, M.J.S., and Wasson, J.S. *J. Am. Chem. Soc. 94, 9095 (1972).*

32. Bauschlicher, C.W., Jr., Haber, K., Schaeffer, H.F. III, and Bender, C.F. *J. Am. Chem. Soc. 99, 3610 (1977) and references therein.*

33. (a) Hoffmann, R. *J. Am. Chem. Soc. 90, 1475 (1968);* (b) Hoffmann, R., Hayes, D.M., and Skell, P.S. *J. Phys. Chem. 76, 664 (1972).*

34. (a) Skell, P.S., and Garner, A.Y. *J. Am. Chem. Soc. 78, 5430 (1956);* (b) Doering, W. von E., and Henderson, W.A., Jr. *J. Am. Chem. Soc. 80, 5274 (1958);* (c) Skell, P.S., and Cholod, M.S. *J. Am. Chem. Soc. 91, 7131 (1969).*

35. Zurawski, B., and Kutzelnigg, W. *J. Am. Chem. Soc. 100, 2654 (1978).*

36. Hehre, W.J., Stewart, R.F., and Pople, J.A. *J. Chem. Phys. 51, 2657 (1969).*

37. Ditchfield, R., Hehre, W.J., and Pople, J.A. *J. Chem. Phys. 54, 724 (1971).*

38. Walling, C., and Thaler, W. *J. Am. Chem. Soc. 38, 3877 (1961).*

39. Hefter, J.H., Hecht, T.A., and Hammond, G.S. *J. Am. Chem. Soc. 94, 2793 (1972).*

40. Bertrand, M.P., and Surzur, J.-M. *Tetrahedron Lett., 3451 (1976).*

41. Garcia, L.A., Williams, J.C., Jr., and Houk, K. N., in preparation.

AN EXERCISE IN COMPUTER MODELING: THE HIGH-TEMPERATURE PYROLYSIS OF HYDROCARBONS

John N. Bradley

Department of Chemistry, University of Essex,
Colchester, Essex, England.

INTRODUCTION

The high-temperature ($>1000^{o}K$) pyrolysis of hydrocarbons is of interest for two reasons. Firstly, the cracking of hydrocarbons to give low molecular weight species is of enormous importance industrially : the temperatures in use already exceed $1000^{o}K$ and there is considerable evidence to suggest that the economics of the process can be further improved by moving to even higher temperatures and shorter reaction times. Secondly, although pyrolysis involves a large number of chemical species and chemical reactions, many of them have closely-related physical and chemical properties which are reasonably well-characterized. Thus there is a very real opportunity for developing a computer model which can adequately account for the experimental observations using only a very simple set of reactions. The value of such a model is that it should provide a satisfactory prediction of the product composition obtained from the pyrolysis of any hydrocarbon or mixture of hydrocarbons under specified conditions of temperature, pressure and reaction time, based solely on the known molecular structure and thermochemistry of the reactants.

Although the choice of the temperature boundary is somewhat arbitary, a very real transition in behaviour occurs as the pyrolysis temperature is raised. The reason is that, at these higher temperatures, alkyl radicals decompose unimolecularly to smaller fragments rather than reacting with other molecules so that "polymerization" cannot occur and the products are all of lower molecular weight than the initial reactants. The transition is not completely sharp because the stability of unsaturated radicals extends to rather higher temperatures: nevertheless, it can be assumed for modelling purposes that the quantity of higher molecular weight products is sufficiently low to be neglected.

ISBN 0-12-566550-4

Several workers have investigated the pyrolysis of individual hydrocarbons but have not been particularly concerned with generalized mechanisms so that inevitably discussion must be restricted largely to the author's own work. There are other justifications: because the current goal is to develop mechanisms suitable for "practical" systems, relatively high conversions have been employed. In most of the earlier investigations, the aim has been to derive accurate rate constants for the initial stages of reaction and so only very low conversions have been studied. Furthermore, it turns out that transferring a rate constant from one experimental system to another is often of very questionable validity, particularly if the temperatures/pressures involved are significantly different. One aspect of the current approach has been to develop self-consistent reaction sets which require an absolute minimum of "external" data.

This review summarizes the current state-of-the-art in the field and highlights the major problems which still have to be solved. The amount of quantitative detail has been kept to an absolute minimum but is available in the original papers.

COMPUTER SIMULATION

This article is primarily concerned with the computer simulation of experimental data and the mathematical techniques and modelling philosophy adopted are therefore of crucial importance.

The experimental results appear in the form of final product compositions corresponding to individual sets of starting conditions and residence times. The first decision required is the extent to which the raw data should be processed prior to optimization. Virtually all previous workers have reduced their data to empirical relationships involving the input parameters. This introduces an element of "smoothing" which may distort or conceal the true behaviour. The present approach has been to use the results of actual experiments, which meet appropriate criteria of reproducibility and mass balance, although in some instances it has proved more satisfactory to work with a smaller number of hypothetical experiments instead. In no case are the data adjusted to fit any specific dependences on concentration, temperature, or time. A very substantial advantage of this approach is that conclusions concerning the mechanisms are not invalidated by appreciable errors in temperature and reaction time, although numerical values of rate constants will naturally differ. This is important because single-pulse shock tube experiments at very high temperatures/conversions are open to criticism(1). Also, none of the data employed in this work displayed the anomalous behaviour reported for certain other systems(2).

The major optimization routine for solving a series of non-linear equations was developed by J. Oliver and J.A. Ford of the University of Essex Computing Centre(3). The user provides a set of trial rate constants, equal in number to the independent products, together with upper and lower bounds for each. The program then adjusts the rate constants, within these bounds, until it locates the best fit to the experimental data. During the process it integrates the whole of the kinetic equations, using a fifth-order Gear predictor-corrector method containing no steady-state assumptions, for each trial. Complete optimization frequently involves several thousand numerical integrations.

Test mechanisms often contain hidden constraints which prohibit solution: unfortunately, due to the high non-linearity of these systems, failure to achieve optimization is no guarantee that a solution does not exist although, in several years' operation, the program has never yet failed to find a known solution.

The modelling 'philosophy' is novel in a number of ways. The conventional approach is to include every possible reaction which is likely to occur, together with the best estimate of its rate constant. In the case of hydrocarbon pyrolysis, this would generate literally hundreds of reactions for which barely a handful of rate constants are known. The present approach has been to search for the *minimum* reaction set, based on roughly-reasonable rate constants, which satisfies the experimental requirements. A major advantage of this approach is that the mechanism eventually adopted is sufficiently brief to allow a genuine understanding of what is taking place. Although this may lead to a scheme which is over-simplified in places, with a single reaction carrying the burden of several, it is invaluable for the purpose of producing a generalized model.

Another feature has been to keep to an absolute minimum the incorporation of literature rate constants. This means that results obtained on one hydrocarbon are as independent as possible of those on any other and hence do not depend on information which may not be applicable under the same conditions. The greater the degree of independence of the different optimizations, the more which can be learnt by comparison.

Various devices can be accepted to keep the number of adjustable parameters low. To begin with, forward and reverse rate constants can usually be related through an equilibrium constant. Secondly, it often happens with competing reactions that it is necessary to consider only rate constant ratios. Thirdly, it is convenient to relate chemically-similar reactions, e.g. methyl radical abstraction reactions, by simple numerical constants. The validity of any computer modelling exercise must always be checked by thorough sensitivity analysis and the dependence of the optimized rate constants on such simplifications is readily ascertained.

A satisfactory generalized model must employ *reasonable* and self-consistent rate constants. The criteria for reasonableness are agreement with (i) *comparable* literature data and (ii) chemical intuition. These criteria are not easy to define and a more objective measure is that of self-consistency, i.e. that a particular rate has the same value when measured in different systems, which is essential to any generalized model.

EXPERIMENTAL

In principle, several experimental techniques are available for use in this temperature regime and devices such as the tubular flow reactor(4) and the "wall-less" reactor(5) have been employed. However, the field is dominated by the single-pulse shock tube, employed in the majority of investigations.

Two versions of the device are currently in use: the conventional two-diaphragm apparatus developed originally by Glick, Squire and Herzberg(6) and the "magic-hole" apparatus introduced by Lifshitz, Bauer and Resler(7). Each geometry has its individual protagonists and it is certainly open to question whether their relative merits would remain unaltered for very different tube dimensions, pressures, etc. However, it is evident that both geometries produce virtually identical results under the conditions employed so far. In either case the reactants are confined to a short length of the experimental section in order to minimize the uncertainty in the reaction time.

Almost all studies use pressure transducers to measure both the shock velocity, from which the temperature is obtained, and the pressure-time history, which leads to the reaction time. The final products are analyzed by gas-liquid chromatography and mass spectrometry. Reaction times are typically in the range 0.5 to 1.5 msec, with pressures between 0.5 and 4.0 atmospheres and temperatures in the range 1000°-1700^oK.

THEORETICAL CONSIDERATIONS

Initiation Reactions

In one of the first papers in the series(8), it was shown theoretically that as carbon-carbon bonds are normally weaker than carbon-hydrogen bonds, it is necessary to consider only the former as possible initiation reactions. Confirmation of this was shown by the work on isobutene (see later), in which it was demonstrated that C-H rupture did not exceed 5% of the total initiation.

There is little likelihood in the near future of preidicting *a priori* the relative rates of rupture of the various C-C bonds in a hydrocarbon with sufficient accuracy. The experimental studies thus far conducted have been restricted mainly to molecules where competition between different C-C bond ruptures does not arise.

Radical Dissociation

Once radicals are formed from the initiation step, a whole range of reactions can in principle occur. However at the present temperatures and concentrations, theoretical considerations demonstrate that apart from two exceptions, radicals can only decompose. The exceptions are, of course, the methyl radical and the hydrogen atom, neither of which has a decomposition route available. Although the case of unsaturated radicals, e.g. C_4H_7, C_3H_5 and C_2H_3, is less clearcut, computer simulations indicate that their principal reaction is that of decomposition.

Radical dissociation is normally governed by the "β-bond breaking" rule i.e. that the bond which breaks is β to the radical centre. Simultaneous isomerization may occur via 1-5, and to a lesser extent 1-4, hydrogen transfer.

Propagation Reactions

Propagation of the pyrolysis chain takes place almost entirely by attack of CH_3 and H on the parent hydrocarbon. Although such reactions have been studied extensively in the past, it has been demonstrated in recent years(9,10) that the methyl radical reactions show non-Arrhenius behaviour at higher temperatures. This may be attributed, at least in part, to low frequency vibrations (CH_2 rocking motions) in the vicinity of the reaction co-ordinate. Whatever the explanation, the disappointing consequence is that the vast body of literature data obtained at lower temperatures cannot be applied to the present problem.

In the case of abstraction by H, the situation proves less serious than might be expected because the stationary concentration of H is low and because the majority of H atoms react via this route.

In all cases so far studied, reactions involving abstraction by methyl have been treated as variables and the problem of non-Arrhenius behaviour has been circumvented. The optimizations have always confirmed this non-Arrhenius temperature dependence although its magnitude has not been entirely consistent. Another complication is that there are often several such reactions and certainly always two - one with the parent and the other with ethane. The current procedure is to assume a simple relationship between the different rate constants; usually that they are proportional to the number of abstractable H atoms in the molecule, although alternatives have been examined. In practice, sensitivity tests show that the choice is unimportant since reaction with the parent is always dominant and treated as a variable.

Radical Recombination

Although H atoms do not decompose, their reactivity is high and hence their stationary concentration is low: recombination can therefore be neglected. However, CH_3 radicals will associate rapidly to give C_2H_6 and a significant concentration is soon built up. The ethane is itself pyrolyzed via the sequence

$$C_2H_6 \rightleftharpoons CH_3 + CH_3 \qquad k_{ethane}\ (1,-1)$$

$$CH_3 + C_2H_6 \rightarrow CH_4 + C_2H_5 \qquad k_{ethane}\ (2)$$

$$H + C_2H_6 \rightarrow H_2 + C_2H_5 \qquad k_{ethane\ (3)}$$

$$C_2H_5 \rightarrow C_2H_4 + H \qquad k_{diss,\ C_2H_5(1)}$$

This reaction sequence has to be included in *all* pyrolysis simulations. As it is impossible completely to avoid specifying one or two rate constants, it has proved convenient to select them from the above. In practice, $k_{ethane\ (1)}$ and $k_{ethane\ (-1)}$ are related by the equilibrium constant, and $k_{ethane\ (2)}$ and $k_{ethane\ (3)}$ are tied to the main propagation reactions, so that it is only necessary to choose numerical values for $k_{ethane\ (1)}$ and $k_{diss,\ C_2H_5\ (1)}$.

The radical concentration in the stationary state, which is attained only late during the reaction if at all, is high. It has been assumed, and confirmed by computer simulation, that radicals are removed solely by recombination during the quenching process.

At this stage, the reactions described are sufficient to form the *basic* mechanism for alkane pyrolysis shown in Table 1. The reactions which have subsequently been added to the basic mechanism are discussed below.

TABLE 1

"BASIC" MECHANISM of ALKANE PYROLYSIS

Initiation (rupture of C–C only)	$R–R' \rightarrow R + R'$
Radical Dissociation (via β-bond rupture + 1,4 and 1,5 shifts)	$R \rightarrow Alkene + R'$
Radical Recombination	$CH_3 + CH_3 \rightarrow C_2H_6$
Propagation Reactions (attack of H and CH on parent)	$CH_3 + RH \rightarrow CH_4 + R$
	$H + RH \rightarrow H_2 + R$
Reactions of Ethane	$C_2H_6 \rightleftharpoons CH_3 + CH_3$
	$CH_3 + C_2H_6 \rightarrow CH_4 + C_2H_5$
	$H + C_2H_6 \rightarrow H_2 + C_2H_5$
	$C_2H_5 \rightarrow C_2H_4 + H$

Hydrogen Atom Addition to Alkenes

Theoretical considerations show that inclusion of hydrogen atom addition to alkenes has only a minimal effect on the other optimized rate constants under most conditions. However, at high conversions and in the pyrolysis of alkenes, these reactions have an appreciable affect. Since the aim is to produce a generalized model, they have been included throughout. At high temperatures, the rate constants for these reactions are remarkably similar: it is likely that, within experimental error, they may be treated as identical although the optimizations have not yet resolved this point.

The "Ethylene Anomaly": Forbidden Isomerizations

In all the hydrocarbon pyrolyses so far studied, the basic mechanism has failed to match the experimental data and it has proved necessary to extend the mechanism further. Now that a series of different investigations has been completed and the results of other work in the literature re-examined, it has become clear that there is a common cause. At higher temperatures, the relative yield of ethylene increases by a much greater amount than any reaction mechanism has so far predicted - the "ethylene anomaly". A quantitative reappraisal of all possible routes to ethylene will be reported shortly: while this is not yet complete, it is already apparent that there is no simple, unequivocal solution.

On the basis of the evidence currently available, radical dissociation involving 1,2 and 1,3 hydrogen shifts, and naturally more complex rearrangements, can be ruled out at lower temperatures but there is conflicting evidence regarding the high-temperature behaviour. In general, it has been found necessary to postulate the occurrence of these so-called *forbidden* reactions, with temperature coefficients which make their role totally insignificant below about 1200°. Examples of such reactions are

$$(CH_3)_2CH\cdot \rightarrow CH_3 + C_2H_4$$

$$(CH_3)_3C\cdot \rightarrow C_2H_5 + C_2H_4$$

$$(CH_3)_2CHCH_2\cdot \rightarrow C_2H_5 + C_2H_4$$

Intuitively, the 1,2 hydrogen shift involved in the first of these reactions does not seem outrageous for a high temperature reaction but the other two are more difficult to accept.

This uncertainty over the decomposition reactions leads to an additional complication in the computer model. If each radical isomer has its own distinctive modes of decomposition, then it is correct to regard each isomer as a distinct species. However, relaxing the constraint on the routes for decomposition leads to an unacceptable increase in the number of chemical reactions. This difficulty can be circumvented by regarding the various isomers as identical. This leads again to a tractable number of variables but means that the optimized rate constants become system - dependent as they represent a summation over a particular isomer distribution. Although the latter approach has generally been followed, recent unpublished work suggests that it may prove more profitable, as understanding grows, to return to the original approach.

The "Ethylene Anomaly": Methyl Radical Addition Reactions

Theoretical considerations based on rate data obtained at low temperatures(11) indicate that methyl radical addition to the product alkene can be neglected. However the recent work on decomposition of C_3 hydrocarbons(12) shows that, in these cases, the anomalous ethylene yield cannot be explained by *forbidden* isomerizations and that the reaction

$$CH_3 + C_3H_6 \rightarrow C_2H_6 + C_2H_5$$

is almost certainly responsible. [It will be noted that even this requires a 1,2 hydrogen transfer]. The rate constant required for the reaction is very high (up to 10^9l mole^{-1}s^{-1} at the highest temperatures) despite the fact that the two reactants are present in the highest concentration, apart from the initial hydrocarbon.

On this basis, therefore, it may prove possible to replace *forbidden* isomerizations by routes involving methyl addition reactions. Even so, it seems that some of these isomerizations e.g.

$$(CH_3)_2CH\cdot \rightarrow C_2H_4 + CH_3$$

must still be included. Furthermore, the high rate constants required for the methyl addition reactions render them only marginally more acceptable than the forbidden isomerizations one is attempting to replace.

Although many other routes to ethylene are available, all of them prove even less acceptable in terms of the rate constants required.

Formation of Alkynes

Small quantities of acetylene are always observed during hydrocarbon pyrolysis. In most cases, these have simply been added to the ethylene yield, which implies that acetylene is formed either from ethylene itself or from its immediate precursor. At higher temperatures and conversions, acetylene, allene and methylacetylene are formed in significant quantities and have to be incorporated in the mechanism. The preceding considerations generally prove adequate: hydrogen abstraction from the alkene yields an alkenyl radical which then decomposes to an alkyne, thus

$$CH_3,H + C_4H_8 \rightarrow CH_4,H_2 + C_4H_7$$

$$C_4H_7 \rightarrow C_3H_4 + CH_3$$

$$CH_3,H + C_3H_6 \rightarrow CH_4,H_2 + C_3H_5$$

$$C_3H_5 \rightarrow C_3H_4 + H$$

$$CH_3,H + C_2H_4 \rightarrow CH_4,H_2 + C_2H_3$$

$$C_2H_3 \rightarrow C_2H_2 + H$$

The decomposition reactions are all *allowed* and hence proceed with high efficiency.

A difficulty which arises is that the rate constants required for abstraction from ethylene are more than two orders of magnitude greater than the literature values and the reaction

$$C_2H_5 \rightarrow C_2H_3 + H_2$$

has therefore been included in the mechanism. The resulting success does not necessarily prove that such a reaction actually occurs although the high activation energy required would not be unreasonable if both H atoms were removed from the same carbon atom.

EXAMPLES OF HYDROCARBON PYROLYSIS

I. The Pyrolysis of Isobutane

The investigation of isobutane pyrolysis(8) provided the first test of the theoretical considerations outlined above. The optimization showed that the *basic* mechanism in Table 1, incorporating literature rate constant data, was inadequate to model the pyrolysis. It proved necessary to incorporate the *forbidden* isomerizations and to invoke non-Arrhenius behaviour for the methyl radical abstraction reactions. The complete mechanism is illustrated in Table 2.

TABLE 2

THE PYROLYSIS OF ISOBUTANE

	Reaction	Rate constant
Initiation	$C_4H_{10} \rightarrow C_3H_7 + CH_3$	$k_{I,init}$
Propagation	$CH_3 + C_4H_{10} \rightarrow CH_4 + C_4H_9$	$k_{I,prop,CH_3}$
	$H + C_4H_{10} \rightarrow H_2 + C_4H_9$	$k_{I,prop,H}$
Ethane Reactions	$C_2H_6 \rightleftharpoons CH_3 + CH_3$	$k_{ethane(1,-1)}$
	$CH_3 + C_2H_6 \rightarrow CH_4 + C_2H_5$	$k_{ethane(2)}$
	$H + C_2H_6 \rightarrow H_2 + C_2H_5$	$k_{ethane(3)}$
Radical Decomposition	$C_4H_9 \rightarrow C_4H_8 + H$	$k_{diss,C_4H_9(1)}$
	$\rightarrow C_3H_6 + CH_3$	$k_{diss,C_4H_9(2)}$
	$\rightarrow C_2H_4 + C_2H_5$	$k_{diss,C_4H_9(3)}$
	$C_3H_7 \rightarrow C_3H_6 + H$	$k_{diss,C_3H_7(1)}$
	$\rightarrow C_2H_4 + CH_3$	$k_{diss,C_3H_7(2)}$
	$C_2H_5 \rightarrow C_2H_4 + H$	$k_{diss,C_2H_5(1)}$
H Addition to Alkenes	$H + C_4H_8 \rightarrow C_4H_9$	k_{add,C_4H_8}
	$H + C_3H_6 \rightarrow C_3H_7$	k_{add,C_3H_6}
	$H + C_2H_4 \rightarrow C_2H_5$	k_{add,C_2H_4}

The experimental data cannot be used both for the formulation of the model and in its evaluation: nevertheless some comparisons with literature data are possible. Sensitivity tests showed that six of the rate constants are critical: as there are only five *independent* products, a value for $k_{ethane,(1)}$ was specified beforehand and the others were obtained by optimization. Variation of the $k_{I,prop,CH_3}/k_{ethane(2)}$ ratio from 1.5 to 2.0 and inclusion of the H + alkene reactions affected the optimized rate constants by less than 25%. The initiation reaction $k_{I,init}$ fell somewhat below the literature value, obtained at lower temperatures. Although $k_{ethane(2)}$ was well above the value predicted by extrapolation from the low-temperature data, due to non-Arrhenius behaviour, it still fell below the result obtained independently by Pacey and Purnell(10). The contentious *forbidden* isomerizations

$$\left.\begin{array}{l} t\text{-}C_4H_9 \searrow \\ i\text{-}C_4H_9 \nearrow \end{array}\right. C_2H_4 + C_2H_5$$

had very high activation energies, relative to the normal dissociation, and only became important above 1300°. The isomerization rates for propyl decomposition also indicated that the effective overall reaction

$$(CH_3)_2CH\cdot \rightarrow CH_3 + C_2H_4$$

is quite rapid.

II. The Pyrolysis of Neopentane

The pyrolysis of neopentane(13) provided a genuine test of the model developed in the preceding section. The most important outcome was that precisely the same additional features - forbidden isomerizations, H atom addition to alkenes, and high rates of methyl radical abstractions - had to be included to match the data. It was also necessary to provide a route to C_3H_4 in order to simulate the results at high conversion.

As before, it proved necessary to specify the value of $k_{ethane(1)}$ at the outset. The value obtained for the initiation reaction $k_{II,init}$ is in satisfactory agreement with a value reported by Tsang(14) but is closer to a subsequent measurement by Pratt and Rogers(15) using a

TABLE 3

THE PYROLYSIS OF NEOPENTANE

	Reaction	Rate constant
Initiation	$C_5H_{12} \rightarrow C_4H_9 + CH_3$	$k_{II,init}$
Propagation	$CH_3 + C_5H_{12} \rightarrow CH_4 + C_5H_{11}$	$k_{II,prop,CH_3}$
	$H + C_5H_{12} \rightarrow H_2 + C_5H_{11}$	$k_{II,prop,H}$
Ethane Reactions	as C_4H_{10}	
Radical Decomposition	$C_5H_{11} \rightarrow C_4H_8 + CH_3$	$k_{diss,C_5H_{11}(2)}$
	$\rightarrow C_3H_6 + C_2H_5$	$k_{diss,C_5H_{11}(3)}$
	$\rightarrow C_2H_4 + C_3H_7$	$k_{diss,C_5H_{11}(4)}$
	C_3H_7 and C_2H_5 as for C_4H_{10}	
	$C_2H_5 \rightarrow C_2H_3 + H_2$	$k_{diss,C_2H_5(2)}$
Addition to Alkenes	as C_4H_{10}	
Alkyne Formation	$CH_3,H + C_4H_8 \rightarrow CH_4,H_2 + C_4H_7$	$k_{IV,prop,CH_3,H}$
	$C_4H_7 \rightarrow C_3H_4 + CH_3$	$k_{diss,C_4H_7,(1)}$
	$CH_3,H + C_2H_4 \rightarrow CH_4,H_2 + C_2H_3$	$k_{VI,prop,CH_3,H}$
	$C_2H_3 \rightarrow C_2H_2 + H$	$k_{diss,C_2H_3(1)}$

"wall-less" reactor. The value for $k_{ethane(2)}$ is also in better agreement with Pacey and Purnell's value. The rate constant ratio for radical isomerization would not be expected to agree with that from isobutane pyrolysis because the method of formation, and hence the energy and isomer distribution, are not the same. However the value agrees well with that obtained from the pyrolysis of isobutene (see later) i.e. 0.6 *cf.* 0.7 at 1200°, where the radical is formed by an identical route.

This investigation showed that a model which is required to cope with both high temperatures and high conversions must incorporate subsequent reactions of alkenes to give alkynes. Such reaction steps are included in the mechanism in Table 3 although, as these are secondary processes, the values of the rate constants obtained by computer simulation are subject to considerable scatter.

III. The Pyrolysis of Propane

The pyrolysis of propane poses two special problems, one concerned with the reaction scheme and the other with the optimization.

The first of these is a manifestation of the ethylene paradox in a particularly difficult form. A C_3 hydrocarbon will decompose to give one C_1 and one C_2 fragment, the former leading to CH_4 and C_2H_6, and the latter to C_2H_4 and C_2H_2. The mechanism proposed in an investigation by Lifshitz and Frenklach(16) contains the implicit constraint that the quantities of C_1 and C_2 hydrocarbons must be equal. The current model allows for a limited degree of "exchange" via the sequence

$$CH_3,H + C_2H_6 \rightarrow CH_4,H_2 + C_2H_5$$

$$C_2H_5 \rightarrow C_2H_4 + H$$

The experimental results(12) demonstrated that the constraint ($\Sigma C_1 = \Sigma C_2$) is not obeyed and the computer simulation shows that the sequence above is inadequate to explain the discrepancy. The problem has eventually been overcome by including the reaction

$$CH_3 + C_3H_6 \rightarrow [C_4H_9] \rightarrow C_2H_4 + C_2H_5$$

The complete mechanism is shown in Table 4.

TABLE 4

THE PYROLYSIS OF PROPANE

Initiation	$C_3H_8 \rightarrow C_2H_5 + CH_3$	$k_{III,init}$
Propagation	$CH_3 + C_3H_8 \rightarrow CH_4 + C_3H_7$	$k_{III,prop,CH_3}$
	$H + C_3H_8 \rightarrow H_2 + C_3H_7$	$k_{III,prop,H}$
Ethane Reactions	$C_2H_6 \rightleftharpoons CH_3 + CH_3$	$k_{ethane(1,-1)}$
	$CH_3 + C_2H_6 \rightarrow CH_4 + C_2H_5$	$k_{ethane(2)}$
	$H + C_2H_6 \rightarrow H_2 + C_2H_5$	$k_{ethane(3)}$
Radical Decomposition	$C_3H_7 \rightarrow C_3H_6 + H$	$k_{diss,C_3H_7(1)}$
	$\rightarrow C_2H_4 + CH_3$	$k_{diss,C_3H_7(2)}$
	$C_2H_5 \rightarrow C_2H_4 + H$	$k_{diss,C_2H_5(1)}$
	$\rightarrow C_2H_3 + H_2$	$k_{diss,C_2H_5(2)}$
	$C_2H_3 \rightarrow C_2H_2 + H$	$k_{diss,C_2H_3(1)}$
H Addition to Alkenes	$H + C_3H_6 \rightarrow C_3H_7$	k_{add,C_3H_6}
	$H + C_2H_4 \rightarrow C_2H_5$	k_{add,C_2H_4}
Transfer	$CH_3 + C_3H_6 \rightarrow C_2H_4 + C_2H_5$	k_{trans}

Unfortunately each time an additional reaction is incorporated, it is necessary to specify another rate constant so that the number of adjustable parameters remains equal to the number of independent products. The sensitivity of the optimized rate constants to the various approximations was shown to be either negligible or entirely predictable. As far as comparisons are possible, the optimized rate constants agree with those from other sources; thus the values of $k_{ethane(2)}$ and of $k_{diss,C_3H_7(1)}/k_{diss,C_3H_7(2)}$ agree with those obtained from the previous investigations.

The *transfer* reaction brings up the ethylene anomaly yet again. The rate constant is higher than would be expected from low-temperature measurements and is comparable with the rates of H-atom addition reactions. As the reaction involves the two species in highest concentration, apart from the reactant itself, any alternative explanation would demand even larger rate constants.

IV. The Pyrolysis of Isobutene

An investigation of isobutene pyrolysis(17) enables a test to be made of the changes required to the model to cope with an alkene reactant.

The first problem concerns initiation. While the argument which leads to C-C rupture as the principal initiation step in alkanes is strong, the situation is rather less clear-cut for alkenes. Rather than adopt the situation without question, it must therefore be tested by computer simulation. Unfortunately the optimization procedure cannot simultaneously handle two competing initiation reactions since the magnitude of one depends critically on that of the other and in any chain sequence the nature of the initiation step has only a secondary influence. The procedure adopted was to carry out independent simulations using each initiation reaction in turn. The behaviour of the other optimized rate constants indicated strongly that C-H rupture was responsible for no more that 5% of the total initiation and probably much less. Both theoretical and experimental considerations suggest that initiation by C-H rupture may be neglected in the pyrolysis of alkenes.

The second problem is that unsaturated radicals turn out to decompose in preference to undergoing other reactions, at least over 1200°K, and the rate constant for reaction with the parent alkene does not differ appreciably from that of CH_3.

As far as individual rate constants are concerned, methyl radical abstraction from the alkene $k_{IV,prop,CH_3}$, which was subject to considerable scatter in the neopentane analysis, now takes a value in agreement with the corresponding reactions with alkanes. The hydrogen atom reaction rate $k_{add,\ C_4H_8}$ also agrees with the previous value.

The complete mechanism is given in Table 5.

TABLE 5

THE PYROLYSIS OF ISOBUTENE

Initiation	$C_4H_8 \rightarrow C_4H_7 + H$	$k_{IV,init(1)}$
	$C_4H_8 \rightarrow C_3H_5 + CH_3$	$k_{IV,init(2)}$
Propagation	$CH_3 + C_4H_8 \rightarrow CH_4 + C_4H_7$	$k_{IV,prop,CH_3}$
	$H + C_4H_8 \rightarrow CH_4 + C_4H_7$	$k_{IV,prop,H}$
	$C_3H_5 + C_4H_8 \rightarrow CH_4 + C_4H_7$	$k_{IV,prop,C_3H_5}$
Ethane Reactions	as C_4H_{10}	
Radical Decomposition	C_4H_9 as for C_4H_{10}	
	C_3H_7, C_2H_5 as for C_3H_8	
	$C_4H_7 \rightarrow C_3H_4 + CH_3$	
	$C_3H_5 \rightarrow C_3H_4 + H$	
Addition to Alkenes	as C_4H_{10}	
Alkyne Formation	as C_5H_{12}	

V. *The Pyrolysis of Propylene*

Although modelling of the pyrolysis of propylene is not yet complete, it raises an interesting problem regarding radical addition reactions. Once again it is a reaction which has been analyzed in the past(18) but which is not adequately represented by the model proposed. The $\Sigma C_1/\Sigma C_2$ ratio is below unity at high temperatures but is actually greater than unity at low temperatures. One possible explanation is the reaction

$$C_2H_3 + C_2H_4 \rightarrow [C_4H_7] \rightarrow C_3H_4 + CH_3$$

TABLE 6

"COMPLETE" MECHANISM FOR HYDROCARBON PYROLYSIS

Initiation (rupture of C–C only)	$R{-}R' \rightarrow R + R'$
Radical Dissociation (including 1,2 and 1,3 H shifts)	$R_s \rightarrow \text{Alkene} + R_s'$
	$R_u \rightarrow \text{Alkyne} + R_s$
Radical Recombination	$R_u + R_u \rightarrow \text{Dialkene}$
	$R_u + CH_3 \rightarrow \text{Alkene}$
	$CH_3 + CH_3 \rightarrow C_2H_6$
Propagation Reactions	$CH_3 + RH \rightarrow CH_4 + R$
	$H + RH \rightarrow H_2 + R$
	$(R_u + RH \rightarrow \text{Alkene} + R)$
Reactions of Ethane	$CH_3 + C_2H_6 \rightarrow CH_4 + C_2H_5$
	$H + C_2H_6 \rightarrow H_2 + C_2H_5$
	$C_2H_5 \rightarrow C_2H_4 + H$
Radical Addition to Alkenes	$CH_3 + \text{Alkene}_1 \rightarrow \text{Alkene}_2 + R$
	$R_u + \text{Alkene}_1 \rightarrow \text{Alkyne}/\text{Dialkene} + R_s$

R denotes an unspecified radical, subscripts s and u denote saturated (alkyl) and unsaturated (alkenyl) radicals respectively.

which shifts the balance in the right direction but involves an intramolecular 1-3 hydrogen transfer. Other reactions, such as

$$C_2H_3 + C_3H_6 \rightarrow [C_4H_9] \rightarrow C_4H_6 + CH_3$$

which will be faster and are *allowed*, may also contribute.

CONCLUSIONS

The latest version of the generalized mechanism, shown in Table 6, is certainly able to explain quantitatively all the major features of both alkane and alkene pyrolysis. Rate constants are reasonably self-consistent from one hydrocarbon to the next, although room for improvement exists. The relative importance of *forbidden* isomerizations and of methyl radical addition reactions, both with unexpectedly high rates, is still not resolved. However, it has not yet proved possible to dispense completely with either.

REFERENCES

1. G.B. Skinner, *Int. J. Chem. Kinetics* 9, 863 (1977).
2. J.N. Bradley and M.A. Frend, *Trans. Faraday Soc.* 67, 1 (1971); *J. Phys. Chem.* 75, 1492 (1971).
3. J.N. Bradley and J.A. Ford to be published.
4. L. Crocco, I. Glassman and I.E. Smith, *J. Chem. Phys.* 31, 506 (1959).
5. J.E. Taylor, D.A. Hutchings and K.J. Frech, *J. Phys. Chem.* 73, 3167 (1969); *J. Amer. Chem. Soc.* 91, 2215 (1969).
   G. Pratt and D. Rogers, *J.C.S. Faraday Trans. I*, 75, 1089 (1979).
6. H.S. Glick, W. Squire and A. Hertzberg, Fifth Symposium (International) on Combustion, Reinhold, New York p. 393 (1955).
7. A. Lifshitz, S.H. Bauer and E.L. Resler, Jnr., *J. Chem. Phys.* 38, 2056 (1963).
8. J.N. Bradley, *Proc. Roy. Soc.* A337, 199 (1974).
9. T.C. Clark, T.P.J. Izod, and G.B. Kistiakowsky, *J. Chem Phys.* 54, 1295 (1971).
10. P.D. Pacey and J.H. Purnell, *J.C.S. Faraday Trans. I*, 68, 1462 (1972).

11. J.A. Kerr and M.J. Parsonage, "Evaluated Kinetic Data on Gas Phase Addition Reactions", Butterworths, London, 1972.
12. J.N. Bradley, *J.C.S. Faraday Trans. I*, 75, 2819 (1979).
13. J.N. Bradley and K.O. West, *J.C.S. Faraday Trans. I*, 72, 1 (1976).
14. W. Tsang, *J. Chem. Phys.* 44, 4283 (1966).
15. G. Pratt and D. Rogers, private communication.
16. A. Lifshitz and M. Frenklach, *J. Phys. Chem.* 79, 686 (1975).
17. J.N. Bradley and K.O. West, *J.C.S. Faraday Trans. I*, 72, 558 (1976).
18. A. Burcat, Fuel 54, 87 (1975).

HOMOGENEOUS ALKANE CRACKING: THE ROUTE TO QUANTITATIVE DESCRIPTION TO VERY HIGH CONVERSION

J. H. PURNELL

Department of Chemistry
University College of Swansea
Swansea, Wales, U.K.

Summary

The current status of our understanding of alkane cracking mechanisms is outlined and outstanding quantitative problems regarding important chain initiation, propagation and termination reactions identified. The direction of studies leading to resolution of these problems is discussed and an early solution is predicted.

Illustrations are given of how present knowledge has allowed an excellent quantitative description of homogeneous n-butane cracking to over 90% conversion at temperatures up to nearly 700°C and pressures to 1 atm, notwithstanding residual problems regarding precise values of certain rate parameters. The success of this exercise has both academic and industrial implications, which are briefly indicated.

The important role of alkane cracking studies of the past decade in, first, revealing serious deficiencies in theories of chemical reactivity and, secondly, defining the direction of future theoretical activity, is stressed.

ISBN 0-12-566550-4

Twenty years ago our real understanding of the mechanism of alkane pyrolyses was little better than rudimentary. Not only was there dispute regarding the relative roles of molecular and radical chain processes but our quantitative knowledge was limited and conflicting. The turning point came with the advent of gas chromatography which allowed a detailed analytical approach providing, rapidly, large volumes of comprehensive data. The almost immediate consequence was that a number of groups (1) working along independent lines were able unequivocally to establish the radical chain nature of these reactions.

Subsequent development was swift and within a decade the general features of the early stages of the pyrolyses of all alkanes in the C_2-C_5 range were well established. Probably the most important studies were those devoted to n-butane (2) because, of all the alkane pyrolyses, this is the least affected by the mechanistic complexities now identified. In contrast to most other alkane pyrolyses it is only significantly self-inhibited at quite large extents of reaction, it is the least surface sensitive, and the one for which the effects of adventitious oxygen are at a minimum. By to-day the detailed analytical product data can be reproduced so well from laboratory to laboratory that the uncatalysed n-butane pyrolysis can be used as a standard to check equipment and personal performance. The mutual reproducibility of their data and the relative mechanistic simplicity of the reaction gave workers in different laboratories their first common ground and a largely agreed starting point.

The intervention of vessel walls and both the homogeneous and heterogeneous effects of traces of oxygen were studied in great detail by Niclause and his colleagues (3) and their eventual observation that isopentane pyrolysis can be almost stopped by appropriate wall treatment constituted the final and most convincing evidence for the totally radical chain nature of these pyrolyses. Meanwhile, our own group at Swansea was quantitatively evaluating the nature of the self-inhibition processes occurring (4). These are of such potency that in ethane, propane and isobutane pyrolyses, for example, the self-inhibition is very marked even at fractions of one percent decomposition. In general terms, these studies established the view that heterogeneity can be traced to hydrogen atom removal while self-inhibition normally involves hydrogen atom addition to ethylene or propylene, where these predominate, and hydrogen abstraction where isobutene is a significant product. The situation was summarised in a set of rules enunciated by Leathard (5) in 1969 and nothing has since emerged to cause us to dispute his account.

The elucidation of the origins of heterogeneity and self-inhibition, in turn, permitted very comprehensive evaluation of rate data for formation of both major and minor products and, in consequence, it became possible to use the pyrolyses, complex though they are, as a means to establish the values of rate constants and of rate parameters for individual reaction steps. It is largely this facility that has opened up some of the areas of current interest which have gone far to expose and will eventually clarify anomalies in alkyl radical reactivities and thermochemistry. Before proceeding to expand on this it is worthwhile to note that current understanding does not extend to methane pyrolysis. The position has been well summarised in relatively recent publications by Chen, Back and Back (6) who emphasise discrepancies in the literature, propose a non-chain reaction in which all the characteristic products are secondary and can offer no firm suggestion as to the origins of the remarkable, homogeneous autocatalysis that occurs. However, given the same initiative as has been displayed in alkane studies in recent years, which has been summarised in two lengthy reviews(7), there seems no reason to doubt that the residual detail, and its interpretation, for methane, can be provided shortly and so allow us to provide, for the first time, an account of the early stages of all homogeneous alkane pyrolyses, the most significant group of organic straight chain processes.

Achievement of the aforestated goal demands resolution of two problems which have arisen as a result of these studies and alluded to earlier. First, the rate parameters of pyrolytic initiation processes, in association with currently accepted thermochemical values for alkyl radicals, have yielded radical recombination (chain termination) rate constants, for all but methyl-methyl at high pressure, that are substantially smaller than was once thought and are also smaller than alternative modern techniques suggest. Secondly, reaction rate data for atom or radical attack on alkanes (chain propagation), measured for reaction at pyrolytic and higher temperatures, yield both higher rates and higher A and E values than have been determined in experiments at lower temperatures. These observations are evidently important in any attempt to provide a quantitative description of alkane pyrolyses but are, perhaps, even more important in their theoretical implications. The problem they pose merits a brief review.

Radical combinations and thermochemistry

Much of the most intensively studied radical combination process is the mutual recombination of methyls. The reaction has been investigated by almost every known relevant technique on more than twenty occasions, and, with only one exception, the high pressure rate constant has been established as close to $10^{13.3}mol^{-1}cm^{3}s^{-1}$. It is particularly noteworthy that all pyrolytic studies have provided such a result as an illustration based on some representative data for

$$C_2H_6 \rightleftarrows 2CH_3 \qquad (1)$$

shows. Using any one of three (8 - 10) pyrolytically derived Arrhenius equations for k_1^∞ in combination with JANAF/API thermochemical data, as listed in Table 1, one finds that, although the Arrhenius equations derive from experiments in very different temperature ranges, covering the region 840 - 1500 K, they all yield $10^{13.3\pm0.1}$, the commonly accepted (11), low temperature value for k_{-1}^∞. There is thus, not only a comprehensive agreement regarding k_{-1}^∞ but it is clear that the JANAF thermochemical data for methyl are excellent. This has received striking confirmation recently in work carried out by Baghal-Vayjooee, Colussi and Benson (12) who have measured directly, for the first time, an equilibrium constant for a reaction involving a highly reactive free radical, viz.

$$Cl + CH_4 \rightleftarrows HCl + CH_3$$

and derive from the result the value $\Delta H_f^\ominus = 34.9 \pm 0.15$ kcal mol^{-1} for the enthalpy of formation (298 K) of CH_3, in remarkable accord with the JANAF/API value.

We can clearly feel confident in the current thermochemistry of methyl and the value of k_{-1}^∞ cited.

TABLE 1

Thermochemical estimation of k_{-1} from pyrolysis data for $C_2H_6 \rightarrow 2CH_3$ (1)

	Temp(K)	log A	E_1	ΔU_1	ΔS_1	log k_{-1}
Trenwith (8)	840-875	16.3	88.0	89.7	18.9	13.2
Pacey and Purnell(9)	930.1030	16.7	88.4	89.1	18.3	13.3
Burcat et al.(10)	1000-1500	16.9	89.5	88.4	17.6	13.3

(units: cal., mol., cm^3, s)

However, the pressure dependence of k_1, and hence of k_{-1} is still quantitatively unresolved. The position is well summarised in the paper of Waage and Rabinovitch (13) where virtually all data available at the time are plotted together and the reported fall off half-pressures cover a range of two-orders of magnitude at any given temperature over a wide range. Some of these data are obviously in error but, even discounting these, significant quantitative difficulty remains. Only for studies in the vicinity of 600°C can we rely upon the data, as has recently been confirmed by Trenwith (39). Definitive study of this problem, particularly in the temperature range 300 - 800 K, is much to be desired since so many reaction rates are determined relative to that of methyl-methyl recombination.

It had long been assumed that the recombination rate constants for ethyl, and of higher alkyls, were essentially identical with that for methyls. The first hint that this might not be true came in 1968 when both we (4*b*) and Niclause and his colleagues (14) pointed out that our studies of neopentane pyrolysis led to results relevant to recombination of t-butyl which were quite incompatible with the current thermochemistry of the radical if a rate constant around 10^{13} was assumed. It was pointed out too that the same conclusion must be derived from earlier shock tube results due to Tsang (15). That this was a more general problem, however, was only brought to light with the development of the radical buffer technique by Hiatt and Benson (16), who derived recombination rate constants for ethyl (16*a*) and the propyls (16*b*) of $10^{11.6} cm^3 mol^{-1} s^{-1}$, and of $10^{9.6} cm^3 mol^{-1} s^{-1}$ for t-butyl (16*c*). Shortly, thereafter, we showed (2*e*) that the widely accepted rate data for n-butane pyrolysis, with current thermochemistry of ethyl, again yielded $10^{11.5} cm^3 mol^{-1} s^{-1}$ while Hase, Johnson and Simons (17) deduced the same values for ethyl and the propyls via a chemical activation technique. Finally, we showed (18) that pyrolytic data plus current thermochemistry gave $10^{9.6} cm^3 mol^{-1} s^{-1}$ for t-butyl recombination. This quite remarkable agreement of results obtained by three different techniques, operating over a 500° temperature span, is a considerable testament to the quality of the data from the several sources, a point totally overlooked in the subsequent debate.

That the new "low" value for the recombination constant for ethyl-ethyl was valid in the context of equilibrium calculations involving ethyl was established by Marshall and Purnell (19) who showed that the previously irreconcil-able literature rate data for the forward and reverse reactions

$$C_2H_5 + H_2 \rightleftarrows C_2H_6 + H$$

and

$$C_2H_5 \rightleftarrows C_2H_4 + H$$

were instantly reconciled by use of this new value.

All three techniques mentioned above suffer the common shortcoming that for evaluation of radical combination rates, thermochemical data must be employed. All three groups recognised this and made the point strongly. Indeed, as long ago as 1968 it had been suggested (4*b*) that the accepted heat of formation of t-butyl must be seriously in error.

The foregoing series of events triggered off efforts to determine the rate constants in an absolute way. The first data of consequence came from very low pressure pyrolysis (VLPP) studies by Golden and his colleagues (20). This technique is free of thermochemical uncertainty, but is very subject to experimental error. Even so, the value for ethyl recombination found (20*a*), close to $10^{13} cm^3 mol^{-1} s^{-1}$, could hardly be subject to a degree of error approaching $10^{1.5}$. Of much more consequence has been the development of the photochemical molecular modulation technique by Parkes and Quinn (11,21) and their associates. Their first study (11) confirmed the accepted high pressure rate constant for methyl-methyl recombination and they subsequently (21) produced values for mutual recombination of ethyl, isopropyl and t-butyl. Their data for 300 K are listed in Table 2.

TABLE 2

Radical recombination rate constants (300 K) determined by molecular modulation (Parkes and Quinn)

Radical	$10^{-12}k$ ($cm^3 mol^{-1} s^{-1}$)
C_2H_5	8.4 ± 3.6
isoC_3H_7	4.8 ± 1.2
tC_4H_9	2.4 ± 0.4

Combining their data with the VLPP results of Golden et al they found k, over the range 300 - 850 K, for ethyl/ethyl to be temperature independent, for isopropyl/isopropyl to be proportional to $T^{-\frac{1}{2}}$ and for t-butyl/t-butyl to be proportional to $T^{-3/2}$. However, the VLPP data are quantitatively unreliable and a more recent study (22) from Quinn's laboratory suggests that the isopropyl recombination reaction is described by the Arrhenius parameters $A = 8.4 \times 10^{12} cm^3 mol^{-1} s^{-1}$, $E = 320$ cal mol^{-1}. The evidence is, in fact, so inconclusive that until better data are available the reaction is probably best regarded as temperature independent, having the value of Table 2. A negative temperature dependence of the recombination constant for t-butyl, however, remains a clear possibility although as Table 2 shows, molecular modulation provides data showing a fair degree of error. Further, Marshall and Purnell (23) have shown that a thermochemistry for t-butyl can be proposed which reconciles all known data covering the temperature range spanned by radical-buffer and the shock-tube techniques. This temperature independent value is $10^{12.2} cm^3 mol^{-1} s^{-1}$, in remarkable accord with that of Table 2, but it requires acceptance of a substantial increase over the accepted values of ΔH_f° for t-butyl. Atri, Baldwin, Evans and Walker (24) in the most recent published study, have derived a value of 10.5 kcal mol^{-1}, a very acceptable result in the light of the current discussion.

These studies have established three things. First, it is clear that all alkyl radical recombinations do not have the same rate constant. Secondly, the rate constant diminishes progressively with increasing size of the radical. Thirdly, heats of formation of ethyl and higher radicals are almost certainly higher, by 1-4 kcal mol^{-1}, than is currently accepted.

The situation has been dealt with in a very detailed and closely reasoned recent paper by Tsang (25). His conclusions are that alkyl heats of formation are indeed too low but, further, that the Arrhenius parameters for the unimolecular dissociation of alkanes viz.

$$R_1R_2 \rightarrow R_1 + R_2$$

must diminish with increasing temperature in accord with Benson's restricted rotor model (26) for radical combination. However, the extent of change almost certainly defies experimental determination. Finally, and interestingly in the light of recent criticism (20, 21), he establishes the considerable value of the radical-buffer technique and pin-points the fact that the whole problem outlined here derives from inadequacy of established values of the enthalpies of formation of alkyl halides, against which background the data for alkyls rests.

Thus, the issues are now clear as to how we may effect a complete reconciliation of these matters that have occupied so prominent place in alkane chemistry for a decade.

Bimolecular chain propagation reactions

The determination of rates and rate constants in this very important area have now been in progress for over thirty years. Until relatively recently, most work involved some photochemical technique or other, and in general, temperatures were sufficiently low that pyrolytic complications were avoided. In a majority of cases, rate parameters for reactions of this type

$$R + R_1H \rightarrow RH + R_1$$

have been determined via product yield comparison with the alternative reaction

$$2R \rightarrow (R)_2$$

Thus, such rate parameters are only as accurate as is the value assumed for the rate constant of the radical recombination. The foregoing discussion of radical recombination rates therefore has considerable relevance in this context too. Since we now know the situation for the high pressure constants of the recombinations with reasonable clarity the remaining problem in this area lies in the uncertain pressure dependence of the rate constant for methyl-methyl recombination alluded to earlier. This problem arises because, of all the recombinations, this is the one most commonly in its fall-off region in the conditions of most experiments. There is no doubt at all that resolution of this problem is a matter of urgency.

Turning now to the more general problem of the Arrhenius parameters of such metathetical processes, the first significant statement can be traced to Purnell and Quinn (*2e*), who, almost twenty years ago, pointed out that low temperature data for radical attack on alkanes implied stability of the alkane at observed pyrolysis temperatures. To a first approximation, the overall pyrolysis activation energy (E) can be set as the sum of a chain propagation step energy (E_p) and one-half the initiation energy (E_i). Since E generally lies between 56 and 64 kcal and $\frac{1}{2}E_i$ between about 40 and 44 kcal, E_p must considerably exceed 10 kcal mol^{-1} and thus much exceed low temperature estimates. They, thus pointed out the probability that the propagation reactions were associated with strongly curved Arrhenius plots.

Subsequently, Kondratiev (27) took up this theme in a paper devoted to the reaction of H with CH_4 and Hay (28) later offered an explanation for this particular reaction based on a proposed equilibrium between planar and non-planar methyl radicals.

The problem came into clear focus in 1972 when Pacey and Purnell (29) established rate parameters for the reaction

$$CH_3 + C_2H_6 \rightarrow CH_4 + C_2H_5$$

in the vicinity of 1000 K and found them to be very substantially greater than the low temperature values of McNesby (30) et al would predict and, themselves predicted a lower rate than that observed by Clark and Izod (31) at higher (shock-tube) temperatures. A plot of all available data showed continuity and very pronounced curvature. Curvature of the Arrhenius plot for CH_3 attack on CH_4 has since been established by Chen, Back and Back (6*c*) while Camilleri, Marshall and Purnell (32) have shown that all known data for

$$CH_3 + C_3H_8 \rightarrow CH_4 + C_3H_7$$

also fit well on a curved Arrhenius plot. Finally, Pacey and Purnell have established (2*e*) that, almost certainly, the plots for both CH_3 and C_2H_5 attack on n-butane must also be strongly curved.

There can now be little doubt of the reality of this phenomenon where alkyl attack on an alkane is concerned and it remains to be seen what theoretical account can be given. In principle, it would seem that Clark and Dove (33), who have carried out a theoretical study using a modified BEBO method have clarified the issue since they obtain a satisfactory reconciliation of the data for methyl attack on ethane. However, they do just as well in describing the assumed curvature of the Arrhenius plots for

$$H + C_2H_6 \rightarrow H_2 + C_2H_5$$

and the corresponding reaction (27, 28)

$$H + CH_4 \rightarrow H_2 + CH_3$$

both of which we have now shown (34) unequivocally to be linear over a range of 1000 K. Clark and Dove's ability to fit a satisfactory curve to data which we now know not to give any detectable curve must clearly cast suspicion on the basic assumptions of the theory.

The position then is that it appears that where H is involved with an alkane, Arrhenius plots are straight over a very wide temperature range. In contrast, where an alkyl radical abstracts hydrogen from an alkane, curvature is so pronounced that activation energies at high temperatures may rise to over 20 kcal mol^{-1}, while *A*-factors approach collision numbers.

The current situation is illustrated in fig.1 where we show the most recently assessed data set out as Arrhenius plots for CH_3 attack on, respectively, CH_4, C_2H_6 and C_3H_8. Their curvature is apparent, and interestingly, for all three, extrapolation to $T^{-1} = 0$ yields close to $10^{14.5} cm^3 mol^{-1} s^{-1}$ with the high temperature range (>1000 K) activation energies, 25.7, 20.9 and 18.6, respectively. For the temperature range < 500 K, *A* for attack on both methane and ethane approximates $10^{11.5} cm^3 mol^{-1} s^{-1}$ while the respective activation energies are 14.1 and 10.9 kcal mol^{-1}. If the *A* value is the same for C_3H_8, continuity of the data of fig.1 gives $E \simeq 9$ kcal mol^{-1}. The summary of data in Table 3 shows that at high and low temperature the differential activation energies (ΔE) closely approximate the known bond dissociation energy differences. The data of Table 3, further identify the scale of the problem; over the range 500 - 100 K, the activation energies drop by 10 kcal or more and *A* factor by as much as 10^3.

TABLE 3

Arrhenius parameters for CH_3 attack on alkanes, (A-$cm^3 mol^{-1} s^{-1}$, E/kcal mol^{-1})

Temperature range	> 1000 K			500 K <		
Alkane	log *A*	*E*	ΔE	log *A*	*E*	ΔE
CH_4	14.5	25.7		11.5	14.1	
			4.8			3.2
C_2H_6	14.5	20.9		11.5	10.9	
			2.3			1.9*
C_3H_8	14.5	18.6		-	-	

*If log *A* for C_3H_8 is also ∿ 11.5, for continuity with experimental data for the mid-temperature range 670 - 910 K, $E \simeq 9.0$ kcal mol^{-1}.

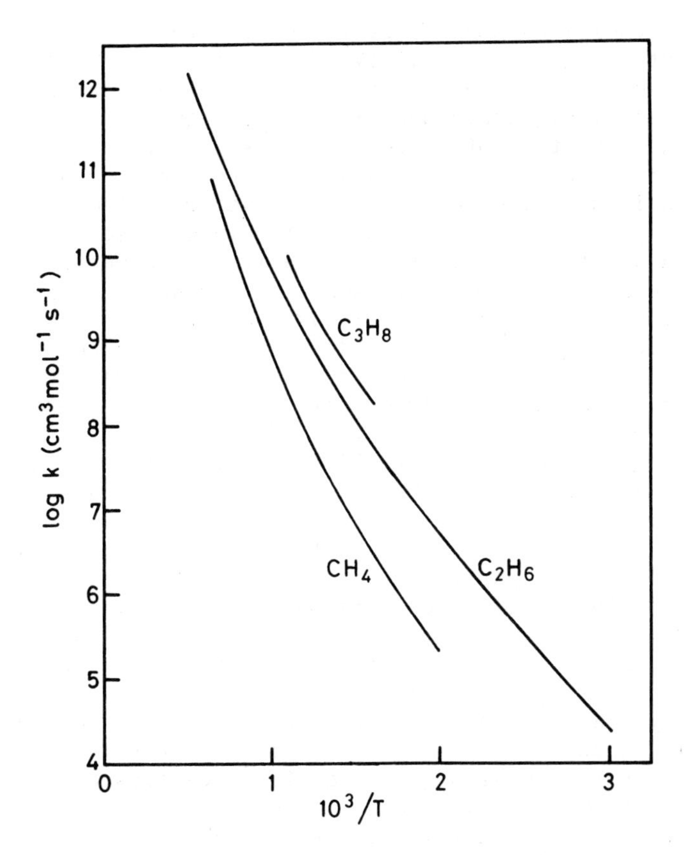

Fig.1. Arrhenius plots for $CH_3 + RH \rightarrow CH_4 + R$ *for* $RH = CH_4$ *(ref. 6c),* C_2H_6 *and* C_3H_8 *(refs. 29, 32). These plots derived from recent studies and assessment of all published data.*

It may well be that the tunnelling approach of Clark and Dove will eventually yield an acceptable description of events but it will be equally worthwhile to turn our minds to the question of the structure and properties of the transition states. Whatever the answer, we are clearly faced with an intriguing problem the resolution of which must advance our knowledge and understanding.

High severity cracking of n-butane

Quantitative studies of alkane cracking have, with one exception (*2d*), so far, been carried out in "academic" conditions, i.e. low conversion, low temperature, low pressure. The several published attempts that have been made to computer model the reactions, have all referred to these conditions and, fundamentally, have advanced our understanding and capability of description very little since such modelling has largely already been achieved by traditional algebraic and arithmetic methods.

The real test of success in this area lies in achieving a description of alkane cracking to very high conversions. The obvious candidate for an introductory attempt at this is the pyrolysis of n-butane, not only because it is the best understood and most quantitatively documented of such reactions but, because, in addition, in some geographical areas, this reaction is an industrial process of some economic importance. In order to proceed, a considerable volume of experimental information is obviously required. The only published study of relevance prior to our own work appears to be that of Illes (*2d*), which provides comprehensive product distribution data for a reasonably large, higher temperature range, for reaction at 1 atm pressure. However, valuable though this is, it provides only a starting point to a really detailed study and so, we have, over a period of five years, investigated the reaction in detail to conversion of over 90%.

Using a flow system, capable of high temperature operation and high conversion the work was initiated by Dr. David Hughes and completed latterly by Dr. Paul Boughey (38). Their data for the region of low conversion totally quantitatively confirmed the lower temperature, static system results (*2a-e*) in flow reactors of S/V ranging from about 1-18 cm^{-1}. Thus, the homogeneous character and essential validity of published rate data has been fully confirmed over the extended temperature range of about 450 - 700°C. At extents of reaction, roughly in excess of 15%, strong self-inhibition appears and, with increasing conversion, there is a marked superimposed fall-off in C_2H_6 and C_3H_6 formation, the former generally reaching a yield plateau, the latter passing through a maximum and, in certain circumstances becoming very small.

It is self-evident that a steady state calculational approach to data handling is precluded on account both of the high conversions involved and of the consequent inconstancy of flow rate and associated reaction time. An appropriate computer programme (FLOWCRAC) was therefore devised. For comparison of experimental and computed data, a common reaction time scale is required. Methods of "real" time calculation such as those of Panchenkov and Baranov (36), of Hirst (37) or such as used by Illes (*2d*), were found, by comparison with computer calculated "real" times, to be so inaccurate on account of approximation that they were discarded. The problem is most simply resolved by recourse to the chemical engineering concept of fictive time, i.e.

the residence time calculated as though no gas expansion occurs. Empirical though this is, it has the merits of being both instantly calculable and a direct measure of throughput.

In order to construct a programme we must first, evidently, set up a mechanism. The major products are H_2, CH_4, C_2H_4, C_2H_6 and C_3H_6; the minor products are C_3H_8, butenes, pentenes, hexenes, hexadienes, octadienes, and depending on reaction conditions, benzene and a trace of toluene. The sum total of these latter never exceeded 2% and so, in the present illustration, will be largely neglected, although in the detailed study a reasonable, quantitative account of production of most was achieved.

The mechanism finally chosen is as listed in Table 4 being based on the findings of static studies supplemented by the results of our own flow studies. And also on published observations regarding the nature, origins and consequences of self-inhibition. The corresponding rate constants are listed in Table 5.

It is important to recognise the following features in the choice of the rate parameters listed.

(1) All data for reactions involving ethyl radicals in either the forward or backward step were made thermochemically self-consistent. Thus, the questions of the ethyl-ethyl recombination rate constant and the enthalpy of formation of the radical, discussed earlier, was avoided.

(2) Rate parameters for the pyrolysis of ethyl (reaction (7)) quoted are for the first order constant. The low conversion flow data for the relevant temperature range yielded an average order for this reaction of 1.7, an entirely acceptable result in light of the lower temperature (static system) value (*2b*, *2c*) of 1.5. This pressure dependence was built in, as was also that for the reverse reaction (22).

(3) Referenced rate data represent our assessment of the most recent and/or generally acceptable and compatible values available.

(4) Estimated rate parameters were based on comparison and compatibility with established data for similar reactions, relevant thermochemical requirements, and, where indicated, best computer fit.

TABLE 4

Mechanism for high conversion cracking of n-butane

Primary reactions

$n\text{-}C_4H_{10} \rightarrow C_2H_5^{\cdot} + C_2H_5^{\cdot}$	(1)
$n\text{-}C_4H_{10} \rightarrow C_3H_7^{\cdot} + CH_3^{\cdot}$	(2)
$C_2H_5^{\cdot} + C_4H_{10} \rightarrow C_2H_6 + C_4H_9^{\cdot}$	(3)
$CH_3^{\cdot} + C_4H_{10} \rightarrow C_4H_9^{\cdot} + CH_4$	(4)
$H^{\cdot} + C_4H_{10} \rightarrow C_4H_9^{\cdot} + H_2$	(5)
$1\text{-}C_4H_9^{\cdot} \rightarrow C_2H_4 + C_2H_5^{\cdot}$	(6a)
$2\text{-}C_4H_9^{\cdot} \rightarrow C_3H_6 + CH_3^{\cdot}$	(6b)
1- (and) $2\text{-}C_4H_9^{\cdot} \rightarrow C_4H_8 + H^{\cdot}$	(6c)
$C_2H_5^{\cdot} \rightarrow C_2H_4 + H^{\cdot}$	(7)
$2C_2H_5^{\cdot} \rightarrow C_4H_{10}$	(8a)
$2C_2H_5^{\cdot} \rightarrow C_2H_4 + C_2H_6$	(8b)
$2CH_3^{\cdot} \rightarrow C_2H_6$	(9)
$CH_3 + C_2H_5 \rightarrow C_3H_8$	(10)

Secondary reactions

$H^{\cdot} + C_3H_6 \rightarrow C_2H_4 + CH_3^{\cdot}$	(11)
$H^{\cdot} + C_2H_6 \rightarrow C_2H_5^{\cdot} + H_2$	(12)
$CH_3^{\cdot} + C_2H_6 \rightarrow C_2H_5^{\cdot} + CH_4$	(13)
$H^{\cdot} + C_3H_6 \rightarrow C_3H_5^{\cdot} + H_2$	(14)
$CH_3^{\cdot} + C_3H_6 \rightarrow C_3H_5^{\cdot} + CH_4$	(15)
$C_2H_5^{\cdot} + C_3H_6 \rightarrow C_2H_6 + C_3H_5^{\cdot}$	(16)
$CH_3^{\cdot} + C_3H_5^{\cdot} \rightarrow C_4H_8$	(17)
$C_3H_5^{\cdot} + C_2H_5^{\cdot} \rightarrow C_5H_{10}$	(18)
$C_3H_5^{\cdot} + C_3H_5^{\cdot} \rightarrow C_6H_{10}$	(19)
$C_3H_5^{\cdot} + C_2H_6 \rightarrow C_3H_6 + C_2H_5^{\cdot}$	(20)
$C_3H_5^{\cdot} + C_4H_{10} \rightarrow C_3H_6 + C_4H_9^{\cdot}$	(21)
$H^{\cdot} + C_2H_4 \rightarrow C_2H_5^{\cdot}$	(22)
$H^{\cdot} + C_2H_4 \rightarrow C_2H_3^{\cdot} + H_2$	(23)

TABLE 5

Arrhenius parameters for the reaction steps in high conversion cracking of n-butane

Reaction	A (s^{-1} or $cm^3 mol^{-1} s^{-1}$)	E (kcal mol^{-1})	Source
(1)	2×10^{15}	77.3	K_1, reaction (8), refs. *2b/c/e*
(2)	$k_2 = k_1/3$		ref. 17
(3)	7.94×10^{13}	22.0	ref. *2e*
(4)	3.16×10^{13}	14.8	ref. *2e*
(5)	5.58×10^{14}	9.8	Based on data for (12)
(6a/6b)	1.30	2.0	ref. *2b*
(6c/6a)	2×10^{-2}	-12.0	Estimate from best fit
(7)	3.18×10^{13}	40.9	ref. 19
(8)	3.98×10^{11}	0	ref. 19
(9)	2×10^{13}	0	Accepted value
(10)	5.6×10^{12}	0	Geometric mean value
(11)	1.95×10^{13}	2.4	ref. *2e*
(12)	1.86×10^{14}	9.8	ref. 35
(13)	5.01×10^{14}	21.5	ref. 29
(14)	1.0×10^{14}	9.0)	Best available estimates and best fit.
(15)	2.0×10^{13}	14.0)	
(16)	4.0×10^{13}	20.0)	
(17)	1.41×10^{13}	0	ref. 35
(18)	1.99×10^{12}	0	Geometric mean
(19)	2.5×10^{12}	0	Best estimate
(20)	2.0×10^{15}	31.8	From K_{20}/k_{16}
(21)	2.0×10^{15}	30.0	Based on reaction (20)
(22)	5.0×10^{13}	2.6	ref. 19
(23)	$k_{22}/12$		ref. 8

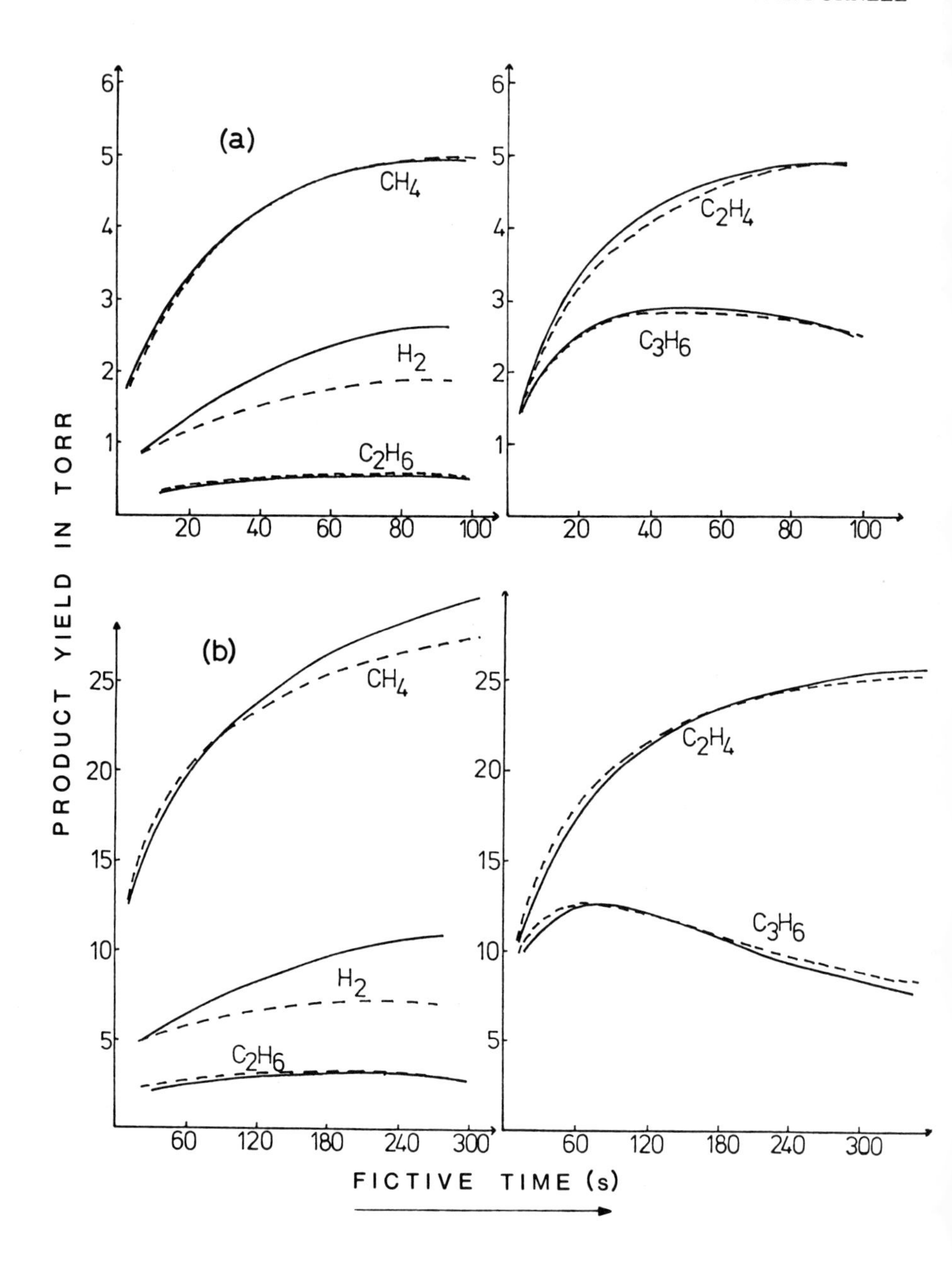

Fig. 2. Comparison of experimental (full lines) and computer calculated yields (broken lines) of the five major products of n-butane cracking at 658°C. Upper diagrams (a) for initial n-butane pressure of 20 Torr, ca. 75% conversion; lower diagrams; (b) for 80 Torr, ca. 92% conversion. Data from ref. 38.

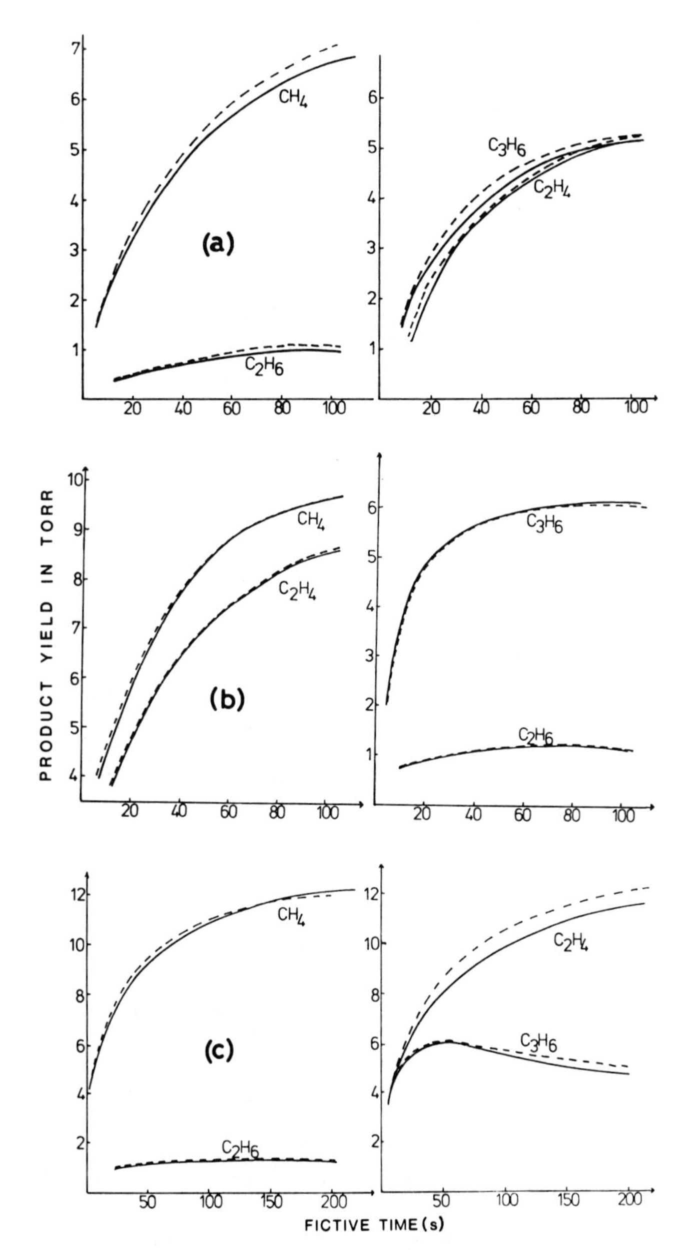

Fig. 3. Comparison of experimental (full lines) and computer calculated yields (broken lines) of the four major hydrocarbon products of pyrolysis of 40 Torr of n-butane at: (a) 613°C, ca. 45% conversion; (b) 642°C, ca. 70% conversion, and (c) 658°C, ca. 80% conversion. Data from ref.38.

The degree of success achieved is illustrated in figs. 2 and 3. The former shows a comparison of measured and computed yields of the five major products in the pyrolysis of 20 and 80 Torr of n-butane, respectively, at 658°C (931 K). The reactions have been carried to more than 90% conversion and the fit of data is obviously remarkably good. The discrepancy in hydrogen yield results from the omission of higher unsaturate forming steps from the mechanism and corresponds to about 2% material loss, as was earlier stated. The latter figure shows comparisons of the data for the major product hydrocarbons arising in the pyrolysis of 40 Torr of n-butane at three temperatures in the range 613 - 658°C (886 - 931 K). Again we see an excellent fit.

The data of the figures clearly establish a high level of success in modelling, to conversions in excess even of those normal even in industrial practice, on the basis of rate parameter data characteristically derived in academic studies for much lower temperatures. The only residual weakness in this demonstration lies in the relatively low initial pressures of n-butane dealt with. To illustrate that the scheme outlined can cope with this too, we show, in figure 4, a comparison of our computed product yields with the experimental data of Illes (*2d*) for pyrolysis of n-butane at 1 atm. Since this exercise was undertaken early in the investigation the computer programme used was truncated in that almost all the secondary reactions were left out. It is for this reason that we see some divergence between computed and experimental yields above 60% conversion.

We can now be confident of having found a total description, for, so far as we are aware, the first time, of the cracking of an n-alkane in conditions of industrial interest. Further, it is evident that data derived from academically orientated studies can be extrapolated far outside the range of conditions in which they were originally generated. Finally, it emerges that, notwithstanding the quantitative mechanistic problems yet to be resolved, as outlined earlier, it is possible to select data in a self-consistent way, even now, so as to get around the difficulties. If, as we have confidently proposed here, these matters are very soon resolved, it will of course be possible to re-work the current model in terms of rate parameters acceptable to all. It goes without saying that not all would accept certain of the data listed in Table 5, but this is, after all, only a matter of time rather than of dispute.

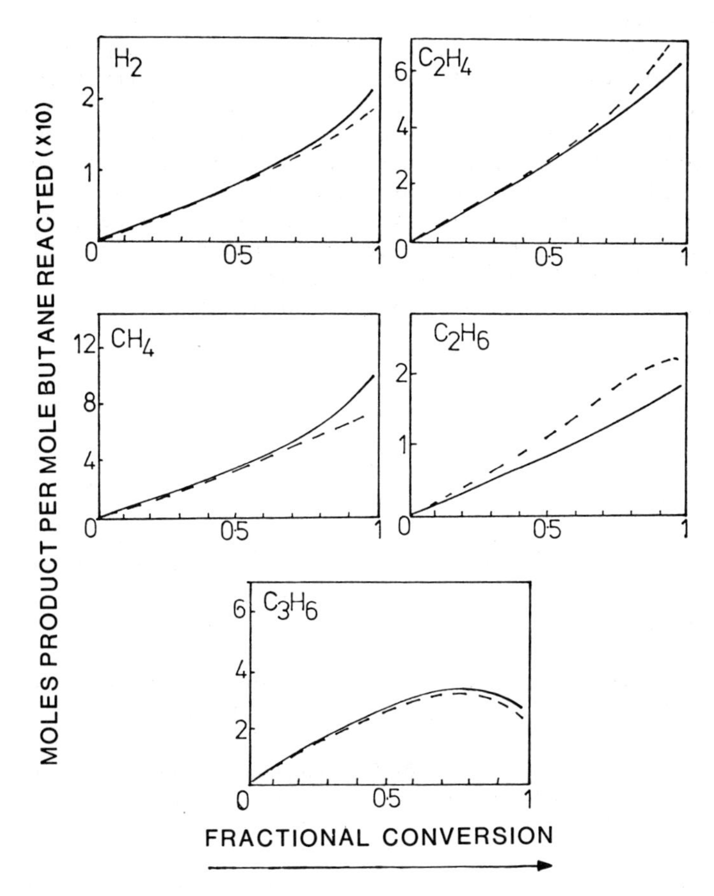

Fig.4. Comparison of computer predictions (broken lines) of current mechanism and rate parameters with experimental data (full lines) of Illes (2d) for pyrolysis of 760 Torr of n-butane at 664°C. Divergence beyond ca. 60% conversion due to use of truncated mechanism. Data from ref.38.

The n-butane study (38) briefly reported here, provides a reasonable basis for projecting high conversion mechanisms for other lower alkanes; the provision of an adequate body of relevant information should then allow test. Whether or not detailed computer modelling of the high severity cracking process for alkane mixtures becomes a reality depends on many factors, not least economic, but it is clearly possible.

We have attempted to chart progress in alkane cracking in this work by bringing together two areas, superficially disparate, in an effort to show how each profits from the other, what remains to be done and, not least, to show how far we have come in little more than a decade of concerted international activity. No single technique can ever reveal all the mysteries surrounding a given chemical problem. Alkane pyrolysis studies have, however, done more than most in bringing to light important aspects of alkyl radical chemistry which may well ultimately lead us towards a satisfactory theory of chemical reactivity.

References

1. See for example: Purnell, J. H., and Quinn, C. P., *Nature, 189*, 656 (1961); Wojciechowski, B. N., and Laidler, K. J., *Proc.Roy.Soc., A260*, 103 (1961); Pratt, G. L., *Nature, 197*, 143 (1963); Quinn, C. P., *Symp.Kinetics of Pyrolytic Reactions*, Chemical Institute of Canada, Ottawa, 1964; Pratt, G. L., and Purnell, J. H., *Trans.Faraday Soc., 60*, 371 (1964); Martin, R., Dzierzynski, M., and Niclause, M., *J.Chim.Phys., 61*, 286 (1964); Norrish, R. G. W., and Pratt, G. L., *Nature, 197*, 143 (1963).

2. [a]Purnell, J. H., and Quinn, C. P., *J.Chem.Soc.*, 4128 (1961); [b]*Proc.Roy.Soc., A270*, 267 (1962); [c]Sagert, N.H., and Laidler, K. J., *Can.J.Chem., 41*, 838 (1963); [d]Illes, V., *Acta Chim.Acad.Sci.Hung., 59(2)*, 35 (1969); [e]Pacey, P. D., and Purnell, J. H., *Ing.Eng.Chem.Fund., 11*, 223 (1972).

3. Martin, R., Niclause, M., and Dzierzynski, M.,*Compt.Rend., 254*, 1786 (1962); *J.Chim.Phys.*, 790 91964); and Martin, R., and Niclause, M., *J.Chim.Phys.*, 802 (1964).

4. [a]Leathard, D. A., and Purnell, J. H., *Proc.Roy.Soc., A305*, 517 (1968); [b]Halstead, M. P., Konar, R. S., Leathard, D. A., Marshall, R. M., and Purnell, J. H., *Proc.Roy.Soc., A310*, 525 (1969); [c]Konar, R. S., Purnell, J. H., and Quinn, C. P., *Trans.Faraday Soc., 64*, 1319 (1968); [d]Konar, R. S., Marshall, R. M., and Purnell, J. H., *Int.J.Chem.Kin., 5*, 1007 (1973); [e]Bull, K. R., Marshall, R. M., and Purnell, J. H., *Proc.Roy.Soc., A342*, 259 (1975). See also: [f]Come, G., Baronnet, F., Scacchi, G., Martin, R., and Niclause, M., *Compt.Rend., 267*, 1192 (1968); *268*, 1971 (1969).

5. Leathard, D. A., *Int.Gas Kin.Symp.*, Szeged, Hungary, Preprints, p.263 (1969).

6. Chen, C-J., Back, M. H., and Back, R. H., *Can.J.Chem.*, *53*, 3580 (1975)[a]; *54*, 3175 (1976)[b]; *55*, 1624 (1977)[c].

7. Purnell, J. H., Quinn, C. P., *"Photochemistry and Reaction Kinetics"*, ed. Ashmore, P. G., Dainton, F. S., and Sugden, T. M., Cambridge University Press, 1966, p.330; Leathard, D. A., and Purnell, J. H., *Ann.Rev. Phys.Chem.*, *21*, 197 (1970).

8. Trenwith, A. B., *Trans.Faraday Soc.*, *62*, 1538, 2486 (1966).

9. Pacey, P. D., and Purnell, J. H., *J.Chem.Soc.Faraday Trans.I*, *68*, 1472 (1972).

10. Burcat, A., Skinner, B. G., Crossley, R. W., and Scheller, K., *Int.J.Chem.Kin.*, *5*, 345 (1973).

11. see e.g. Parkes, D. A., Paul, D. M., and Quinn, C. P., *J.Chem.Soc.Faraday Trans.I*, *72*, 1935 (1976).

12. Baghal-Vayjooee, M. H., Colussi, A. J., and Benson, S.W., *J.Amer.Chem.Soc.*, *100*, 3214 (1978).

13. Waage, E. V., and Rabinovitch, B. S., *Int.J.Chem.Kin.*, *3*, 105 (1971).

14. Baronnet, F., Dzierzynski, M., Come, G. M., Martin, R., and Niclause, M., *Int.Gas Kin.Symp.*, Szeged, Hungary, Preprints p.73 (1969); *Int.J.Chem.Kin.*, *3*, 197 (1971).

15. Tsang, W., *J.Chem.Phys.*, *44*, 4283 (1966); see also *Int.J.Chem.Kin.*, *5*, 651 (1973).

16. Hiatt, R., and Benson, S. W., *J.Amer.Chem.Soc.*, *94*, 25, 6886 (1972)[a]; *Int.J.Chem.Kin.*, *41*, 151 (1972)[b]; *Int.J.Chem.Kin.*, *5*, 385 (1973)[c].

17. Hase, W. L., Johnson, R. L., and Simons, J. W., *Int.J.Chem.Kin.*, *4*, 1 (1972).

18. Konar, R. S., Marshall, R. M., and Purnell, J. H., *Int.J.Chem.Kin.*, *5*, 1007 (1973).

19. Marshall, R. M., and Purnell, J. H., *Chem.Comm.*, 764 (1972).

20. [a]Golden, D. M., Choo, K. Y., Perona, M. J., and Piskiewicz, L. W., *Int.J.Chem.Kin.*, *8*, 381 (1976); [b]Golden, D. M., Piskiewicz, L. W., Perona, M. J., and Beadle, P. C., *J.Amer.Chem.Soc.*, *96*, 1645 (1974); [c]Choo, K. Y., Beadle, P. C., Piskiewicz, L. W., and Golden, D. M., *Int.J.Chem.Kin.*, *8*, 45 (1976).

21. Parkes, D. A., Quinn, C. P., *J.Chem.Soc.Faraday Trans.I*, *72*, 1952 (1976).

22. Arrowsmith, P., Kirsch, J. L., *J.Chem.Soc.Faraday Trans. I*, *74*, 3016 (1978).

23. Marshall, R. M., Purnell, J. H., Storey, P. D., *J.Chem. Soc.Faraday Trans.I*, *72*, 85 (1976).

24. Atri, G. M., Baldwin, R. R., Evans, G. A., and Walker, R. W., *J.Chem.Soc.Faraday Trans.I*, *74*, 366 (1978).

25. Tsang, W., *Int.J.Chem.Kin.*, *10*, 821 (1978).

26. Benson, S. W., *"Thermochemical Kinetics"*, 2nd ed., Wiley, New York, 1976.

27. Kondratiev, V. N., *Russ.Chem.Rev.*, *34*, 893 (1965).

28. Hay, J. M., *J.Chem.Soc.B.*, 1174 (1967).

29. Pacey, P. D., and Purnell, J. H., *J.Chem.Soc.Faraday Trans.I*, *68*, 1462 (1972).

30. McNesby, J. R., and Gordon, A. S., *J.Amer.Chem.Soc.*, *77*, 4719 (1955).

31. Clark, T. C., Izod, T. P. J., Kistiakowsky, G. B., *J.Chem.Phys.*, *54*, 1295 (1971).

32. Camilleri, P., Marshall, R. M., and Purnell, J. H., *J.Chem.Soc.Faraday Trans.I*, *71*, 1491 (1975).

33. Clark, T. C., and Dove, J. E., *Can.J.Chem.*, *51*, 2147 (1973).

34. Camilleri, P., Marshall, R. M., and Purnell, J. H., *J.Chem.Soc.Faraday Trans.I*, *70*, 1434 (1974); Jones, D., Morgan, P. A., and Purnell, J. H., *J.Chem.Soc.Faraday Trans.I*, *73*, 1311 (1977); Sepehrad, A., Marshall, R.M., and Purnell, J. H., *J.Chem.Soc.Faraday Trans.I*, *75*, 835 (1979).

35. Throssell, J. J., *Int.J.Chem.Kin.*, *4*, 273 (1972).

36. Panchenkov, G. M., and Baranov, V. Y., *Kin.i.Kat.Akad. Nuak, S.S.S.R.*, *1*, 188 (1960).

37. Hirsch, I. H., Crawford, C. L., and Holloway, C., *Ind.Eng.Chem.*, *33*, 885 (1946).

38. Hughes, D. G., Ph.D. Thesis, Wales (1974); Boughey, P. J., Ph.D. Thesis, Wales (1978); Boughey, P. J., Hughes, D. G., and Purnell, J. H., in course of publication.

39. Trenwith, A. B., *J.Chem.Soc.Faraday Trans.I*, *75*, 614 (1979).

HYDROGEN TRANSFER CATALYSIS IN HYDROCARBON PYROLYSIS

CHARLES REBICK

Corporate Research Science Laboratories
Exxon Research and Engineering Company
Linden, New Jersey 07036

SUMMARY

Inorganic hydrides such as H_2S, HCl and HBr(HX) are known to alter cracking rates and selectivities in hydrocarbon pyrolysis. They do this by catalyzing the transfer of hydrogen between hydrocarbon reactant and free radicals. Such "hydrogen transfer catalysis" may have several effects: (1) An increase in reaction rate due to catalysis of a rate determining hydrogen transfer step; (2) A change in product distribution due to "healing" of unstable radicals by HX before decomposition occurs; and (3) A change in product distribution due to changes in the selectivity of hydrogen atom abstraction when HX is present. Such effects have been observed in the H_2S catalyzed pyrolyses of n-butane, n-hexadecene, 1-dodecene, n-butylbenzene and tetralin.

Hydrocarbon pyrolyses are complex chain reactions proceeding via free radical intermediates. Such reactions are typically non-selective and even at low conversions produce large numbers of primary products. One goal of current pyrolysis research is the control of reaction rates and product distributions in cracking reactions. A method of achieving this is the selective catalysis of one of the steps in the free radical chain process. If the reaction catalyzed is rate limiting, an acceleration in rate will result. If the catalyzed reaction is one of several competing steps or if its inherent selectivity is changed by catalysis, the product distribution may be changed. Inorganic hydrides such as HCl, HBr and H_2S(HX) possess this ability. They catalyze hydrogen transfer reactions between free radicals and hydrocarbons---reactions of the type $R\cdot + R'H \rightarrow RH + R'$. Such "hydrogen transfer catalysis" was first observed in the pyrolysis of dimethyl ether, by Anderson and Benson using HCl (1) and Imai and Toyama using H_2S (2). Since then, numerous studies of the effects of inorganic hydrides on thermal cracking reactions have been carried out (3-12), a comprehensive model

ISBN 0-12-566550-4

has been proposed to explain their effects on cracking rates (13), and at least one commercial process has employed hydride catalysts (11,12). A partial listing of these studies is given in Table I.

From this work has emerged a mechanistic description that qualitatively and to some extent quantitatively describes the effects of hydrogen transfer catalysis on the rates and selectivities of a wide variety of compound types. This mechanism is reviewed in the next section and then is illustrated for catalysis by H_2S by series of examples from recent work in our laboratory.

THE MECHANISM OF HOMOGENEOUS HYDROGEN TRANSFER CATALYSIS

To understand the effects of inorganic hydrides such as HCl, HBr and H_2S (subsequently referred to collectively as HX) on a pyrolysis reaction, one must first examine the details of

TABLE I

Hydrogen Transfer Catalysis of Hydrocarbon Pyrolysis

Compound Studied	Catalyst	Temp(°C)	Reference
Ethane	HCl, HBr	540	(3,6)
	H_2S	540	(4)
	H_2S	630	(5)
Propane	HCl,HBr,H_2S	490	(6)
n-Butane	H_2S	500	(7)
	H_2S	550-650	This work
Isobutane	HCl,HBr,H_2S	480	(6)
Neopentane	HCl	480	(8)
	HCl,HBr	480	(3,6)
	H_2S	512	(9)
n-Hexadecane	H_2S	475-550	This work(10)
1-Dodecene	H_2S	500	This work
n-Butylbenzene	H_2S	500-595	This work
Tetralin	H_2S	540-595	This work
2-Methyl-2-Pentene	H_2S	516-587	(11)
Piperylene	H_2S	650	(12)

the cracking mechanism. Most cracking reactions involve four types of reactions: (a) initiation - usually carbon-carbon bond cleavage creating free radicals, (b) termination - radical combination or disproportionation, removing radicals from the reaction, (c) decomposition - the degradation of larger radicals into stable products and smaller radicals, (d) hydrogen transfer - the abstraction of hydrogen from the reactant by radicals formed in the initiation or decomposition reactions. Reactions (c) and (d) do not change the overall radical concentration and together make up the propagation step of the reaction.

HX will potentially affect only two steps in the mechanism, termination and hydrogen transfer. The bond dissociation energies of the species discussed here are six to twenty kilocalories greater than typical C-C BDE's, making their participation in initiation unlikely (19). However, they can increase the termination rate by introducing new termination reactions. In general, this will have an inhibiting effect on the overall cracking rate. By contrast, HX can increase the cracking rate by catalyzing the hydrogen transfer reaction. This occurs by the replacement of a slow hydrogen transfer step:

$$R\cdot + R'H \rightarrow RH + R'\cdot \qquad [1]$$

by two faster ones:

$$R\cdot + HX \rightarrow RH + X\cdot \qquad [2]$$

$$X + R'H \rightarrow HX + R'\cdot \qquad [3]$$

HX is not consumed in [2] and [3] making it a true hydrogen transfer catalyst.

Whether or not the net effect of HX will be to accelerate or inhibit the overall cracking rate will depend on the extent to which reactions of the type [1] are rate limiting and the rate of the new termination processes introduced by HX, relative to the rates of existing termination reactions. The latter will depend in part on the extent to which X· replaces hydrocarbon radicals such as $CH_3\cdot$, $C_2H_5\cdot$ and H· as the predominant chain carrier in the system, i.e. on the relative rates of reactions [2] and [3]. Rate acceleration versus inhibition by HX in cracking reactions has been discussed in detail by Niclause, et al. (13) and will only be summarized here.

In general, pyrolyses in which the primary radical decomposition reaction involves carbon-carbon bond scission are rate limited by the hydrogen transfer step and are accelerated by HX addition. Most paraffins, C_5^+ olefins and long chain alkyl aromatics fall in this category. The effectiveness of HX addition is in the order HCl > $H_2S \sim$ HBr.

If, however, the radical decomposition step involves carbon-hydrogen scission, decomposition rather than hydrogen transfer may be rate limiting and no acceleration is observed. In such cases, HCl usually has no effect while H_2S and HBr inhibit the overall cracking rate. One example of this is ethane pyrolysis. Niclause, et al. have been able to fit rate data for the cracking of several compounds to a generalized formula (3,13).

In addition to changing pyrolysis rates, HX can alter product selectivities. It does this in two ways: (a) Stabilization of reactive radicals: A free radical R· may often have two reaction paths to open to it---decomposition or stabilization by hydrogen transfer. In general, at pyrolysis conditions (>500°C) decomposition processes will dominate. However, in the presence of HX, hydrogen transfer may become competitive with decomposition and some radicals may be stabilized before they can decompose. Consider the hypothetical chain:

$$R_1\cdot \rightarrow R_2 + O_2 \qquad [4]$$

$$R_2\cdot \rightarrow R_3\cdot + O_3 \qquad [5]$$

$$R_3\cdot + R_1H \rightarrow R_3H + R_1\cdot \qquad [6]$$

where O_2 and O_3 are olefins and the $R_i\cdot$ are alkyl radicals.

In the presence of HX, the reaction:

$$R_2\cdot + HX \rightarrow R_2H + X\cdot \qquad [7]$$

may be in competition with [5]. Then the relative yields of O_3 and R_3H would decrease and R_2H would appear in the products. Butane and n-hexadecane display this type of selectivity modification.

(b) Alteration of hydrogen atom abstraction selectivity: Consider a reactant R_mH from which hydrogen abstraction yields two different radicals $R_m\cdot$ and R_m'. One can postulate a hypothetical reaction scheme as follows:

$$R_n\cdot + R_mH \begin{matrix} \nearrow R_m\cdot \quad [8a] \\ \searrow R_m'\cdot \quad [8b] \end{matrix} + R_nH$$

$$R_n'\cdot + R_mH \begin{matrix} \nearrow R_m\cdot \quad [9a] \\ \searrow R_m'\cdot \quad [9b] \end{matrix} + R_n'H$$

$$R_m\cdot \rightarrow O_m + R_n\cdot \qquad [10a]$$

$$R'_m\cdot \rightarrow O'_m + R'_n\cdot \qquad [10b]$$

At steady state the ratio of O_m to O'_m in the products is given by

$$\frac{O_m}{O'_m} = \frac{k_{9a}\ (k_{8a} + k_{9b})}{k_{8b}\ (k_{9a} + k_{9b})}$$

The selectivity is determined by the relative rates of hydrogen abstraction by the chain carriers R'_m and $R'_n\cdot$. HX addition will modify the hydrogen transfer process by replacing reactions [8] and [9] with [11] and [12]:

$$\left.\begin{matrix} R_n\cdot \\ R'_n\cdot \end{matrix}\right\} + HX \rightarrow X\cdot + \left\{\begin{matrix} R_nH \\ R'_nH \end{matrix}\right. \qquad \begin{matrix} [11a] \\ [11b] \end{matrix}$$

$$X\cdot + R_mH \rightleftharpoons \left.\begin{matrix} R_m\cdot \\ R'_m\cdot \end{matrix}\right\} + HX \qquad \begin{matrix} [12a] \\ [12b] \end{matrix}$$

The steady state ratio of O_m and O'_m is now:

$$\frac{O_m}{O'_m} = \frac{k_{10a}}{k_{10b}}\ \frac{(k_{12a})}{(k_{12b})}\ \frac{(k_{10b} + k_{-12b}\ HX)}{(k_{10a} + k_{-12a}\ HX)} \qquad [13]$$

At low HX concentrations, the reverse processes k_{-12a} and k_{-12b} will be unimportant and [13] will reduce to

$$\frac{O_m}{O'_m} = \frac{k_{12a}}{k_{12b}} \qquad [14]$$

The product selectivity will depend only on the selectivity for hydrogen abstraction by X·. If X· abstracts differently than $R_n\cdot$ and $R'_n\cdot$, the selectivity will be altered.

At high HX concentrations, [13] reduces to

$$\frac{O_m}{O'_m} = \frac{k_{10a}}{k_{10b}}\ \frac{k_{12a}\ k_{-12b}}{k_{12b}\ k_{-12a}} \qquad [15]$$

At such HX concentrations, hydrogen transfer is no longer the rate limiting step and one might expect the selectivity to depend on relative radical decomposition rates as is the case in [15].

Thus, for a reactant which decomposes by two different hydrogen abstraction paths, as the HX concentration is increased, the product selectivity may change first in one direction reflecting differences in hydrogen abstraction rates by X·, and then in another due to acceleration of hydrogen transfer to point where radical decomposition is rate limiting. Such behavior has been observed by Lyon in the HCl catalyzed thermal alkylation of isobutane with ethylene (21).

n-BUTANE

All of the effects discussed in the previous section are displayed in the cracking of butane modified by H_2S. An examination of the cracking mechanism indicates why:

Initiation: $$C_4H_{10} \rightarrow 2C_2H_5^{\cdot} \quad [16]$$

H· Transfer: $$C_2H_5^{\cdot} + C_4H_{10} \rightarrow C_2H_6 + \begin{cases} CH_3CH_2\dot{C}HCH_3 & [17a] \\ CH_3CH_2CH_2\dot{C}H_2 & [17b] \end{cases}$$

$$CH_3^{\cdot} + C_4H_{10} \rightarrow CH_4 + \begin{cases} CH_3CH_2\dot{C}HCH_3 & [18a] \\ CH_3CH_2CH_2\dot{C}H_2 & [18b] \end{cases}$$

$$H^{\cdot} + C_4H_{10} \rightarrow H_2 + \begin{cases} CH_3CH_2\dot{C}HCH_3 & [19a] \\ CH_3CH_2CH_2\dot{C}H_2 & [19b] \end{cases}$$

Decomposition: $$CH_3CH_2\dot{C}HCH_3 \rightarrow C_3H_6 + CH_3^{\cdot} \quad [20]$$

$$CH_3CH_2CH_2\dot{C}H_2 \rightarrow C_2H_4 + C_2H_5^{\cdot} \quad [21]$$

$$C_2H_5^{\cdot} \rightarrow C_2H_4 + H^{\cdot} \quad [21a]$$

Termination: $$\left.\begin{array}{l} C_2H_5^{\cdot} + C_2H_5^{\cdot} \rightarrow C_4H_{10} \text{ or } C_2H_4 + C_2H_6 \\ C_2H_5^{\cdot} + CH_3^{\cdot} \rightarrow C_3H_8 \\ CH_3^{\cdot} + CH_3^{\cdot} \rightarrow C_2H_6 \end{array}\right\} [22]$$

In the presence of H_2S, new hydrogen transfer paths are possible:

$$R^{\cdot} + H_2S \rightarrow RH + HS^{\cdot} \quad [23]$$

where $R^{\cdot} = C_2H_5^{\cdot}$, $CH_3^{\cdot}$ or $H^{\cdot}$.

$$HS^{\cdot} + C_4H_{10} \rightarrow H_2S + \begin{cases} CH_3CH_2\dot{C}HCH_3 & [24] \\ CH_2CH_2CH_2\dot{C}H_2 & [25] \end{cases}$$

New termination paths may also open up:

$$\left.\begin{matrix} CH_3\cdot \\ C_2H_5\cdot \end{matrix}\right\} + HS\cdot \rightarrow \text{Products} \qquad [26]$$

Since the primary decomposition reactions [20,21] involve C-C bond scission, the hydrogen transfer reaction is rate limiting. H_2S addition would be expected to accelerate the rate of butane disappearance. Our data at 650°C confirm this as Figure 1 indicates. If steady state radical concentrations

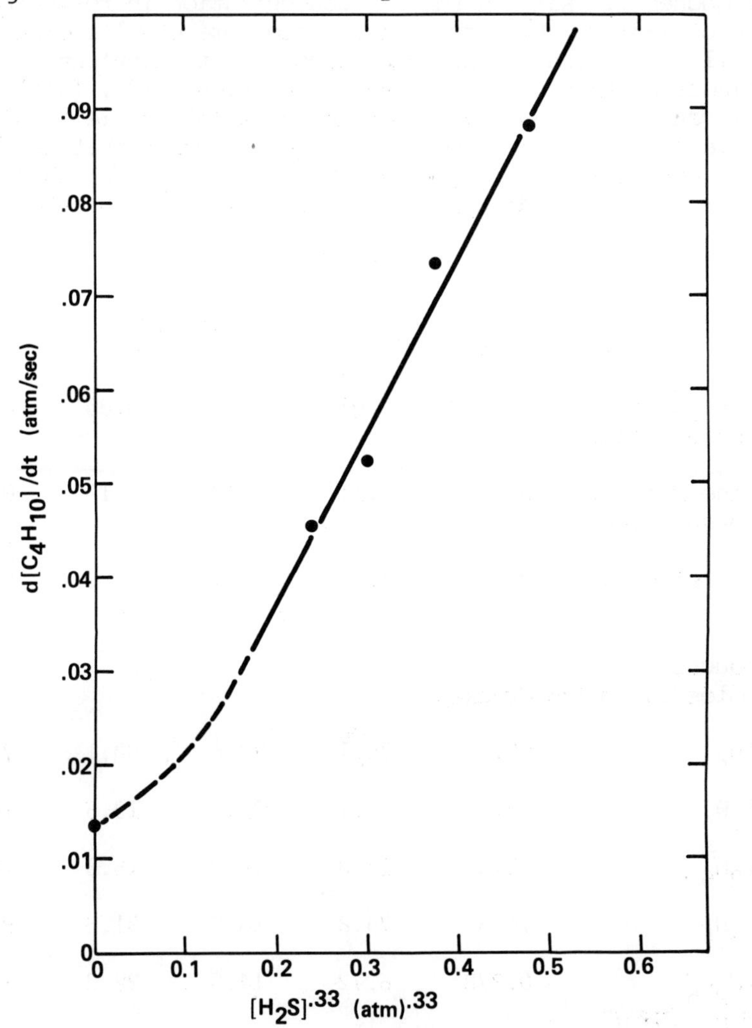

Fig. 1. Variation of the Initial Cracking of 0.1 atm of n-Butane at 650°C and 0.07 Seconds Residence Times with H_2S Partial Pressure to the 0.33 Power.

are assumed and H_2S introduces only reactions [23,24,25], the overall decomposition of butane can be shown to be between half and first order in H_2S. However, as Figure 1 indicates, the rate is proportional to $[H_2S]^{0.33}$ suggesting that termination reactions [26] are significant.

A comparison of product yields for butane cracking at 650°C with and without H_2S addition is shown in Table II. The changes in product distributions reflect both modes of selectivity modification discussed in the last section. As the H_2S pressure is increased, the ratio of ethane to ethylene approaches 1. Raising the H_2S concentration increases the rate of [23] relative to [21a]; hence more ethyl radicals are stabilized to ethane and fewer decompose to ethylene. An estimate of the relative rates of reactions [17], [21a] and [23] can be obtained by examining the relative yields of ethylene and ethane. The percentage of butane that decomposes via reaction [21] is given by $([C_2H_4] + [C_2H_6])/2$; this is the number of ethyl radicals produced by [21] per 100 butane

TABLE II

H_2S Catalysis of n-Butane Pyrolysis at 650°C and 0.07 sec Residence Time

H_2S Partial Pressure (atm)	-	0.013	0.026	0.051	0.106
Butane Partial Pressure (atm)	0.11	0.11	0.10	0.10	0.11
Conversion (%)	0.88	2.96	3.41	4.80	5.76
Products Moles/100 Moles Cracked					
CH_4	68.2	79.1	79.6	80.1	79.8
C_2H_4	53.5	23.1	21.5	19.6	19.4
C_2H_6	8.5	17.8	18.9	18.3	18.8
C_3H_6	70.3	79.8	80.5	81.5	81.3
$\frac{2C_2H_6}{C_2H_4 - C_2H_6}$	0.378	6.72	14.5	28.2	67.7

molecules decomposed. Then the ratio of ethyl radicals stabilized to ethane, to those that decompose to ethylene is given by:

$$\frac{(k_9[C_4H_{10}] + k_{15}[H_2S])[C_2H_5]}{k_{13a}[C_2H_5\cdot]} = \frac{[C_2H_6]}{([C_2H_4] + [C_2H_6])/2-[C_2H_6]}$$

$$= \frac{2[C_2H_6]}{[C_2H_4]-[C_2H_6]}$$

This ratio is included in Table II. A plot of $2[C_2H_6]/[C_2H_4]-[C_2H_6]$ versus $[H_2S]$ should yield a straight line with slope $\frac{k_{15}}{k_{13a}}$ and intercept $\frac{k_9\ [C_4H_{10}]}{k_{13a}}$. Such a plot is shown in Figure 2 for cracking at 650°C, yielding values of

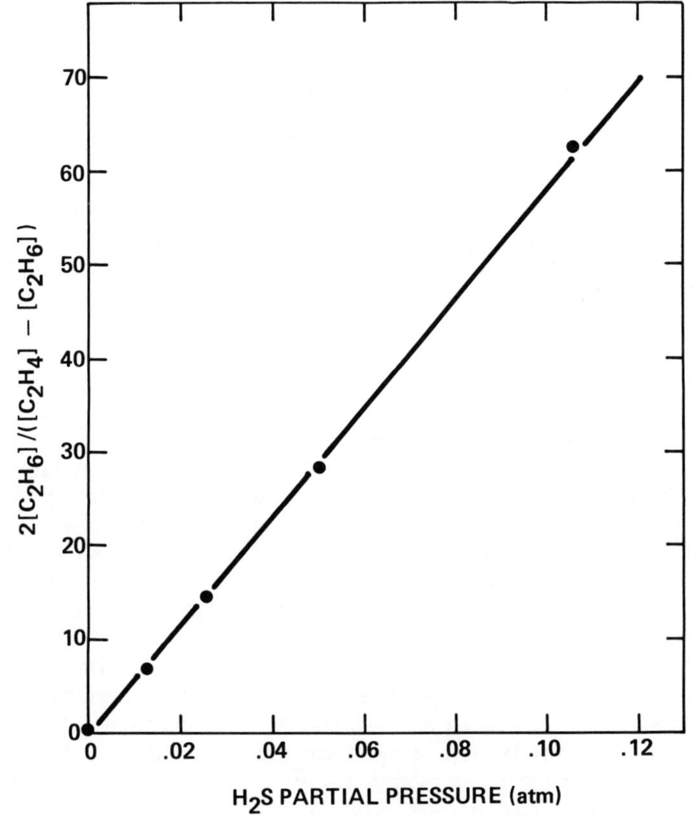

Fig. 2. Determination of the Relative Rate of Hydrogen Abstraction of $C_2H_5\cdot$ from C_4H_{10} and H_2S (see text).

1:3.5:575 for $k_{13a}:k_9:k_{15}$ (k_9 and k_{15} in [atm sec]$^{-1}$). The ratio of the rates of hydrogen transfer to ethyl radicals by H_2S and n-butane is about 160. The only literature values available for comparison are those for hydrogen transfer to methyl radicals by H_2S and methane (14), where the ratio extrapolated to 650°C is 110.

The other selectivity change shown by Table II when H_2S is present is an increase in the yields of propylene and methane, relative to those of ethylene and ethane. This suggests that reaction [20] is increasing in importance relative to [21] or that hydrogen abstraction by HS· is more selective to secondary butyl radicals than hydrogen abstraction by $CH_3\cdot$, $C_2H_5\cdot$ or H·.

n-HEXADECANE

As with n-butane, H_2S addition to n-hexadecane pyrolysis results in a cracking rate increase and a change in product distribution. The mechanism of cetane cracking is significantly more complicated than that of butane but the uncatalyzed product distribution can be predicted by the method of Rice and Kossiakoff (15). This mechanism assumes decomposition of $C_{16}H_{33}\cdot$ radicals in several steps to $CH_3\cdot$, $C_2H_5\cdot$ and H· which abstract hydrogen from cetane to complete the chain. Intermediate length radicals formed by decomposition of $C_{16}H_{33}\cdot$ isomerize by 5, 6 and 7 membered ring transition states before themselves decomposing. In this mechanism, the hydrogen abstraction step is assumed to be rate limiting. Therefore, H_2S addition should increase the rate by catalyzing the hydrogen transfer step. The experimental data show that this is indeed the case. The rate increases with the square root of the H_2S pressure as Figure 3 indicates. The higher order in H_2S with cetane than with n-butane suggests that H_2S participation in termination is less important in the former.

The uncatalyzed pyrolysis of hexadecane gives ethylene, propylene, ethane and methane as major products, with smaller amounts of C_4 to C_{15} alpha olefins completing the product distribution. In the presence of H_2S, the selectivity to methane, ethane and ethylene decreases and C_3^+ paraffins appear in the products. Table III shows typical product distributions with and withou H_2S present. This can be explained by the first of the selectivity modifying mechanisms discussed previously: stabilization of reactive radicals. When H_2S is added to the pyrolysis, some intermediate length ($C_3\cdot$-$C_{14}\cdot$) radicals are converted by hydrogen transfer to paraffins before they can decompose. Hence the decrease in light products and the increase in heavier paraffins shown in Figures 4 and 5 respectively.

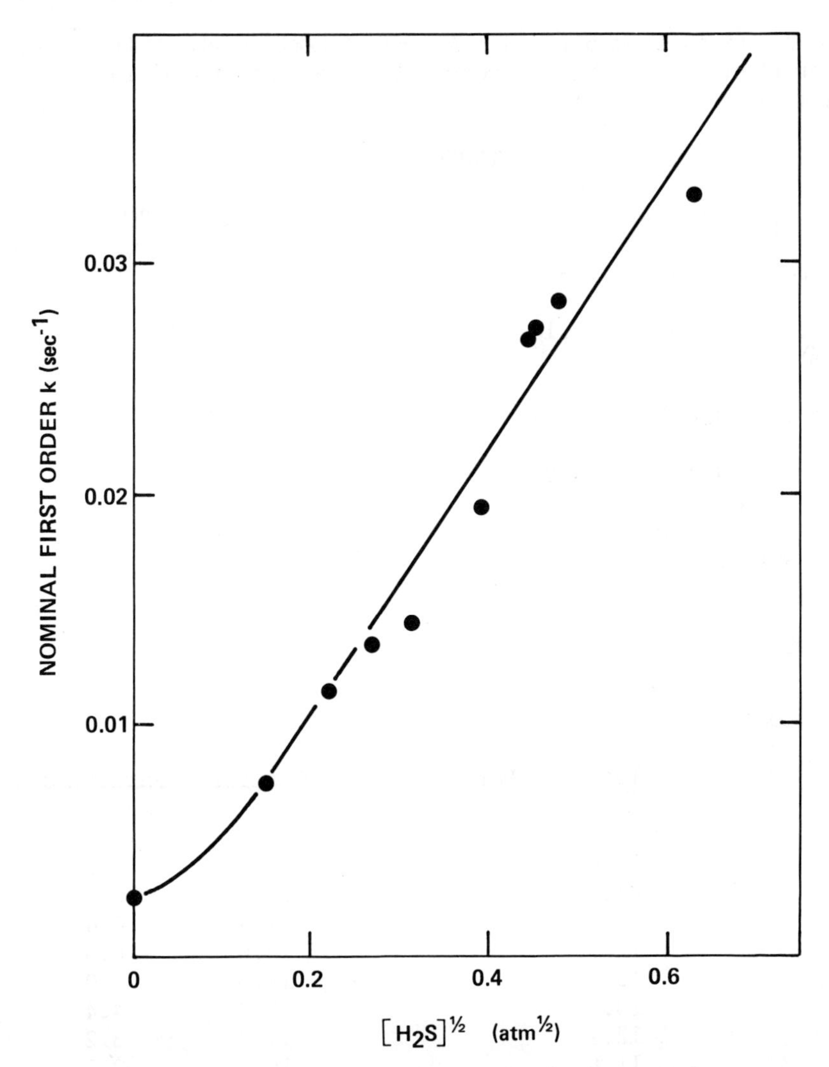

Fig. 3. Variation in the Nominal First Order Rate Constant for n-Hexadecane Cracking at 502°C with $\sqrt{H_2S}$

A similar change in product distribution has been reported for pure n-hexadecane pyrolysis at high pressures (16). By comparing the cetane and H_2S pressures necessary to achieve the same selectivity modifications, one can estimate the average relative rate constants of the reactions:

$$C_nH_{2n+1}^{\cdot} + C_{16}H_{34} \rightarrow C_nH_{2n+2} + C_{16}H_{33}^{\cdot} \qquad [27]$$

and $$C_nH_{2n+1}^{\cdot} + H_2S \rightarrow C_nH_{2n+2} + HS\cdot \qquad [28]$$

where n ranges from 5 to 14. At 500°C [28] is about 140 times faster than [27]. This procedure has been discussed in more detail previously (10).

TABLE III

H_2S Catalysis of n-Hexadecane Pyrolysis at 502°C

H_2S Pressure (atm)	-		0.16	
$C_{16}H_{34}$ Pressure (atm)	0.18		0.18	
Space Time (sec)	20.2		8.9	
Conversion (%)	4.9		13.6	
First Order Rate Constant (sec^{-1})	0.0025		0.0168	
Products (Moles/100 Moles Cracked)				
	Olefins	Paraffins	Olefins	Paraffins
C_1		48.8		12.0
C_2	83.6	48.1	25.6	28.3
C_3	43.6	4.4	25.3	22.9
C_4	23.5		11.4	6.4
C_5	12.0		13.0	6.3
C_6	13.2		18.5	5.0
C_7	13.0		13.3	4.4
C_8	12.2		11.4	3.2
C_9	11.6		10.6	2.8
C_{10}	11.7		10.6	2.7
C_{11}	10.3		9.6	2.3
C_{12}	9.4		8.8	2.0
C_{13}	8.2		7.6	2.0
C_{14}	8.9		7.8	
C_{15}	2.8		2.2	
Total	370.5		275.6	

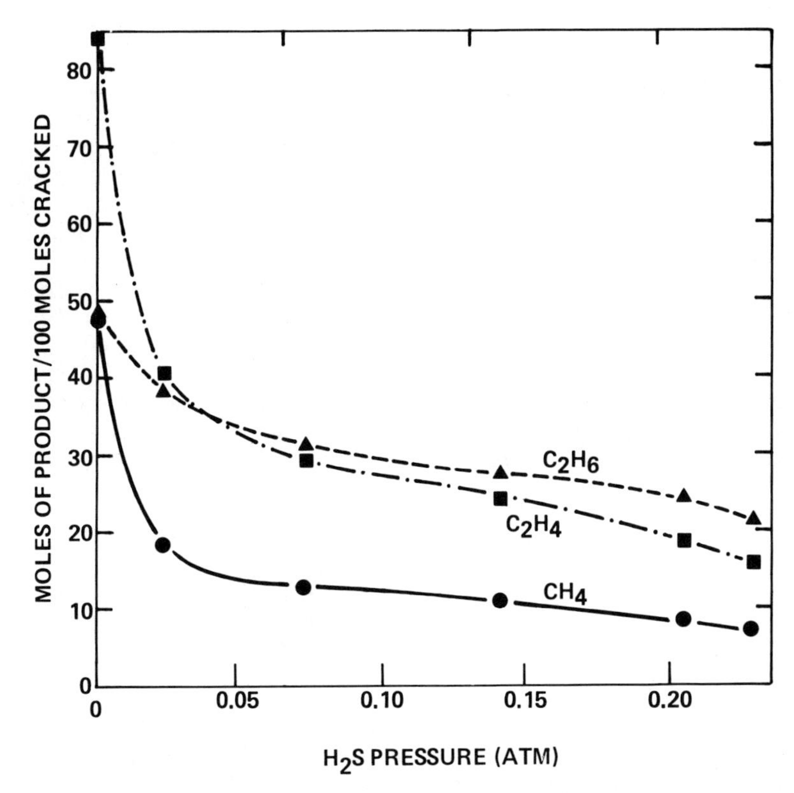

Fig. 4. Effect of H_2S Pressure on Light Gas Yields in n-Hexadecane Pyrolysis at 502°C

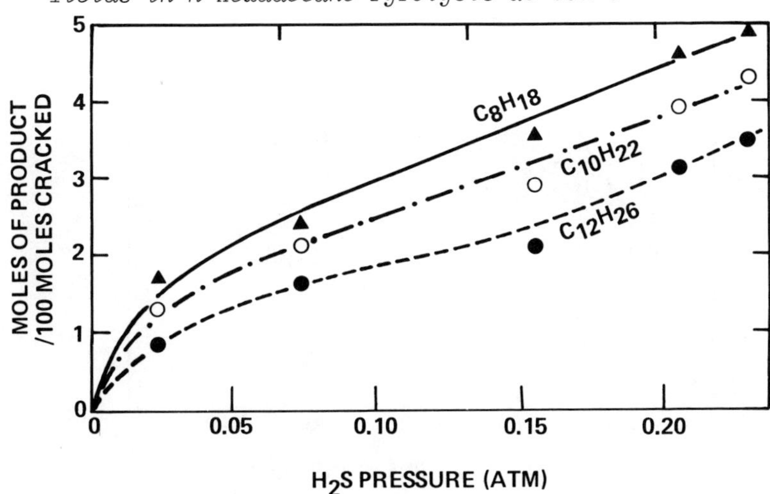

Fig. 5. Effect of H_2S Pressure on n-Paraffin Yields in n-Hexadecane Pyrolysis at 502°C

1-DODECENE

The effects of H_2S addition on dodecene pyrolysis are similar to these for n-hexadecane--a significant rate increase and a change in product distribution. Figure 6 shows the change in nominal first order decomposition rate constant with H_2S pressure at 500°C. The major changes in product distribution are the same as those for cetane--a reduction in methane and ethylene and the appearance of C_3^+ paraffins. Figures 7 and 8 illustrate this for the pyrolysis of 0.1 atm of dodecene at 500°C. These trends are presumably due to the stabilization of intermediate length alkyl radicals before they decompose further, as in the cetane case.

Uncatalyzed dodecene pyrolysis gives small amounts of diolefins and C_5, C_6 and C_7 cyclic compounds as primary products. These compounds are not produced in the initial

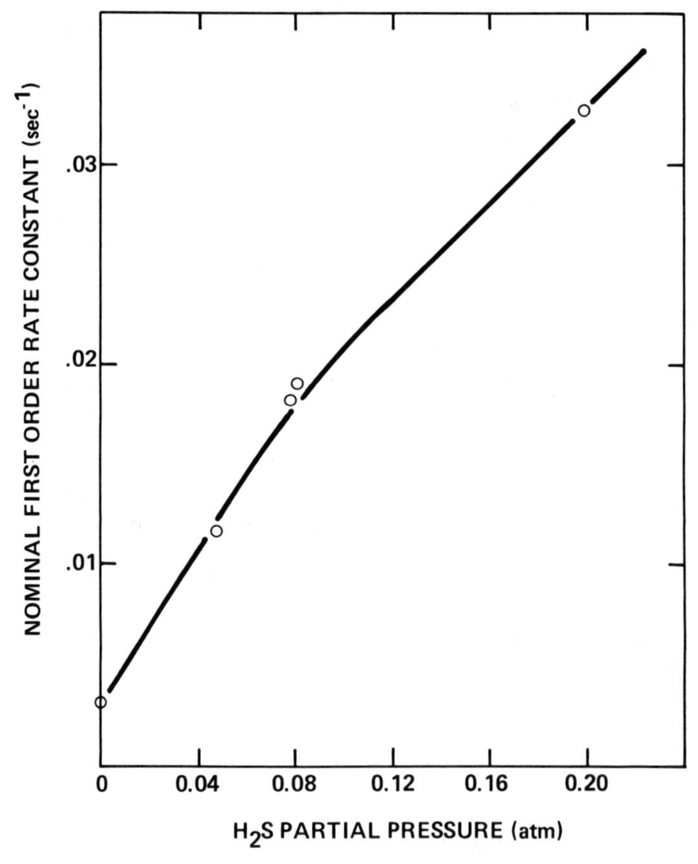

Fig. 6. Variation in Nominal First Order Rate Constant for 1-Dodecene Cracking at 500°C with H_2S Pressure

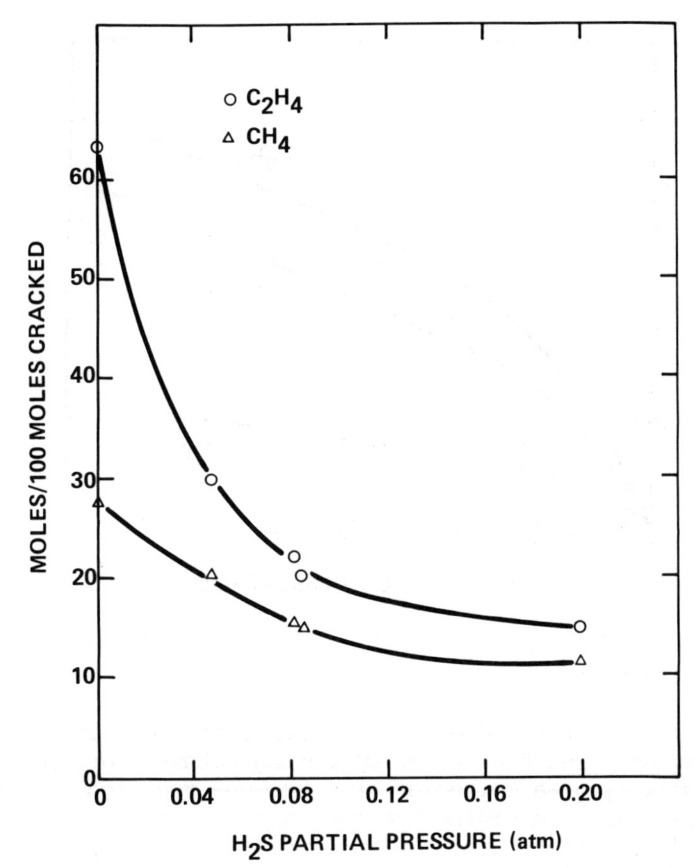

Fig. 7. Effect of H_2S Pressure on Methane and Ethylene Yields in Dodecene Pyrolysis at 500°C

steps of paraffin cracking. The addition of H_2S significantly increases the yields of both diolefins and cyclic compounds, particularly cycloparaffins. Similar results have been reported by Frech, et al. (12).

The increase in cycloparaffin yield possibly is due to the stabilization of cyclic radicals by H_2S. These cyclic species are formed from olefinic radicals. For example the sequence

$$\rightleftharpoons \quad \rightleftharpoons \qquad [29]$$

$$+ H_2S \longrightarrow \quad + HS\cdot \qquad [30]$$

might explain the increase in methylcyclopentane yield when H_2S is present.

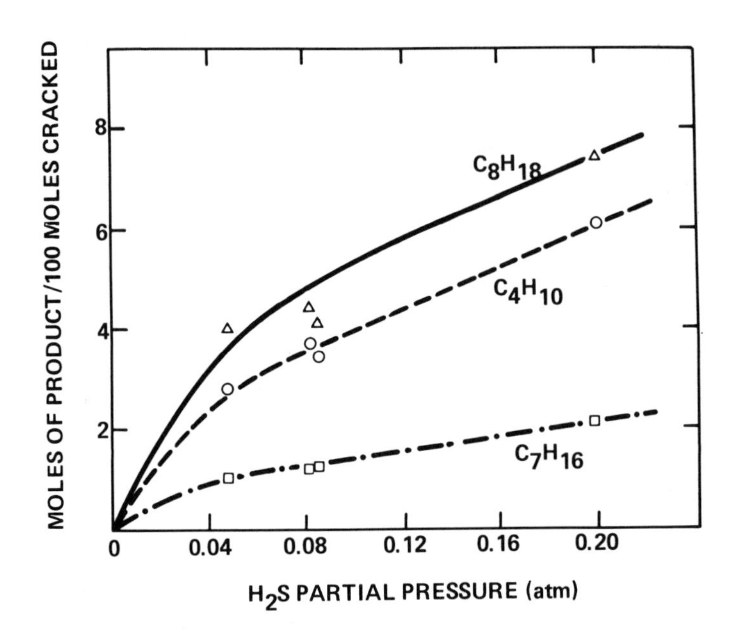

Fig. 8. Effect of H_2S Pressure on n-Paraffin Yields in Dodecene Pyrolysis at 500°C

The reasons for the diolefins increase are more subtle. It has been proposed that three competing pathways are present in olefin cracking--a molecular or retro-ene reaction, and two free radical paths involving hydrogen abstraction from or radical addition to the reactant olefin (17,18). Diolefins are produced only by the H-abstraction route. If H_2S replaces the hydrocarbon chain carriers in the system with HS·, the radical addition path will be inhibited and the abstraction path enhanced. This will lead to the observed increase in selectivity to diolefins.

n-BUTYLBENZENE

The major effect of H_2S addition on n-butylbenzene pyrolysis is a significant rate increase. Table IV compares typical runs with and without H_2S at 595°C. The small changes in the product distribution are due to two factors: (a) There are differences in H-abstraction selectivity between the hydrocarbon chain carriers present in the uncatalyzed reaction and HS·, (b) H_2S inhibits the two secondary decomposition reactions:

$$C_2H_5\cdot \rightarrow C_2H_4 + H\cdot \qquad [31]$$

and $C_6H_5CH_2CH_2\cdot \longrightarrow C_6H_5\cdot + C_2H_4$ [32]

TABLE IV

H_2S Catalysis of n-Butylbenzene Pyrolysis at 595°C

n-Butylbenzene Pressure (atm)	0.46	0.47
H_2S Pressure (atm)	-	0.19
Space Time (sec)	5.5	1.1
Conversion (%)	21.9	22.6
Nominal First Order Rate Constant	0.047	0.246
Products (Moles/100 Moles Cracked)		
CH_4	15.5	7.0
C_2H_4	28.8	4.9
C_2H_6	41.4	67.3
C_3H_6	19.7	23.1
C_3H_8	0.9	1.6
C_4S	0.7	0.1
Benzene	2.3	0.4
Toluene	19.8	21.0
Ethyl Benzene	4.5	3.0
Styrene	57.1	67.3
Alkyl Benzene	5.3	3.7
Others	7.4	2.7
Total	203.4	202.1

decreasing the yields of benzene and ethylene and increasing those of ethane and ethylbenzene.

TETRALIN

In contrast to the pyrolysis of n-butylbenzene, the tetralin cracking rate is unchanged by the addition of H_2S, but the product distribution is significantly altered. Table V compares conversion and product selectivities for experiments with and without added H_2S at 540°C. The primary mode of tetralin decomposition at this temperature is dehydrogenation to dehydronaphthalene and naphthalene. Although the chain length is unknown, the primary propagation steps are assumed to be:

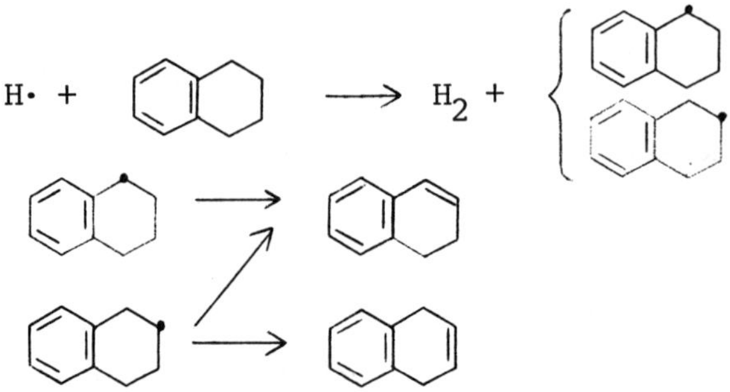

$$H\cdot + \text{tetralin} \longrightarrow H_2 + \left\{ \begin{array}{l} \text{1-tetralyl radical} \\ \text{2-tetralyl radical} \end{array} \right. \qquad [33]$$

and

$$\text{1-tetralyl radical} \longrightarrow \text{1,2-dihydronaphthalene} \qquad [34]$$

$$\text{2-tetralyl radical} \longrightarrow \text{1,2-dihydronaphthalene (and 1,4-dihydronaphthalene)} \qquad [35]$$

In this scheme, the hydrogen transfer reactions are exothermic and are unlikely to be the rate limiting steps. Hence H_2S has little effect on rate.

However, H_2S has two effects on the product distribution. (a) Enhancement of C_4-ring opened product yield at the expense of dehydrogenated products and C_3 and lower alkyl aromatics. This effect is easily explained if it is assumed that the radicals formed by reactions [33] can also decompose by ring-opening:

e.g. $$\text{1-tetralyl radical} \rightleftharpoons \text{ring-opened radical} \qquad [36]$$

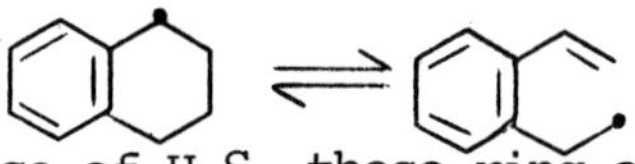

In the absence of H_2S, these ring-opened radicals will re-cyclize or further fragment. In the presence of H_2S they will be stabilized to C_4 alkylbenzenes.

TABLE V

H_2S Catalysis of Tetralin Pyrolysis at 540°C

Tetralin Pressure (atm)	0.42	0.43
H_2S Pressure (atm)	-	0.18
Space Time (sec)	24	24
Conversion (%)	3.4	3.5
Products (Mole/100 Moles Cracked)		
Ethyl Benzene and Styrene	4.2	1.1
Indene	7.3	4.3
C_4 Alkylbenzenes	4.7	39.9
Dihydronaphthalene	73.1	37.2
Naphthalene	10.6	17.5
Total	99.9	100.0

The second effect of H_2S on the product selectivity is the promotion of the secondary dehydrogenation of dihydronaphthalene to naphthalene. Presumably the ratio of primary to secondary dehydrogenation is determined by the relative ease with which the chain carriers abstract hydrogen from tetralin and dihydronaphthalene. One would expect H· to be nonselective since for it both reactions are exothermic, while HS· should prefer to attack dihydronaphthalene.

CONCLUSIONS

The results presented in the previous sections indicate that H_2S can modify pyrolyses in various ways. The common element in all cases though is the catalysis of hydrogen transfer from reactants to free radicals by H_2S. Similar results are observed with HCl and HBr. This hydrogen transfer catalysis can result in a cracking rate increase and changes in the product distribution. Selectivity modifications come from the stabilization of free radicals by hydrogen transfer from H_2S before they can further decompose, or from differences in hydrogen abstraction selectivity between HS· and hydrocarbon free radical chain carriers. Thus homogeneous hydrogen transfer catalysis represents an interesting approach to the problem of rate and selectivity control in hydrocarbon pyrolysis.

ACKNOWLEDGMENTS

The author would like to thank Drs. W. H. Davis, Jr. and R. K. Lyon for helpful discussions. The excellent technical assistance of Mr. J. C. Dowling is also gratefully appreciated.

REFERENCES

1. Benson, S. W. and Anderson, K. H., *J. Chem. Phys.* *39*, 1677 (1963).

2. Imai, N. and Toyama, O., *Bull. Chem. Soc. Japan* *34*, 328 (1961).

3. Muller, J., Baronnet, F., Scacchi, G., Dzierzynski, M. and Niclause, M., *Int. J. Chem. Kin.* *9*, 425 (1977).

4. Scacchi, G., Dzierzynski, M., Martin, R. and Niclause, M. *Int. J. Chem. Kin.* *2*, 115 (1970).

5. McLean, P. R. and McKenney, D. J., *Can. J. Chem.* *48*, 1782 (1970).

6. Niclause, M., Baronnet, F., Scacchi, G., Muller, J. and Jezequel, J. Y., Industrial and Laboratory Pyrolyses, ACS Symposium Series, 32, Washington, DC (1976), pp. 17-36.

7. Large, J. F., Martin, R. and Niclause, M., *C. R. Acad. Sci*, Paris, 274C, 322 (1972).

8. Anderson, K. H. and Benson, S. W., *J. Chem. Phys. 40*, 3747 (1964).

9. Scacchi, G., Baronnet, F., Martin, R. and Niclause, M., *J. Chem. Phys. 65*, 1671 (1968).

10. Rebick, C., "H_2S Catalysis of n-Hexadecane Pyrolysis", presented at Joint ACS-CIC Conference, Physical Chemistry Division, Montreal, Canada, June 2, 1977.

11. Hutchings, D. A., Frech, K. J. and Hoppstock, F. H., Industrial and Laboratory Pyrolyses, ACS Symposium Series, 32, Washington, DC (1976), pp. 178-196.

12. Frech, K. J., Hoppstock, F. H. and Hutchings, D. A., Ibid., pp. 197-217.

13. Niclause, M., Martin, R., Baronnet, F. and Scacchi, G., *Rev. Inst. Fr. Petrole, 21*, 1724 (1966).

14. Kerr, J. A. and Parsonage, M. J., Evaluated Kinetics Data on Gas Phase Hydrogen Transfer Reactions of Methyl Radicals, Butterworths, London (1976).

15. Kossiakoff, A. and Rice, F. O., *J. Am. Chem. Soc. 65*, 590 (1943).

16. Doue, F. and Guiochon, G., *J. Chim. Phys. 64*, 395 (1969).

17. Rumyanstev, A. N., et al., Proceedings of the Ninth World Petroleum Congress, Tokyo (1975) Vol. 5, p. 155.

18. Rebick, C., "Pyrolysis of Alpha Olefins - A Mechanistic Study", ACS Advances in Chemistry Series, 183, Washington, DC (1979), pp. 1-19.

19. Cottrell, T. L., The Strengths of Chemical Bonds, pp. 270-274, Butterworths, London (1958).

20. Benson, S. W., Thermochemical Kinetics, John Wiley and Sons, Inc., New York (1968).

21. Lyon, R. K. and Mitchell, J. E., "Modifications of Thermal Alkylation with HCl", ACS Advances in Chemistry Series, 183, Washington, DC (1979), pp. 289-296.

THE GAS PHASE CHEMISTRY OF CARBYNES[1]

F.C. JAMES, H.K.J. CHOI, B. RUZSICSKA AND O.P. STRAUSZ

Chemistry Department
University of Alberta
Edmonton, Alberta, Canada

and

T.N. BELL

Chemistry Department
Simon Fraser University
Burnaby, British Columbia, Canada

Introduction

Carbynes, the monovalent carbon radical family comprising CH and its derivatives, constitute the least investigated and consequently the least understood variety of carbon radical species. Our present day knowledge of these radicals is based upon perhaps only three dozen articles published during the last two decades - a mere trifle when compared to the vast accumulation of literature on carbene chemistry [1]; and yet, the investigations indicate that the chemistry of carbynes is just as interesting and varied as that of the analogous carbenes. Undoubtedly, the prime reason for the lack of data on carbynes is the unavailability of suitable, clean, general sources for the generation of carbynes under conditions applicable to kinetic, mechanistic studies. Consequently, each of the carbyne species - CH, CCl, CBr, CCO_2Et - for which the chemistry has been explored to date, has been generated in an individual way. Unfortunately, each method is complicated by the formation of not only carbyne, but a large number of other reactive radicals. Consequently, it is necessary to isolate the specific carbyne reaction under investigation from a maze

[1]*The authors acknowledge the continuing financial support of the Natural Sciences and Engineering Research Council of Canada.*

ISBN 0-12-566550-4

of complex parallel reactions involving the other radicals and secondary products.

In gas phase studies, the carbynes CH, CCl and CBr have been generated by the flash or laser photolysis, or pulsed radiolysis, of C_1-C_2 hydrocarbons, or by nuclear recoil atom reactions in the presence of selected substrates, and their reactions monitored by kinetic absorption spectroscopy or by laser induced fluorescence. In this manner it was possible to measure absolute rate constants at room temperature for a series of reactions of CH, CCl and CBr and to determine Arrhenius parameters for some reactions of CBr over the range 298-423°K. This approach, however, does not allow the establishment of the mechanistic details of the reactions, which have to be inferred on the basis of presumed analogies.

On the other hand, the chemistry of carbethoxymethylidyne has been investigated in the solution phase [2-5]. By means of kinetic studies and product analysis, it was possible to show that the ground state ($\tilde{X}^2\Pi$) carbethoxymethylidyne generated *via* the *in situ* photolysis of mercury bisdiazoacetate ($EtO_2CCN_2HgCN_2CO_2Et$) undergoes concerted, stereospecific cyclo-addition to olefins, regioselective insertion into the C-H bonds of paraffins, olefins and alcohols, polar addition to the O-H bonds of alcohols and cyclo-addition to aromatics resulting in ring expanded products [2-5]. Hydrogen abstraction by the carbyne could not be observed in these studies, neither could rate constant determinations be carried out.

CH is isoelectronic with the nitrogen atom, but unlike nitrogen which obeys Hund's rule and accordingly has an electronic configuration $1s^2 2s^2 2p^3$ and a ^4S spectroscopic ground state, the electron configuration of CH is $1\sigma^2 2\sigma^2 3\sigma^2 1\pi$ [6] which gives rise to only a single spectroscopic state, namely $^2\Pi$; the ground state of all known carbynes conforms to this configuration. It would therefore appear that the single bond of the carbyne carbon does not sufficiently influence the atomic orbital energies of the carbon to make sp^3 hybridization possible. The result is an orbital occupancy that is unique in carbon radical chemistry, in which there is a doubly occupied σ orbital, a singly occupied π orbital and an empty π orbital localized on the carbon atom:

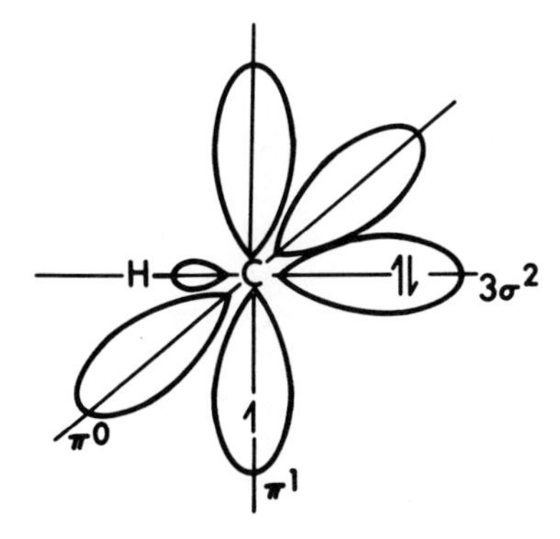

This electron deficiency endows carbynes with a high reactivity and a distinct electrophilic character. In the case of halocarbynes and carbethoxymethylidyne, overlap between the P_π orbitals of carbon and those of the halogen or the π orbital of CO, decreases the electron deficiency of the carbon and hence results in a reduction of the carbyne reactivity.

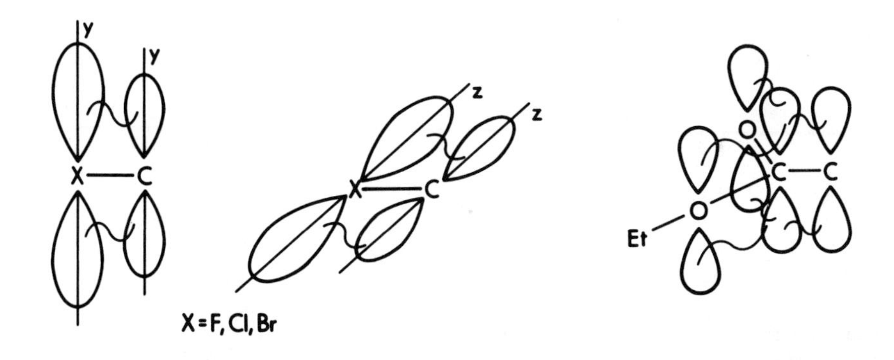

The effect may be stronger with halogens, where both P_π atomic orbitals of the halogen may overlap with the corresponding P_π orbitals of the carbyne carbon, than with the carbethoxy group, where the CO π molecular orbital can overlap with only one of the P_π atomic orbitals of the carbyne carbon. This electronic picture of carbynes is in agreement with the results of experimental studies which indicate that $CH(\tilde{X}^2\Pi)$ undergoes indiscriminate insertion and addition reactions with paraffins, olefins and acetylenes at rates close to collision frequency [7-9]. $CCl(\tilde{X}^2\Pi)$ and $CBr(\tilde{X}^2\Pi)$ on the other hand, are less reactive and more selective reagents; they add readily to olefins and acetylenes but insert only very slowly, if at all, into the C-H bonds of paraffins and olefins [10-12].

$EtO_2CC(\tilde{X}^2\Pi)$ displays an intermediate reactivity, inserting into the C-H bonds of paraffins and olefins, but in a regio-selective manner [2-5].

All of the excited spectroscopic states of carbynes can be deduced from the excited electron configurations. The lowest excited electronic configuration is $1\sigma^2 2\sigma^2 3\sigma 1\pi^2$, and this gives rise to four excited states, $^4\Sigma^-$, $^2\Sigma^+$, $^2\Sigma^-$ and $^2\Delta$ [6]. The lowest of these is the $^4\Sigma^-$ state which has not been observed by conventional spectroscopic techniques, but in the case of CH has recently been detected by laser photo-electron spectroscopy [13]. It lies 17 kcal/mole above the ground state, in reasonable agreement with earlier *ab initio* CI computational values [14].

For the specific case of CH, it can be predicted on the basis of the spin conservation rule and orbital symmetry considerations deduced from detailed *ab initio* m.o. calculations, that the reactivity of ground state $^2\Pi$ CH would be analogous to that of the lowest excited state of CH_2, $(\tilde{A}^1A_1)$, and that the reactivity of the lowest excited state of $CH(\tilde{a}^4\Sigma^-)$, would parallel that of $CH_2(\tilde{X}^3B_1)$:

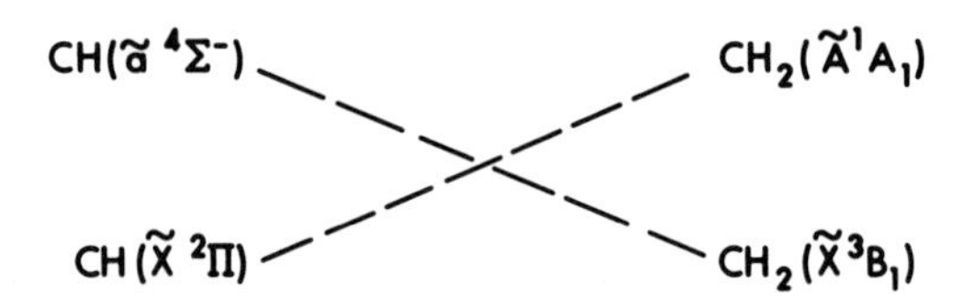

Thus, on this basis alone, the $^2\Pi$ ground state of CH may be predicted to undergo insertion and concerted stereospecific cyclo-addition reactions with hydrocarbons: this is in accord with experimental observations. The $^4\Sigma^-$ excited state of CH should react with hydrocarbons *via* hydrogen abstraction and non-stereospecific addition reactions, although no experimental evidence is available to confirm or deny this prediction.

As mentioned above, CH is a highly reactive reagent and in fact in the few cases where comparative data exist, $CH(\tilde{X}^2\Pi)$ appears to react faster than $CH_2(\tilde{A}^1A_1)$ *(vide infra)*. *Ab initio* calculations by Brooks and Schaefer [15] yielded very low energy barriers for insertion by asymmetric attack of CH $(\tilde{X}^2\Pi)$ on H_2:

$$CH(\tilde{X}^2\Pi) + H_2 \rightarrow CH_3(\tilde{X}^2A_2'')$$

and

$$CH_2(\tilde{A}^1A_1) + H_2 \rightarrow CH_4(\tilde{X}^1A_1)$$

On the other hand the computed energy barriers for the abstraction reactions

$$CH_2(\tilde{X}^3B_1) + H_2 \rightarrow CH_3(\tilde{X}^2A_2'') + H(^2S)$$

and

$$CH(\tilde{a}^4\Sigma^-) + H_2 \rightarrow CH_2(\tilde{X}^3B_1) + H(^2S)$$

are substantial and the value is somewhat higher for the carbene than the carbyne reaction [15]. *Ab initio* m.o. calculations on the asymmetric cyclo-addition of $CH(\tilde{X}^2\Pi)$ to ethylene predict a zero activation energy for the process [16]. All of the chemistry of the carbynes which has been studied to date applies to the $^2\Pi$ ground state species, and no report exists on the chemistry of an excited state carbyne. The observed chemistry of CBr and CCl shows considerable deviation from that of CH in terms of their specific reactivity with respect to insertion, addition, etc.

The near uv banded spectra of CH and its halogenated analogues have almost atomic line-like characteristics under low resolution, making them particularly suitable for kinetic spectroscopic measurements. The absorption spectrum of CH has been known for almost a century. It was first detected in the spectra of the sun and stars, and the emission spectrum, a component of the bunsen flame, was the subject of early analysis [17]. The absorption spectrum consists of two weak ($\tilde{A}^2\Delta \leftarrow \tilde{X}^2\Pi$, 430 nm and $\tilde{B}^2\Sigma^- \leftarrow \tilde{X}^2\Pi$, 390 nm) and one strong ($\tilde{C}^2\Sigma^+ \leftarrow \tilde{X}^2\Pi$, 314.5 nm) transitions in the near uv [18] with other stronger bands in the vacuum uv [19]. The attenuation with time of the intense $\tilde{C}\leftarrow\tilde{X}$ band has been used for conventional kinetic absorption spectroscopic measurements of CH decay [7,8]. The relative intensities of the bands are reversed in emission [20], and more recent kinetic measurements have utilized the laser induced fluorescence (LIF) of the $\tilde{A}^2\Delta \rightarrow \tilde{X}^2\Pi$ transition for monitoring CH concentrations [9,21]. All kinetic measurements of CBr and CCl reactions have been made using the kinetic absorption spectroscopy technique, monitoring the $(Q_1,0,0)$ bands of the $\tilde{A}^2\Delta \leftarrow \tilde{X}^2\Pi$ system at 277.7 nm (CCl) [22] and 301.4 nm (CBr) [23]. To date, there has been no report on direct kinetic studies of CF reactions, although the $\tilde{A}^2\Sigma^+ \leftarrow \tilde{X}^2\Pi$ system centered at 224 nm [24] is suitable for kinetic absorption spectroscopy and studies on this system are currently underway in the authors' laboratory. CI has not been observed spectroscopically, but its existence has been inferred from the fact that O^- ions were detected in an ion-molecule collision study of I^- with CO, presumably from the reaction [25]:

$$I^- + CO \rightarrow O^- + CI$$

Generation of Carbynes

Simons and Yarwood [26,27] have made the most comprehensive investigation of potential sources of halocarbynes, Table 1. CF, CCl and CBr were observed following the flash photolysis of a variety of halogenated molecules, primarily halomethanes, and their generation was ascribed to the unimolecular decay of vibrationally excited halomethyl radicals [26], e.g.

$$CHCl_2Br \xrightarrow{h\nu} CHCl_2^{\dagger} + Br$$

$$(CHCl_2^{\dagger})_{E>E^*} \longrightarrow CCl + HCl$$

$$\xrightarrow{+M} (CHCl_2)_{E<E^*}$$

The major products in decreasing order of importance, $CHCl_2CHCl_2$, $CHCl_3$ and CH_2Cl_2 were attributed to combination and disproportionation reactions of the thermalized $CHCl_2$ radicals, with small quantities of HCl also detected [27]. By varying the wavelength of the photolytic flash, Simons and Yarwood were able to show that almost 100% of the excess photonic energy resides in the $CHCl_2^{\ddagger}$ radical. Vibrational relaxation of the radical reduces the yield of CCl and consequently helium is used as diluent gas to maximize the yields of carbyne in kinetic experiments. Tyerman [30,31] used haloethylenes of the form $CCl_2{:}CXY$ as sources of CCl and attributed its generation to the sequential absorption of two photons,

$$\text{Olefin} + h\nu \rightarrow CCl_2 + CXY$$

$$CCl_2 + h\nu \rightarrow CCl + Cl$$

Interestingly, Tyerman observed vibrationally hot CCl ($\nu = 1$) from the flash photolysis of CF_2CCl_2, but not from C_2Cl_4, $CFCl{:}CCl_2$, $CHCl{:}CCl_2$ or $CH_2{:}CCl_2$. This was attributed [30] to the rapid relaxation of CCl ($\nu = 1$) by parent molecules having fundamental frequencies close to that of CCl at 864 cm^{-1} [22]. CF_2CCl_2 has no fundamental between 640 and 993 cm^{-1}.

CH has been generated by a number of techniques. For early kinetic measurements the vacuum uv ($\lambda > 105$ nm) flash photolysis [7,37] or ns pulse radiolysis [8] of CH was employed as the source of CH. In the photolysis, radiation of $\lambda < 136$ nm is sufficiently energetic to generate CH *via* the unimolecular decomposition of vibrationally excited methylene radicals,

TABLE 1

Sources of halomethylidynes in gas phase mechanistic/kinetic studies

CCl	CBr	CF	Technique[a]	Reference
Direct Observation				
$CHCl_2Br$, $CHClBr_2$	$CHBr_3$	$CHFBr_2$[b]	f.p.	[26,27]
CCl_2Br_2, CCl_3CHO	CBr_3CHO	$CFBr_3$[b]		
CH_2ClBr[b], CCl_3Br[b]	CBr_3NO_2			
CCl_3NO_2, CCl_3CN[b]	$CHClBr_2$			
$CHCl_3$[b]	CH_2Br_2[b]			
CF_2CCl_2			f.p.	[28]
CCl_3NO			f.p.	[29]
CF_2CCl_2, $CFClCCl_2$			f.p.	[30,31]
C_2Cl_4, $CHClCCl_2$				
CH_2CCl_2				
	$CHBr_3$		f.p.	[10,11]
$CHClBr_2$			f.p.	[12,32]
Inferred				
		CF_4/Ar	s.t.	[33]
		$CFBr_3$/NO[b]	f.p.	[34]
		$CFBr_3/SO_2$[b]	f.p.	[35]
		$CHFCl_2$[b]	s.s.p.	[36]

[a] f.p.: flash photolysis; s.t.: shock tube; s.s.p.: steady state photolysis.

[b] photolysis in the vacuum u.v.

$$CH_4 \xrightarrow[\lambda < 136 \text{ nm}]{h\nu} CH_2^{\dagger} + H_2$$
$$CH_2^{\dagger} \rightarrow CH(\tilde{X}^2\Pi) + H$$

However, the broad spectral output of the photoflash will generate substantial quantities of reactive CH_2 radicals having energies below the threshold for unimolecular decomposition. A vacuum uv steady state photolysis study of CH_4 (λ = 123.6 and 104.8-106.7 nm) indicates that CH could also be generated from the unimolecular decomposition of hot methyl radicals, and that C atoms may also be generated at the shortest wavelengths [38].

$$CH_4 \xrightarrow{h\nu} C(^1D) + 2H_2(^1\Sigma_g^+)$$

The presence of so many reactive species could lead to complications in the reaction mechanism and to uncertainties in any kinetic measurements of CH reactions. Similarly in the 600 kV pulse radiolysis of CH_4 [8] it is not clear whether the CH is formed *via* an ionic mechanism or by the fragmentation of neutral excited CH_4 formed by electron impact [39].

The presence of CH was invoked to explain the observation of 5 μm CO laser emission following the vacuum uv ($\lambda > 165$ nm) flash photolysis of $CHBr_3/M/SF_6$ systems, where M = O_2 [40], SO_2 [41] or NO, N_2O, NO_2 [42]. Lin proposed that the pumping mechanism for CO laser emission involves the reactions of CH, e.g.,

$$CH(^2\Pi) + O(^3P) \rightarrow CO^{\dagger}(^1\Sigma^+) + H(^2S)$$

and attributed the generation of CH to the sequential absorption of three photons by $CHBr_3$:

$$CHBr_3 \xrightarrow{h\nu} \xrightarrow[\lambda > 165 \text{ nm}]{h\nu} \xrightarrow{h\nu} CH + 3Br,$$

a scheme at variance with that of Simons and Yarwood. Although CH was not observed spectroscopically in these systems, the mass spectrometric determination of C_2H_2 among the reaction products was taken as confirmation of its presence.

The generation of CH from a variety of molecules by pulsed lasers has stimulated renewed interest in its reaction kinetics. Butler *et al.* [9] photolyzed $CHBr_3$ using a high powered pulsed ArF laser (193 nm) under mildly focussed conditions that would allow two-photon absorption [43], a photodecomposition scheme that is different to the sequential absorption of three photons proposed in the vacuum uv flash photolysis of $CHBr_3$ [40]. The photolysis of CH_3NH_2 by a TEA CO_2 laser was used as the source of CH in the kinetic investigation of Messing *et al.* [21]. In this case the multiphoton dissociation was believed to involve secondary photolysis of primary fragments such as CH_2NH_2. These authors also showed that multiphoton dissociation of CH_3CN, CH_3OH and C_2H_5OH also leads to the generation of CH.

Reactions of Carbynes: Chemistry and Kinetics

1. Methylidyne, CH

Wolff [44] invoked CH as an intermediate in the reactions of ^{11}C atoms with hydrocarbons, and by analogy with $CH_2(\tilde{A}^1A_1)$, postulated that CH would undergo insertion reactions with the C-H bonds of paraffins,

$$^{11}C + HR \rightarrow {}^{11}CH + R\cdot$$

$$^{11}CH + CH_3R \rightarrow [^{11}CH_2CH_2R]^{\dagger} \rightarrow {}^{11}CH_2CH_2 + R$$

and olefins

$$^{11}CH + C_2H_4 \rightarrow [^{11}CH_2\text{-}CH\text{-}CH_2]^{\dagger}$$

$$[^{11}CH_2\text{-}CH\text{-}CH_2]^{\dagger} + C_2H_4 \rightarrow C_5H_9 \xrightarrow{RH} C_5H_{10}$$

The first definitive kinetic measurement by Braun *et al.* [7] confirmed that insertion into the C-H bonds of CH_4 did in fact occur, and at a very rapid rate. CH was generated by the vacuum uv ($\lambda >$ 105 nm) flash photolysis of methane and its decay monitored by following the attenuation of the $\tilde{C} \leftarrow \tilde{X}$ Q branch at 314.3 nm. With the aid of product analysis and an estimated value of $1.2 \times 10^{11}\ M^{-1}s^{-1}$ for the rate of recombination of CH [37], they were able to determine rate constants for the following reactions:

$$CH + CH_4 \rightarrow CH_2\text{-}CH_3^{\dagger}$$

$$CH + H_2 \rightarrow CH_3^{\dagger}$$

$$CH + N_2 \rightarrow HCN + N$$

The results are summarized in Table 2. Bosnali and Perner [8] in the most comprehensive examination of the kinetics of CH to date, have determined rate constants for the reaction of CH with eleven substrates, and these results are also listed in Table 2. $CH(\tilde{X}^2\Pi)$ was generated by the 600 kV, 3 ns pulse radiolysis of CH_4, and the decay was again monitored by kinetic absorption spectroscopy of the $\tilde{C} \leftarrow \tilde{X}$ band. Comparison among the three substrates common to both investigations shows that Bosnali and Perner's rate constant values are a factor of 10 larger than those of Braun *et al.* [7]. Rate constant values for the reactions with CH_4, C_3H_8 and n-C_4H_{10} indicate that insertion into C-H bonds occurs at rates approaching collision

TABLE 2

Bimolecular rate constants for the reactions of $CH(\tilde{X}^2\Pi)$ *and* $CH_2(\tilde{A}^1A_1)$

Substrate	$k \times 10^{-9}$ $(M^{-1}s^{-1})$					
	$CH(\tilde{X}^2\Pi)$					$CH_2(\tilde{A}^1A_1)$
	Early Estimates	Braun... [7]	Bosnali... [8]	Butler... [9]	Messing... [21]	
H_2		0.62	10.5 ± 1.2	13.8 ± 3		4.2 [47]
N_2		0.043	0.61 ± 0.3	0.46 ± 0.12		3.0 [47]
CH_4		1.5	20.1 ± 0.5	181 ± 6		1.1 [47]
O_2	∿0.06 [45]		≲24		19.8 ± 2.4	<18 [48]
NH_3	>0.06 [46]		59 ± 9			
H_2O			27 ± 5			
CO			2.9			<5.4 [48]
C_3H_8			82 ± 20			3.8 [49]
$\underline{n}$-C_4H_{10}			78 ± 7			4.9 [49]
C_2H_4			69 ± 6			
C_2H_2			45 ± 9			

frequency. The higher values for C_2H_2 and C_2H_4 when compared to CH_4 (Table 2) suggest that addition to the unsaturated bonds also occurs.

The rate constants measured for reaction with inorganic substrates are also high, but one can only speculate about the nature of the reactions:

$$CH + NH_3 \rightarrow HCN + H_2 + N$$

$$CH + O_2 \rightarrow CO(^1\Sigma^+) + OH(^2\Sigma^+)$$

$$\rightarrow CO(^1\Sigma^+) + OH(^2\Pi)$$

$$\rightarrow CO_2(^1\Sigma_g^+) + H(^2S)$$

$$CH + CO \rightarrow ?$$

$$CH + H_2O \rightarrow CH_2\text{-}OH$$

Butler *et al.* [9] generated CH by the pulsed laser (ArF at 193 nm) photolysis of $CHBr_3$ and monitored its concentration by the LIF of the $\tilde{A}^2\Delta \rightarrow \tilde{X}^2\Pi(0,0)$ vibronic transition at 429.8 nm. The reactions of CH with CH_4, H_2 and N_2 were investigated for comparison with earlier work. From Table 2 it is apparent that for H_2 and N_2 there is good agreement with the rate constant values determined by Bosnali and Perner [8], but that the value for CH_4 is an order of magnitude larger (and two orders of magnitude larger than the value of Braun *et al.* [7]). The i.r. multiphoton dissociation of CH_2NH_2 was used as the source of CH by Messing *et al.* [21] and the kinetics of reaction with O_2 were monitored by the LIF of the Q branch of the $\tilde{A} \rightarrow \tilde{X}$ band at 431.4 nm. One of the reaction channels leads to the formation of OH $(\tilde{A}^2\Sigma^+)$

$$CH(^2\Pi) + O_2(^3\Sigma_g^-) \rightarrow CO(^1\Sigma^+) + OH(^2\Sigma^+) \quad \Delta H = -56 \text{ kcal/mole}$$

and the kinetic results determined from the chemiluminescent signal of OH* were found to correspond with the LIF measurement of the CH decay.

For comparison, Table 2 also contains kinetic data on the reactions of $CH_2(\tilde{A}^1A_1)$ and as can be seen, the reactivity of $CH(\tilde{X}^2\Pi)$ exceeds that of $CH_2(\tilde{A}^1A_1)$ in most instances, making $CH(\tilde{X}^2\Pi)$ one of the most reactive known organic radicals.

From Lin's results on the CO 5 μm laser emission following the flash photolysis of $CHBr_3/SO_2$, O_2, NO/SF_6 mixtures [40-42], the following reactions appear to explain the emission:

$$CHBr_3 \xrightarrow{h\nu} \xrightarrow[\lambda > 165\ nm]{h\nu} \xrightarrow{h\nu} CH(^2\Pi) + 3Br$$

$$SO_2 \xrightarrow[\lambda > 165\ nm]{h\nu} SO(^3\Sigma^-) + O(^3P)$$

$$CH(^2\Pi) + O(^3P) \longrightarrow CO^{\dagger}(^1\Sigma^+) + H(^2S)$$

$$CH(^2\Pi) + O_2(^3\Sigma_g^-) \longrightarrow CO^{\dagger}(^1\Sigma^+) + OH(^2\Pi)$$

$$\longrightarrow CO^{\dagger}(^1\Sigma^+) + OH(^2\Sigma^+)$$

$$\longrightarrow CO_2^{\dagger}(^1\Sigma_g^+) + H(^2S)$$

$$CH(^2\Pi) + NO(^2\Pi) \longrightarrow CO^{\dagger}(^1\Sigma^+) + NH(^3\Sigma^-)$$

with the side reactions

$$CH + CH\ (+ M) \longrightarrow C_2H_2$$

$$CH + SO \longrightarrow \text{products}$$

$$CH + SO_2 \longrightarrow HCSO_2,\ CHO + SO$$

limiting the emission yield.

2. Fluoromethylidyne, CF

Little work has been done on the chemistry of CF [50] but on the basis of CO 5 μm laser emission studies the following reaction has been suggested in the $CFBr_3/SO_2/SF_6$ system [35]:

$$CF(^2\Pi) + O(^3P) \longrightarrow CO^{\dagger}(^1\Sigma^+) + F(^2P)$$

complicated by the side reactions

$$CF + CF \rightarrow C_2F_2$$

$$CF + SO \rightarrow \text{products}\ (FCO + S,\ \text{etc.})$$

$$CF + SO_2 \rightarrow \text{products}\ (FCO + SO,\ \text{etc.})$$

For the $CFBr_3/NO/SF_6$ system [34], the interactions with $NO(^2\Pi)$, $O(^3P)$ and $N(^2S)$ are probably:

$$CF(^2\Pi) + NO(^2\Pi) \rightarrow FCN + O \text{ and/or } FCO + N$$

$$CF(^2\Pi) + O(^3P) \rightarrow CO^{\dagger}(^1\Sigma^+) + F(^2P)$$

$$CF(^2\Pi) + N(^2S) \rightarrow CN(^2\Pi) + F(^2P)$$

However, there have been no rate coefficient determinations apart from a shock tube study of the reaction CF + F + M which yielded a value of $k = 6.57 \times 10^{26}\ T^{-2.85}\ ml^2\ mol^{-2}\ s^{-1}$ when M = Ar [33].

3. Bromo- and Chloromethylidyne CBr and CCl

Of all the carbynes, $CBr(\tilde{X}^2\Pi)$ and $CCl(\tilde{X}^2\Pi)$ have been the most thoroughly investigated. Systematic kinetic studies on CCl have been reported by Tyerman [30,31] and on CCl and CBr by the authors [10-12,32]. Tyerman used CF_2Cl_2 as the source of CCl, while the authors used $CHClBr_2$ for CCl and $CHBr_3$ for CBr. The yield of CBr from $CHClBr_2$ was insufficient for parallel kinetic measurements. In no instance was $CH(\tilde{X}^2\Pi)$ detected in these experiments even at the shortest time delays. In view of the very low percentage conversions, no attempt at product analysis was made by either group. The results of these studies are summarized in Tables 3 to 7.

3a. Reaction with saturated hydrocarbons and inorganic molecules: The rate constant values for the reactions of CCl and CBr with saturates and inorganic molecules are given in Table 3. The low values of the rate constants for reactions with paraffinic hydrocarbons suggest that insertion into primary or secondary C-H bonds is an inefficient process. The values reported by Tyerman for saturated compounds are all higher by factors ranging from 2 to 50 than those obtained in more recent measurements, probably because of the presence of olefinic impurities in the substrates used. The values for *iso*butane suggest that insertion may occur into the weaker tertiary C-H bond. This behaviour is in marked contrast to that of CH, which inserts even into primary C-H bonds at rates close to the collision frequency.

Both CCl and CBr react readily with O_2 and NO. The following four center reactions with O_2 are highly exothermic [53],

$$CBr(^2\Pi) + O_2(^3\Sigma_g^-) \rightarrow CO^{\dagger}(^1\Sigma^+) + BrO(^2\Pi), \quad \Delta H \approx 117 \text{ kcal/mole}$$

and

$$CBr(^2\Pi) + O_2(^3\Sigma_g^-) \rightarrow CO_2^{\dagger}(^1\Sigma_g^+) + Br(^2P) \quad \Delta H \approx 188 \text{ kcal/mole}$$

and probably contribute to the CO and CO_2 laser emission observed by Lin. The four center reaction with NO

$$CBr(^2\Pi) + NO(^2\Pi) \rightarrow CN(^2\Pi) + BrO(^2\Pi), \quad \Delta H \approx 11 \text{ kcal/mole}$$

is only slightly exothermic [53]. Although the BrN radical has never been detected, the reaction

$$CBr(^2\Pi) + NO(^2\Pi) \rightarrow CO^\dagger(^1\Sigma^+) + BrN$$

could occur and may contribute to the occurrence of CO 5 μm laser emission in the flash photolysis of $CHBr_3$/NO mixtures [42].

TABLE 3

Bimolecular rate constants for the reactions of CBr($\tilde{X}^2\Pi$) and CCl($\tilde{X}^2\Pi$) with saturates and inorganic compounds

Substrate	$k \times 10^{-9}$ ($M^{-1}s^{-1}$)	
	CBr [10]	CCl
CH_4	<0.003	
C_3H_8		<0.004 [12]
iso-C_4H_{10}	0.018 ± 0.009	0.0045 ± 0.0004 [32]
cyclo-C_6H_{12}		<0.004 [51]
CF_3Cl		0.075 ± 0.008 [31]
CH_3Cl		<0.004 [52]
C_2H_5Cl		0.22 ± 0.01 [31]
H_2	<0.003	<0.004 [51]
O_2	1.3 ± 0.9	2.5 ± 0.03 [31]
N_2	<0.004	<0.0015 [31]
NO	13 ± 2	
NH_3		<0.01 [51]

3b. Reaction with Olefins: The available rate constant values for the reactions of CCl and CBr with olefins are collected in Table 4 along with comparative data for $S(^3P)$, $O(^3P)$ and $OH(^2\Pi)$. As seen from the data, the rate constant values for CBr and CCl are very similar to each other and to those of $O(^3P)$ and $S(^3P)$, and in the case of the ethylene

TABLE 4

Bimolecular rate constants for the reactions of $CCl(\tilde{X}^2\Pi)$, $CBr(\tilde{X}^2\Pi)$, $O(^3P)$, $S(^3P)$ *and* $OH(\tilde{X}^2\Pi)$ *with olefins*

Substrate	$k \times 10^{-9}$ $(M^{-1}s^{-1})$				
	$CBr(\tilde{X}^2\Pi)$ [10]	$CCl(\tilde{X}^2\Pi)$ [31]	$O(^3P)$ [55]	$S(^3P)$ [56]	$OH(\tilde{X}^2\Pi)$
C_2H_4	0.46 ± 0.07	0.33 ± 0.04	0.42 ± 0.04	0.9 ± 0.1	4.7 ± 0.5 [57]
C_3H_6	5.5 ± 0.2	2.50 ± 0.25	2.7 ± 0.3	6 ± 1	15.0 ± 1.5 [58]
$1\text{-}C_4H_8$	9 ± 1		2.7 ± 0.3	9 ± 1	21.0 ± 2.1 [58]
$t\text{-}C_4H_8$	6 ± 1	9.6 ± 0.8	14.1 ± 1.4	12 ± 2	42.0 ± 4.2 [58]
$iso\text{-}C_4H_8$		15.5 ± 1.1	9.2 ± 0.9	36 ± 5	30.0 ± 3.0 [58]
$(CH_3)_2CC(CH_3)_2$	12 ± 1			>62 ± 8	
CF_2CCl_2		1.90 ± 1.5			
$CFClCCl_2$		8 ± 4			
$CHClCCl_2$		4 ± 2			
C_2Cl_4		10 ± 5			
C_2H_3F	0.18 ± 0.01				
$c,t\text{-}C_2H_2F_2$	0.12 ± 0.01				
C_2HF_3	side reactions				
C_2F_4	side reactions				
Benzene	0.22 ± 0.02 [54]				

reaction, are about two orders of magnitude smaller than that of CH. The rate constants for the reactions of $OH(^2\Pi)$ are generally two to three times larger than for the other species, and an order of magnitude larger for the reaction with ethylene. For CBr the trend in the rate constant values with alkyl and fluorine substitution on the olefinic carbon reveals the electrophilic nature of the radical, and the difference between the fastest reaction (with $(CH_3)_2CC(CH_3)_2$) and the slowest (with CFHCFH) is about 100-fold. Thus, the decrease in reactivity, relative to CH, causes an increase in selectivity. The rate constant values reported by Tyerman for the CCl + haloethylene systems do not seem to conform to the trend expected for an electrophilic reagent and should be remeasured. The variation of the magnitude of the rate constant for the addition of $O(^3P)$ [59], $S(^3P)$ [60] and other electrophilic reagents to olefins, was attributed primarily to changes in the activation energy, resulting in a linear correlation between log k and the ionization potential of the olefin [11]. In the case of CBr reactions, some downward curvature appears in the plot for the higher substituted olefins, and this was interpreted as due to a lowering of the pre-exponential factor because of steric hindrance [11]. However, temperature studies over the range 298-423°K indicate an increase in the pre-exponential factor with increasing alkyl substitution [61], necessitating a re-evaluation of the influence of steric hindrance in these reactions.

The change from positive to negative activation energies with increasing alkyl substitution of the olefin was first recognized for the addition of $Te(^3P)$ to a series of olefins [56], and is now well established for other group VIa elements [62,63] and $OH(^2\Pi)$ [57,58]. The unusual aspect of the CBr addition is that the activation energy is already negative for addition to ethylene, a characteristic only previously observed with the $OH(^2\Pi)$ reaction [57], and becomes increasingly negative with alkyl substitution of the olefin [61].

Since neither CCl nor CBr react with methane or propane and the rate constant for reaction with *iso*butane is almost four orders of magnitude smaller than for *iso*butene, it may be assumed that the data in Table 4 represent the rate constants of the cyclo-addition reaction to give the corresponding cyclopropyl radical.

3c. Reaction with Acetylenes. Rate constant data for the addition reactions of CBr and CCl with acetylenes are listed in Table 5, along with comparative data for $O(^3P)$ and $S(^3P)$. The main features of the reactions with acetylenes are similar to those with olefins. Thus both CBr and CCl add to

TABLE 5

Bimolecular rate constants for the reactions of $CBr(\tilde{X}^2\Pi)$, $CCl(\tilde{X}^2\Pi)$, $O(^3P)$ *and* $S(^3P)$ *with alkynes*

Substrate	$k \times 10^{-9}$ $(M^{-1}s^{-1})$			
	CBr [11]	CCl [12]	$O(^3P)$	$S(^3P)$ [68]
C_2H_2	0.081 ± 0.007	0.035 ± 0.004	0.079 [64]	0.23 ± 0.05
		0.11 ± 0.01 [31]		
C_2D_2	0.072 ± 0.007	0.042 ± 0.013	0.079 [64]	0.23 ± 0.05
C_3H_6	4.8 ± 0.6	2.2 ± 0.2	0.4 ± 0.1 [65]	4.8 ± 0.2
		2.4 ± 0.2 [31]		
but-1-yne	6.2 ± 1.1	3.7 ± 0.5	1.2 [66]	3.3 ± 0.2
but-2-yne	24 ± 5	18 ± 3	2.9 [67]	16 ± 2
pent-1-yne	3.6 ± 0.8	4.3 ± 0.7		
pent-2-yne	20 ± 3	25 ± 4		18 ± 3
3,3 dimethyl but-1-yne		2.4 ± 0.3		
2,2,5,5, hexamethyl hex-3-yne	9.7 ± 1.6	7.6 ± 0.7		
phenyl acetylene		4.3 ± 0.7		
but-2-yne F_6	0.020 ± 0.002			0.21 ± 0.04

acetylene at a rate that is about three orders of magnitude slower than the CH reaction with acetylene. Also, their rate of addition to acetylene is about an order of magnitude slower than that to ethylene. The selectivity with respect to acetylenes is somewhat higher than with olefins, ranging over a factor of 1000 from pent-2-yne to but-2-yne-F_6, and is comparable to that exhibited by $O(^3P)$ and $S(^3P)$ with a similar range of acetylenes.

The linear correlation observed between log k and the ionization potential, Figure 1, for the addition of CBr and CCl to acetylenes substituted with straight chain alkyl groups,

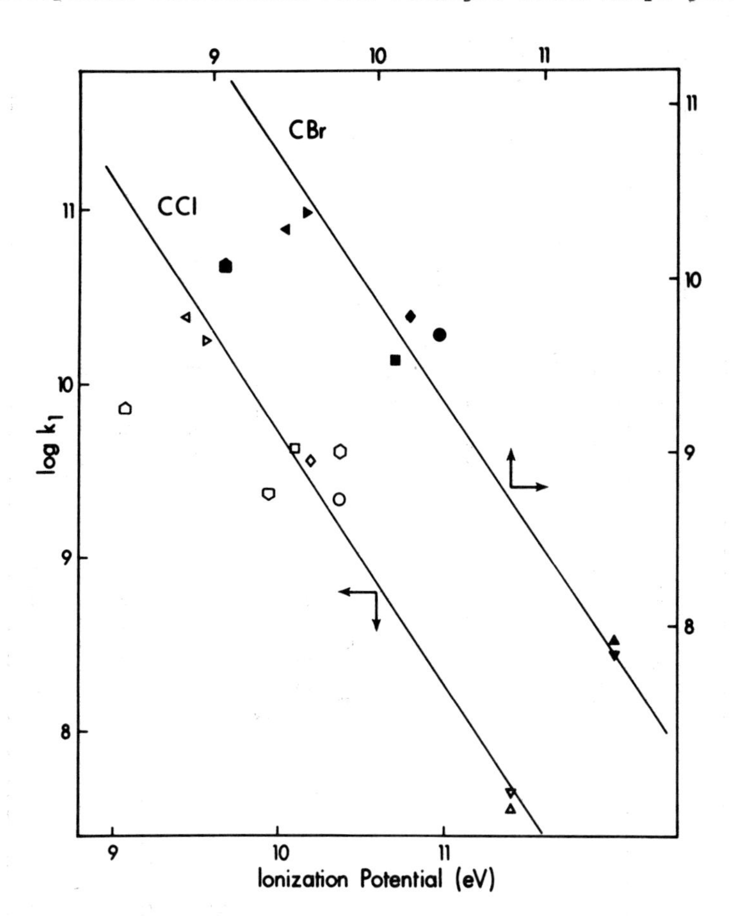

Figure 1. Plot of log k_1 versus ionization potential for $CCl(\tilde{X}^2\Pi)$ (open symbols) and $CBr(\tilde{X}^2\Pi)$ (closed symbols) reactions with acetylenes. △ ≡; ▽ D≡D; ○ ≡-; ◇ ≡--; □ ≡---; ▷ -≡-; ◁ -≡--; ⬡ ≡+; ⌂ +≡+; ⬡ ≡ϕ.

is indicative of only a small change in the preexponential factor, despite the bulkiness of the halocarbyne radical. The marked deviation from linearity in the log k vs. I.P. plot when branched chain alkyl groups are substituted is attributed to increased steric hindrance resulting in a substantial lowering of the A-factor [12].

3d. Reaction with Silanes. For the case of silanes the only carbyne which has been investigated to date is CCl [32]. Rate constant values for silanes are given in Table 6, along with comparative data for $CH_2(\tilde{A}^1A_1)$, $SiH_2(\tilde{X}^1A_1)$, H, D and CH_3.

The data indicate that the reactions of CCl with silanes are very rapid, of the same order as SiH_2 and CH_2, slightly faster than H or D, and up to six orders of magnitude faster than CH_3. Since CH_2 and SiH_2 react primarily by insertion and H, D and CH_3 by abstraction (although CH_2 is known to abstract [78]),it was assumed that the reaction of CCl was also by insertion, although this was not confirmed by product analysis [32]. The lack of reaction with tetramethylsilane (TMS) and with the series $F_{4-x}SiMe_x$ suggests that insertion does not occur into 1° C-H bonds [32] nor into Si-C bonds. Reaction with monosilanes was believed to be due solely to insertion into the Si-H bond. This behaviour parallels that of $SiH_2(\tilde{X}^1A_1)$ which inserts readily into Si-H bonds but does not react with C-H [79,80] or Si-C [79-81] bonds. $CH_2(\tilde{A}^1A_1)$ reacts readily with TMS [72] by insertion into the 1° C-H bonds, but like CCl and SiH_2, it does not insert into Si-C bonds [82]. Interestingly, CCl_2 is more reactive towards Si-C, since it has been shown to insert into the Si-C bond of silacyclobutanes [83]. A comparison between Me_3SiH (Table 6) and *iso*butane (Table 3) shows that the rate of insertion of CCl into 3° Si-H bonds is 600-fold larger than into 3° C-H bonds. Although CH_2 also shows a selectivity, the insertion ratio Si-H/C-H, ∿7-9, observed for CH_2 is much smaller [71, 84]. The electrophilic character of CCl was once again manifested by the linear correlations between log k and the ionization potentials of homologous series of silanes [32], Figure 2.

The CCl insertion reaction exhibits an isotope effect, varying from 1.9 for 1° Si-H bonds to 1.0 for 3° Si-H bonds [32] which can be compared to 1.15 for CH_2 insertion into Si-H [72] and ∿2.7 for methyl abstraction reactions [74]. The slow reaction with Me_6Si_2 was attributed to weak interaction with the Si-Si bond [32]. SiH_2 inserts into Si-Si bonds [81,85,86] but no products resulting from Si-Si insertion could be detected from the reaction of CH_2 with Me_6Si_2 [82]. Halogen substitution on the silicon atom effects a substantial reduction in the rate constant [52], but for alkyl

TABLE 6

Bimolecular rate constants for the reactions of $CCl(\tilde{X}^2\Pi)$, $CH_2(^1A_1)$, $SiH_2(^1A_1)$, CH_3, *H and D atoms with silanes*

Substrate	$k \times 10^{-9}$ $(M^{-1}s^{-1})$					
	CCl [32]	SiH_2	CH_2 [c]	CH_3 [e]	H [76] [g]	D [77] [h]
SiH_4	0.48 ± 0.05	0.57 [69] [a]		6.0×10^{-6} [73]	0.26	0.28
$MeSiH_3$	1.7 ± 0.2	6.44 [70] [b]	8.4 [71]	2.1×10^{-6} [74]	0.37	0.31
Me_2SiH_2	2.8 ± 0.1	3.03 [70] [b]	8.1 [72]	9.6×10^{-7} [74]	0.4	0.30
Me_3SiH	2.8 ± 0.2	4.46 [70] [b]	7.1 [72]	4.0×10^{-7} [74]	0.34	0.23
Me_4Si	No reaction	No reaction [70] [b]	6.9 [72]	1.9×10^{-8} [74]	<0.006	
Et_2SiH_2	2.9 ± 0.3				0.78	
Et_3SiH	4.5 ± 0.1				0.60	
SiD_4	0.25 ± 0.03			9.5×10^{-7} [75]	0.09	
$MeSiD_3$	0.98 ± 0.04		7.3 [71] [d]	3.1×10^{-7} [74]	0.14	
Me_2SiD_2	1.8 ± 0.1			2.8×10^{-7} [74] [f]		
Me_3SiD	2.8 ± 0.2					
Si_2H_6	6.47 ± 0.34	3.69 [69] [a]		6.6×10^{-5} [75]	2.23	2.65
Si_2D_6	5.61 ± 0.32			1.2×10^{-5} [75]	0.90	

Continued...

TABLE 6 Continued

Me_6Si_2	0.025 ± 0.003
Cl_6Si_2	No reaction [52]
Cl_3SiH	<0.004 [52]
Cl_2MeSiH	0.029 ± 0.003 [52]
$ClMe_2SiH$	0.39 ± 0.02 [52]
$F_{4-x}SiMe_x$	No reaction

[a] Arrhenius parameters for silene insertion calculated from co-pyrolysis data and thermochemical information.

[b] Based on co-pyrolysis data with Si_2H_6 and the calculated value for SiH_2 insertion into Si_2H_6 [71].

[c] Values for overall rate constant for CH_2 insertion into both Si-H and C-H bonds. Absolute values calculated from data for CH_2 insertion into silanes relative to CH_2 insertion into CH_4. $CH_2(^1A_1) + CH_4 \rightarrow C_2H_6 = 1.1 \times 10^9$ ℓ. $mole^{-1}s^{-1}$ [47].

[d] Calculated from the isotope effect of 1.15 [71].

[e] Room temperature data calculated from the Arrhenius parameters given in [73-75].

[f] Abstraction by CD_3 radicals.

[g] Values for H abstraction relative to $H + C_2H_4 \xrightarrow{k_H} C_2H_5$. Value for k_H taken as 5.1×10^8 ℓ. $mole^{-1}s^{-1}$ [76].

[h] Values for D abstraction relative to $D + C_2D_4 \xrightarrow{k_D} C_2D_5$. Value for k_D assumed to be 5.3 x 10 ℓ. $mole^{-1}s^{-1}$ [77].

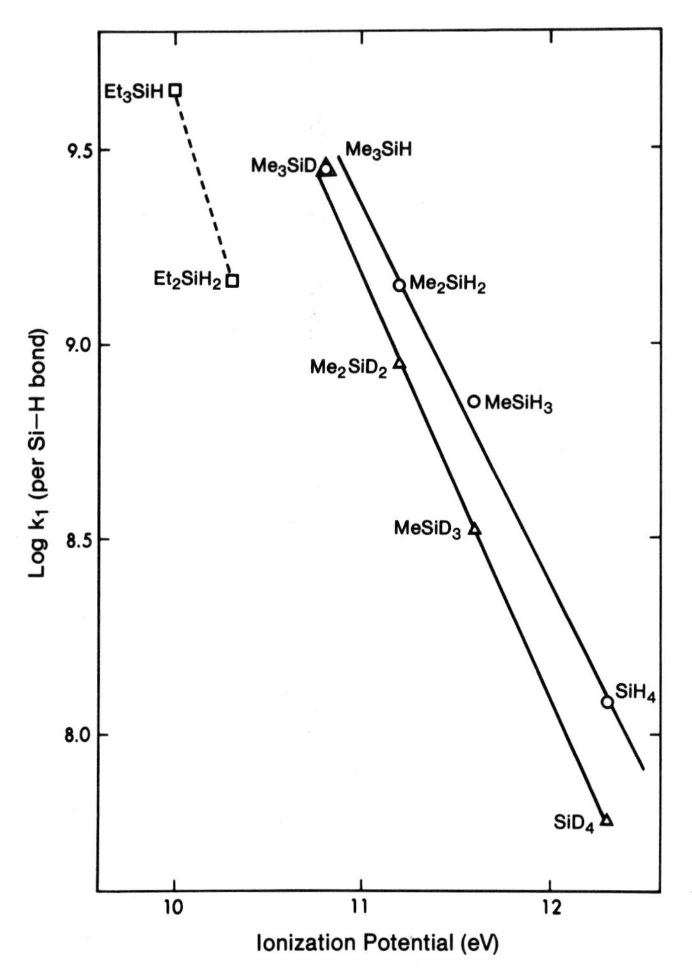

Figure 2. Plot of log (k_1 per Si-H bond) versus ionization potential for CCl($\tilde{X}^2\Pi$) reactions with silanes. (Reproduced with permission from North-Holland Pub. Co., Amsterdam [32]).

substitution the data in Table 6 do not indicate any recognizable trend for Si-H insertion. However, if a comparison is made on a per Si-H bond basis, Table 7, then it is apparent that the reactivities of CCl and CH_2 increase with increasing alkylation. The value for the SiH_2 + $MeSiH_3$ reaction does not follow this trend and may be erroneous. The values for CH_2 insertion in Table 7 were determined by converting the C-H bonds to the equivalent number of Si-H bonds using the known values for the relative rates of C-H and Si-H insertion [71, 84].

The apparent anomaly for the case of SiH_2 is removed if the relative data of Sefcik and Ring [85] are used for

TABLE 7

Bimolecular rate constants for the insertion of $CCl(\tilde{X}^2\Pi)$, $SiH_2(^1A_1)$ and $CH_2(^1A_1)$ into silanes, per Si-H bond

Substrate	$k \times 10^{-9}$ $(M^{-1}s^{-1})$		
	CCl [32]	SiH_2 [a]	CH_2 [b]
SiH_4	0.12	0.14	
$MeSiH_3$	0.85	2.15	2.51
Me_2SiH_2	1.4	1.52	2.82
Me_3SiH	2.8	4.46	3.13
Si_2H_6	1.1	0.61	
SiD_4	0.06		
$MeSiD_3$	0.33		
Me_2SiD_2	0.9		
Me_3SiD	2.8		
Si_2D_6	0.93		

[a] From the calculated data of Purnell and coworkers [69,70].

[b] Per Si-H bond values calculated from the data given in Table 6 (from [71,72]) and the relative rates of $CH_2(^1A_1)$ insertion into C-H and Si-H bonds. SiH/CH = 8.9 for $MeSiH_3$ [71], 6.9 for Me_2SiH_2 and 7.1 for Me_3SiH [84]. Hence the equivalent number of Si-H bonds are: $MeSiH_3$ $3 + 3 \times 1/8.8 = 3.34$; Me_2SiH_2 $2 + 6 \times 1/6.9 = 2.87$; Me_3SiH $1 + 9 \times 1/7.1 = 2.27$.

comparison. Table 8 lists relative rate data for CCl, SiH_2, CH_2, CTCl, H, D, T and $Hg(^3P_1)$ with the values for each series arbitrarily normalized to 1.0 for Me_2SiH_2, the only substrate common to all investigations. In all instances, except for the values for SiH_2 insertion by Purnell and coworkers [69,70], the reactivity on a per Si-H bond basis is $3° > 2° > 1°$.

The much faster rate at which CCl inserts into Si-H bonds compared to C-H bonds may be due to the ease with which a transition state is formed with silanes through the utilization of the 3d orbitals of silicon.

When the measured rate constant values for the CCl + silane reactions are compared to the negative charge densities residing on the H atom attached to silicon, computed

TABLE 8

Relative gas phase insertion rates of CCl, SiH_2, CH_2 and CTCl, abstraction rates of H, D and T atoms and quenching across sections of $Hg(^3P_1)$ per Si-H bond of silanes[a]

Silane	CCl[b]	SiH_2[c]	SiH_2[d]	CH_2[f]	CTCl[g]	H[h]	D[i]	T[j]	$Hg(^3P_1)$[k]
SiH_4	0.08	0.09	0.43[e]		<0.1	0.33	0.47		0.41
$MeSiH_3$	0.61	1.41	0.57	0.89		0.63	0.68	0.87	0.67
Me_2SiH_2	1.0	1.0	1.0	1.0	1.0	1.0	1.0	1.0	1.0
Me_3SiH	2.0	2.93	2.28	1.1	6.2	1.7	1.54	1.44	1.88
Si_2H_6	0.79	0.40	1.43			1.87	2.92		

[a] Each series arbitrarily normalized to 1.0 for Me_2SiH_2, the only common species in all experiments.
[b] From [32].
[c] Data from Table 1 of [70] converted to room temperature.
[d] Values from Table 1 of [85] at 623K.
[e] Determined from fast-neutron irradiation of phosphine-silane-disilane mixtures [86].
[f] Standardized from Table 7.
[g] CTCl produced by the hot atom excitation method. Data of [87].
[h] From Table 1 of [76]. H atoms produced by $Hg(^3P_1)$ photosensitization of H_2.
[i] From Table IX of [77]. D atoms produced by photolysis of C_2D_4.
[j] Value for D atom abstraction from Me_3SiD, Me_2SiD_2 and Me_3SiD by recoil T atoms in excess nitrogen [88].
[k] From [89].

from m.o. theory, an excellent correlation is obtained between reactivity and hydridic character [90,91] and between reactivity and Si-H bond index [91]. This suggests that the rate constant for CCl insertion into Si-H bonds is determined primarily by the properties of the Si-H bond, and it was concluded that interaction with the Si-Si bond plays little part in determining the reactivity of CCl with disilanes [32]. In contrast, Si_2H_6 was displaced in the hydridic series for the SiH_2 reactions, due to the significant interaction of SiH_2 with the Si-Si bond [85].

Summary

From our limited knowledge of carbyne chemistry, it appears that the chemical behaviour of carbynes bears a great deal of resemblance to that of carbenes. CH($\tilde{X}^2\Pi$) may be the most reactive ground state radical in organic chemistry with respect to the rates at which it reacts with a variety of substrates. Halogen substitution, owing to electronic and resonance effects, decreases the general reactivity and increases the selectivity of methylidyne.

One of the main reasons underlying the paucity of reports on carbyne chemistry is the fact that the precursors used to date also generate other radical species upon photolysis, and the ensuing secondary reactions can make kinetic and mechanistic interpretation difficult. It is therefore necessary to explore alternative sources of carbynes and establish optimal photochemical conditions whereby secondary reactions can be minimized.

The chemical behaviour of the lowest excited state of methylidyne, the ($\tilde{a}^4\Sigma^-$) state, is predicted to parallel that of the ground ($^3B_1$) state of methylene. It would be a challenging task to put this prediction to the test of experiment.

References

1. See for example: Laufer, A.H. *Rev. Chem. Intermed.* *4*, (1980) In Press; Gaspar, P.P. and Hammond, G.S. "Carbenes" Vol. II, Chapter 6, Edited by Moss, R.A. and Jones, M. Jr., John Wiley and Sons Inc. (1975); W. Kirmse, "Carbene Chemistry", 2nd Edition, Chapter 1, Academic Press Inc. (1971).

2. Dominh, T., Gunning, H.E. and Strausz, O.P., *J. Amer. Chem. Soc.*, *89*, 6785 (1967).

3. Strausz, O.P., DoMinh, T. and Font, J., *J. Amer. Chem. Soc.*, *90*, 1930 (1968).

4. Strausz, O.P., Kennepohl, G.J.A., Garneau, F.X., DoMinh, T., Kim, B., Valenty, S. and Skell, P.S., *J. Amer. Chem. Soc.*, *96*, 5723 (1974).

5. Kennepohl, G.J.A., Ph.D. Thesis, University of Alberta, (1972).

6. Herzberg, G., "Spectra of Diatomic Molecules", 2nd Ed., D. van Norstrand Company Inc. (1950).

7. Braun, W., McNesby, J.R. and Bass, A.M. *J. Chem. Phys.* *46*, 2071 (1967).

8. Bosnali, M.W. and Perner, D. *Z. Naturforsch.*, *26a*, 1768 (1971).

9. Butler, J.E., Goss, L.P., Lin, M.C. and Hudgens, J.W. *Chem. Phys. Letters*, *63*, 104 (1979).

10. McDaniel, R.S., Dickson, R., James, F.C., Strausz, O.P. and Bell, T.N. *Chem. Phys. Letters*, *43*, 130 (1976).

11. James, F.C., Ruzsicska, B., McDaniel, R.S., Dickson, R., Strausz, O.P. and Bell, T.N. *Chem. Phys. Letters*, *45*, 449 (1977).

12. James, F.C., Choi, J., Strausz, O.P. and Bell, T.N. *Chem. Phys. Letters*, *53*, 206 (1978).

13. Kasdan, A., Herbst, E. and Lineberger, W.C. *Chem. Phys. Letters*, *31*, 78 (1975).

14. Lie, G.C., Hinze, J. and Liu, B. *J. Chem. Phys.*, *57*, 625 (1972).

15. Brooks, B.R. and Schaefer, H.F. III. *J. Chem. Phys.*, *67*, 5146 (1977).

16. Gosavi, R.K., Gunning, H.E. and Strausz, O.P. Unpublished data.

17. Huerlinger, T. (1918) *Dissertation Lund*; Hulthén, E. *Z. f. Phys.*, *11*, 284 (1922); Kratzer, A. *Z. f. Phys.*, *23*, 298 (1924); Mulliken, R.S. *Phys. Rev.*, *30*, 785 (1927)

18. Gerö, L. *Z. f. Phys.*, *118*, 27 (1941).

19. Herzberg, G. and Johns, J.W.C. *Astrophys. J.*, *158*, 399 (1969).

20. Norrish, R.G.W., Porter, G. and Thrush, B.A. *Proc. Roy. Soc. London, 216A,* 165 (1953).

21. Messing, I., Sadowski, C.M. and Filseth, S.V. *Chem. Phys. Letters, 66,* 95 (1979).

22. Verma, R.D. and Mulliken, R.S. *J. Mol. Spectry., 6,* 419 (1961).

23. Dixon, R.N. and Kroto, H.W. *Trans. Faraday Soc., 59,* 1484 (1963).

24. Andrews, E.B. and Barrow, R.F. *Proc. Phys. Soc., A64,* 481 (1951).

25. Refaey, K.M.A. and Franklin, J.L. *Int. J. Mass Spectry. & Ion Phys., 20,* 19 (1976).

26. Simons, J.P. and Yarwood, A.J. *Trans. Faraday Soc., 57,* 2167 (1961).

27. Simons, J.P. and Yarwood, A.J. *Trans. Faraday Soc., 59* 90 (1963).

28. Majer, J.R. and Simons, J.P. *Adv. Photochem., 2,* 137 (1964).

29. Hussain, D. *Nature, 195,* 796 (1962).

30. Tyerman, W.J.R. *Trans. Faraday Soc., 65,* 2948 (1969).

31. Tyerman, W.J.R. *J. Chem. Soc. A.,* 2483 (1969).

32. James, F.C., Choi, H.K.J., Strausz, O.P. and Bell, T.N. *Chem. Phys. Letters, 68,* 131 (1979).

33. Modica, A.P. *J. Chem. Phys., 44,* 1585 (1966); Modica, A.P. and Sillers, S.J. *J. Chem. Phys., 48,* 3283 (1968).

34. Burks, T.L. and Lin, M.C. *J. Chem. Phys., 64,* 4235 (1976).

35. Hsu, D.S.Y. and Lin, M.C. *Int. J. Chem. Kinetics, 10,* 839 (1978).

36. Rebbert, R.E., Lias, S.G. and Ausloos, P. *J. Photochem., 8,* 363 (1978).

37. Braun, W., Welge, K.H. and McNesby, J.R. *J. Chem. Phys.*, *45*, 2650 (1966).

38. Gordon, R., Jr. and Ausloos, P. *J. Chem. Phys.*, *46*, 4823 (1967).

39. Rebbert, R.E. and Ausloos, P. *J. Photochem.*, *1*, 171 (1972/73).

40. Lin, M.C. *J. Chem. Phys.*, *61*, 1835 (1974).

41. Lin, M.C. *Int. J. Chem. Kinetics*, *5*, 173 (1973); *ibid*, *6*, 1 (1974).

42. Lin, M.C. *J. Phys. Chem.*, *77*, 2726 (1973).

43. Lin, M.C. Private Communication.

44. Wolf, A.P. *Adv. Phys. Org. Chem.*, *2*, 201 (1964).

45. Porter, R.P., Clark, A.H., Kaskan, W.E. and Brown, W.G. Eleventh Symposium (International) on Combustion, Berkeley, California, 907 (1966), Combustion Institute (1967)),

46. Safrany, D.R., Reeves, R.R. and Harteck, P. *J. Amer. Chem. Soc.*, *86*, 3160 (1964).

47. Braun, W., Bass A.M. and Pilling, M.J. *J. Chem. Phys.*, *52*, 5131 (1970).

48. Laufer, A.H. and Bass, A.M. *J. Phys. Chem.*, *78*, 1344 (1974).

49. Halberstadt, M.L. and Crump, J. *J. Photochem.*, *1*, 295 (1972/73).

50. Hsu, D.S.Y., Umstead, M.E. and Lin, M.C. *ACS Symp. Ser.* *66*, 128 (1978).

51. Choi, H.K.J. Unpublished results.

52. James, F.C., Choi, H.K.J., Strausz, O.P. and Bell, T.N. To be published.

53. Benson, S.W. "Thermochemical Kinetics", 2nd Ed., J. Wiley and Sons (1976).

54. Ruzsicska, B.P. Unpublished results.

55. Atkinson, R. and Pitts, J.N. Jr. *J. Chem. Phys.*, *67*, 38 (1977).

56. Connor, J., van Roodselaar, A., Fair, R.W. and Strausz, O.P. *J. Amer. Chem. Soc.*, *93*, 560 (1971).

57. Atkinson, R., Perry, R.A. and Pitts, J.N. Jr. *J. Chem. Phys.*, *66*, 1197 (1977).

58. Atkinson, R. and Pitts, J.N. Jr. *J. Chem. Phys.*, *63*, 3591 (1975).

59. Cvetanovic, R.J. *Adv. Photochem.*, *1*, 115 (1963).

60. Strausz, O.P., O'Callaghan, W.B., Lown, E.M. and Gunning, H.E. *J. Amer. Chem. Soc.*, *93*, 559 (1971).

61. James, F.C., Ruzsicska, B.P., Strausz, O.P. and Bell, T.N. Submitted for publication.

62. Singleton, D.L. and Cvetanovic, R.J. *J. Amer. Chem. Soc.*, *98*, 6812 (1976).

63. Davis, D.D. and Klemm, R.B. *Int. J. Chem. Kinetics*, *5*, 841 (1973).

64. Stuhl, F. and Niki, H. *J. Chem. Phys.*, *55*, 3954 (1971).

65. Brown, J.M. and Thrush, B.A. *Trans. Faraday Soc.*, *63*, 630 (1967).

66. Herbrechtsmeier and Wagner, H.Gg. *Ber. Bunsenges. Physik. Chem.*, *79*, 461 (1975).

67. Herbrechtsmeier and Wagner, H.Gg. *Ber. Bunsenges. Physik. Chem.*, *79*, 673 (1975).

68. van Roodselaar, A., Safarik, I., Strausz, O.P. and Gunning, H.E. *J. Amer. Chem. Soc.*, *100*, 4068 (1978).

69. John, P. and Purnell, J.H. *J. Chem. Soc. Faraday I*, *69*, 1455 (1973).

70. Cox, B. and Purnell, J.H. *J. Chem. Soc. Faraday I*, *71*, 859 (1975).

71. Mazac, C.J. and Simons, J.W. *J. Amer. Chem. Soc.*, *90*, 2484 (1968).

72. Hase, W.L. and Simons, J.W. *J. Chem. Phys.*, *54*, 1277 (1971).

73. Berkley, R.E., Safarik, I., Strausz, O.P. and Gunning, H.E. *J. Phys. Chem.*, *77*, 1741 (1973).

74. Berkley, R.E., Safarik, I., Gunning, H.E. and Strausz, O.P. *J. Phys. Chem.*, *77*, 1734 (1973).

75. Strausz, O.P., Jakubowski, E., Sandhu, H.S. and Gunning, H.E. *J. Chem. Phys.*, *51*, 552 (1969).

76. Austin, E.R. and Lampe, F.W. *J. Phys. Chem.*, *81*, 1134 (1977).

77. Obi, K., Sandhu, H.S., Gunning, H.E. and Strausz, O.P. *J. Phys. Chem.*, *76*, 3911 (1972).

78. Simons, J.W. and Mazac, C.J. *Canad. J. Chem.*, *45*, 1717 (1967).

79. Bowrey, M. and Purnell, J.H. *Proc. Roy. Soc. London*, *A 321*, 341 (1971).

80. Bowrey, M. and Purnell, J.H. *J. Amer. Chem. Soc.*, *92*, 2594 (1970).

81. Estacio, P., Sefcik, M.D., Chan, E.K. and Ring, M.A. *Inorg. Chem.*, *9*, 1068 (1970).

82. Conlin, R.T., Gaspar, P.P., Levin, R.H. and Jones, M. Jr. *J. Amer. Chem. Soc.*, *94*, 7165 (1972).

83. Seyferth, D., Damrauer, R. and Washburne, S.S. *J. Amer. Chem. Soc.*, *89*, 1538 (1967).

84. Hase, W.L., Brieland, W.G. and Simons, J.W. *J. Phys. Chem.*, *73*, 4401 (1969).

85. Sefcik, M.D. and Ring, M.A. *J. Amer. Chem. Soc.*, *95*, 5168 (1973).

86. Gaspar, P.P. and Markusch, P. *Chem. Commun.*, 1331 (1970).

87. Tang, Y.N., Daniel, S.H. and Wong, N.-B. *J. Phys. Chem.*, *74*, 3148 (1970).

88. Hosaka, A. and Rowland, F.S. *J. Phys. Chem.*, *77*, 705 (1973).

89. Yarwood, A.J., Strausz, O.P. and Gunning, H.E. *J. Chem. Phys.*, *41*, 1705 (1964).

90. Hollandsworth, R.P. and Ring, M.A. *Inorg. Chem.*, *7*, 1635 (1968).

91. Perkins, K.A. and Perkins, P.G. Personal communication.

THE ROLE OF FREE RADICALS IN ATMOSPHERIC CHEMISTRY

J. ALISTAIR KERR

Department of Chemistry
University of Birmingham
Birmingham, B15 2TT, England

Summary

The reactions of free atoms and radicals in the chemistry of the stratosphere and natural troposphere are described with emphasis on the ozone layer and its potential depletion by man-made pollutants. The reactions are classified into groups consisting of O_x, HO_x, NO_x, CH_4, SO_x, ClO_x and BrO_x species, and their evaluated rate constants at 298 K are presented and discussed.

INTRODUCTION

This article is aimed at giving a broad outline of the role of free atoms and radicals in the chemistry of the atmosphere. Interest in atmospheric chemistry has undoubtedly been stimulated by the occurrence of air pollution, such as photochemical smog in the lower atmosphere and the potential depletion of the ozone layer in the upper atmosphere. It is now widely accepted that free radicals play a paramount role in the chemistry of natural and polluted atmospheres.

ISBN 0-12-566550-4

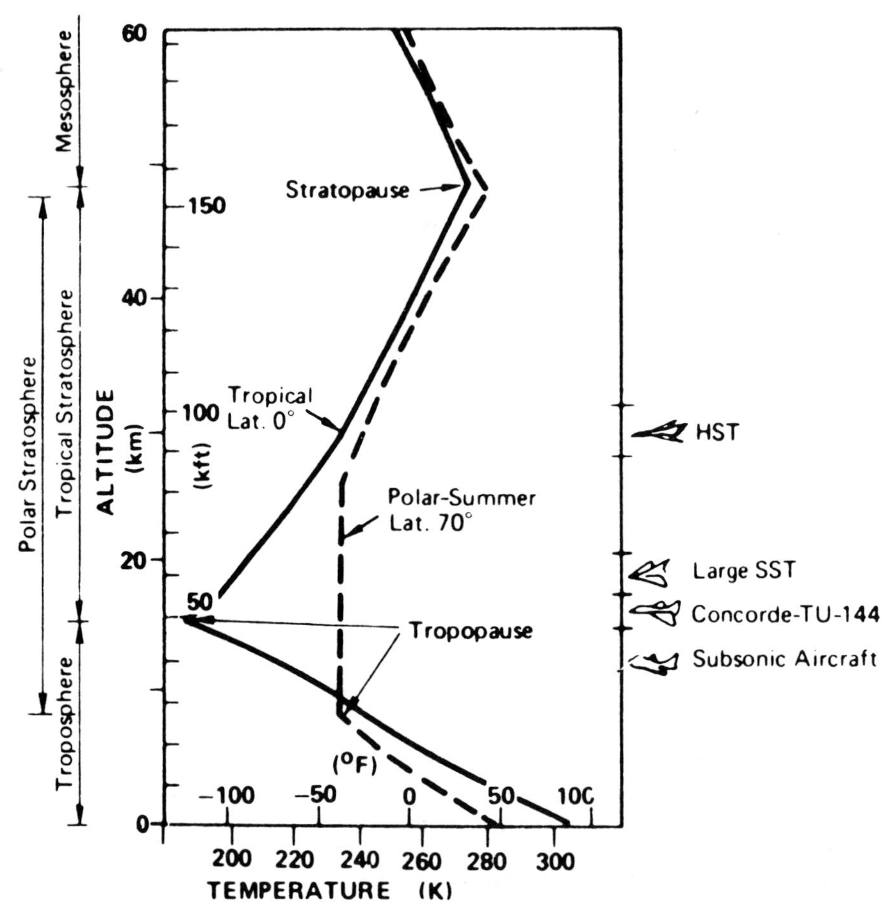

Fig. 1. Regions of the earth's atmosphere as defined by the variation of temperature with altitude (1).

The chemistry involving neutral species takes place in the first two regions of the earth's atmosphere. These regions are illustrated in Figure 1 and are based on temperature minima or maxima. Thus the troposphere or lower atmosphere extends from ground level to an altitude of about 16 km in the tropics and to about 8 km near the poles. It is bounded by the tropopause by which the temperature has fallen to about 210 K. Beyond this is the stratosphere or upper atmosphere in which the temperature rises to a maximum of about 280 K at the stratopause.

At the same time, the pressure of the atmosphere decreases approximately exponentially with increasing altitude, falling to about 1/10 atm at the tropopause and 1/1000 atm at the top of the stratosphere. Up to this altitude gases are transported mainly by eddy mixing rather than molecular diffusion, and gravitational settling-out is negligible.

In the troposphere mixing is rapid because warmer air tends to rise and be replaced by cooler air from above, whereas vertical circulation in the stratosphere is much slower because the cooler, denser air is below. This situation resembles the temperature inversions near ground level which are associated with fogs and smogs. The tropopause is an effective barrier where the rate of vertical transport drops by more than an order of magnitude. In addition, the stratosphere is virtually cloudless, unlike the troposphere where rain rapidly removes water soluble substances.

In order to restrict the number of chemical reactions in this article to manageable proportions we shall focus upon the chemistry of the stratosphere and the unpolluted troposphere. This limits our considerations to C_1 hydrocarbons and considerably reduces the complexity of the free radical reactions. The same basic approach applies to the polluted troposphere but the photo-oxidation sequences of reactions involving the higher hydrocarbons have yet to be established with any certainty. It turns out that about 150 homogeneous elementary gas phase reactions of atoms and radicals are needed to present the initial framework of chemical understanding of the earth's atmosphere. Fortunately the *CODATA Task Group on Chemical Kinetics* has just completed a detailed evaluation of the available kinetic data for these reactions (2,3) and we draw upon their conclusions here.

The driving force for initiating chemical change in the atmosphere is photochemical, i.e. absorption of solar radiation by molecules which dissociate to produce atoms or free radicals. In the stratosphere this involves the chemical dynamics of the ozone layer. On the other hand, in the troposphere photochemistry is also a key feature in the scavenging chemical mechanisms whereby gaseous materials, natural and man-made, emitted at the earth's surface are removed from the atmosphere.

CHEMISTRY OF THE STRATOSPHERE

The ozone layer

The bulk of the ozone in the earth's atmosphere is to be found in the stratosphere, and although it has become customary to use the term 'ozone layer', the concentration of ozone never exceeds a few parts per million. Thus the total amount of ozone in a vertical column corresponds to about 3 mm of pure ozone at NTP.

The basic mechanism for the formation of ozone in the stratosphere was suggested (4) by Chapman in 1930. Molecular oxygen absorbs radiation of wavelength less than 245 nm to yield oxygen atoms; the oxygen atoms then combine rapidly with molecular oxygen to form ozone in a third-body reaction:

$$O_2 + h\nu \rightarrow O + O \quad (\lambda \leqslant 245 \text{ nm}) \qquad [1]$$

$$O + O_2 + M \rightarrow O_3 + M \qquad [2]$$

Ozone is a strong absorber of light in the stratosphere and is rapidly photolyzed, thus preventing ultraviolet light below 290 nm from reaching ground level:

$$O_3 + h\nu \rightarrow O + O_2 \qquad [3]$$

Ozone is also removed by reaction with atomic oxygen which regenerates molecular oxygen:

$$O + O_3 \rightarrow 2O_2 \qquad [4]$$

O and O_3 are referred to as 'odd oxygen' in this scheme. Reactions [2] and [3], which are rapid, merely interconvert odd oxygen, whereas reactions [1] and [4], which are much slower, form and remove odd oxygen. Under stratospheric conditions reactions [2] and [3] reach a steady-state within minutes, while the time scale for reactions [1] and [4] is typically of the order of months. The solar energy absorbed by O_3 in reaction [3] heats the stratosphere and gives rise to the temperature inversion responsible for its stability.

The rate of destruction of 'odd oxygen' in reaction [4] is, however, insufficient to balance the rate of production of O by photodissociation of O_2 in reaction [1] and to maintain a steady-state concentration of O_3 which is consistent with observations of the total column density of ozone. Typical calculations indicate that reaction [4] can account for only 1/5 of the 'odd oxygen' removal in the atmosphere (5).

The remaining 4/5 is largely attributed to catalytic cycles in which a species X abstracts an oxygen atom from O_3 and is then regenerated in a reaction with atomic oxygen:

$$X + O_3 \rightarrow XO + O_2 \qquad [5]$$

$$XO + O \rightarrow X + O_2 \qquad [6]$$

$$O + O_3 \rightarrow O_2 + O_2 \qquad \text{[net reaction]}$$

Such catalytic cycles have been proposed for X = H,HO,(6);

NO,(7); Cl,(8); and Br,(9), according to current chemical knowledge and available kinetic data. These O_3 destruction cycles achieve the same result as reaction [4] but are considerably faster since they have smaller activation energies.

As well as these pairs of relatively simple reactions, there are a large number of elementary photochemical and thermal reactions which are required to define the concentrations of active O_x, HO_x, NO_x, ClO_x and other species at different altitudes.

Reactions of HO_x species

In 1950 Bates and Nicolet (10,11) recognized the importance of H, HO, and HO_2 radicals in atmospheric chemistry. The primary source of these species throughout the atmosphere is from the reaction of excited oxygen atoms, $O(^1D)$, with hydrogen-containing compounds, principally H_2O. The $O(^1D)$ atoms are produced by the photolysis of ozone:

$$O_3 + h\nu \rightarrow O(^1D) + O_2 \quad (\lambda \leqslant 310 \text{ nm}) \qquad [7]$$

$$O(^1D) + H_2O \rightarrow 2HO \qquad [8]$$

Stratospheric water vapour arises chiefly from the oxidation of methane which is transported from the troposphere:

$$HO + CH_4 \rightarrow H_2O + CH_3 \qquad [9]$$

The resulting methyl radicals are oxidized to CO and CO_2 in the following reactions:

$$CH_3 + O_2 + M \rightarrow CH_3O_2 + M \qquad [10]$$

$$CH_3O_2 + NO \rightarrow CH_3O + NO_2 \qquad [11]$$

$$CH_3O + O_2 \rightarrow HCHO + HO_2 \qquad [12]$$

$$HO + HCHO \rightarrow H_2O + HCO \qquad [13]$$

$$HCHO + h\nu \rightarrow H_2 + CO \qquad [14]$$

$$\rightarrow H + HCO \qquad [15]$$

$$H + O_2 + M \rightarrow HO_2 + M \qquad [16]$$

$$HCO + O_2 \rightarrow CO + HO_2 \qquad [17]$$

$$HO_2 + NO \rightarrow HO + NO_2 \qquad [18]$$

$$HO + CO \rightarrow H + CO_2 \qquad [19]$$

Other minor sources of HO radicals involve the photolyses of H_2O or H_2O_2:

$$H_2O + h\nu \rightarrow H + HO \qquad (\lambda \leqslant 190\ nm) \qquad [20]$$

$$H_2O_2 + h\nu \rightarrow 2HO \qquad (\lambda \leqslant 350\ nm) \qquad [21]$$

The major interactions of HO and HO_2 radicals are

$$HO + HO \rightarrow H_2O + O \qquad [22]$$

$$HO + HO + M \rightarrow H_2O_2 + M \qquad [23]$$

$$HO + HO_2 \rightarrow H_2O + O_2 \qquad [24]$$

$$HO_2 + HO_2 \rightarrow H_2O_2 + O_2 \qquad [25]$$

The removal of HO_x species from the stratosphere is by downward transport of H_2O and H_2O_2 to the troposphere where they are involved in the precipitation elements (cloud droplets and aerosol particles).

HO_x species account for about 15% of the O_3 removal from the stratosphere.

Reactions of NO_x species

In 1970 Crutzen (7) first highlighted the importance of natural oxides of nitrogen in the destruction of 'odd oxygen' (X = NO):

$$NO + O_3 \rightarrow NO_2 + O_2 \qquad [26]$$

$$NO_2 + O \rightarrow NO + O_2 \qquad [27]$$

$$O + O_3 \rightarrow 2O_2 \qquad [net reaction]$$

This effect is reduced in daylight since NO_2 is rapidly photolyzed giving the 'do nothing' cycle:

$$NO_2 + h\nu \rightarrow NO + O \qquad [28]$$

$$NO + O_3 \rightarrow NO_2 + O_2 \qquad [29]$$

$$O + O_2 + M \rightarrow O_3 + M \qquad [2]$$

The main natural source of stratospheric NO_X comes from the reaction of $O(^1D)$ atoms with nitrous oxide:

$$O(^1D) + N_2O \rightarrow 2NO \qquad [30]$$

N_2O is produced at the earth's surface by denitrifying bacteria; it is long-lived and well mixed in the troposphere (12). The transport rate of N_2O to the stratosphere and the NO_X source can be estimated reasonably accurately.

In addition to NO, NO_2 and N_2O, stratospheric NO_X species include NO_3, N_2O_5, HNO_3 and HO_2NO_2, which are involved in the following reactions:

$$O + NO_2 + M \rightarrow NO_3 + M \qquad [31]$$

$$O + NO_3 \rightarrow O_2 + NO_2 \qquad [32]$$

$$NO_2 + NO_3 \rightleftarrows N_2O_5 + M \qquad [33]$$

$$HO + NO_2 + M \rightarrow HNO_3 + M \qquad [34]$$

$$HNO_3 + h\nu \rightarrow HO + NO_2 \qquad [35]$$

$$HO_2 + NO_2 + M \rightleftarrows HO_2NO_2 + M \qquad [36]$$

as well as other reactions listed in the summary of reactions. Nitric acid is transported slowly downwards towards the troposphere where it is removed by rain, thereby returning the nitrate to the soil.

The participation of the major NO_X species in the chemistry of the stratosphere is illustrated in Figure 2, which presents a simplified picture of O_3 formation and removal processes in the unpolluted atmosphere.

Present estimates indicate that NO_X species remove about 60% of stratospheric O_3. This could be increased by the NO and NO_2 released in the exhaust gases from supersonic and some subsonic aircraft using polar routes in winter. The Climatic Impact Assessment Program in the U.S. and the Committee on the Meteorological Effects of Supersonic Aircraft in the U.K. examined the consequences of such flights. They deduced respectively that 100 Concordes flying 7 hr/day would deplete the ozone layer by 1 and 0.2 per cent.

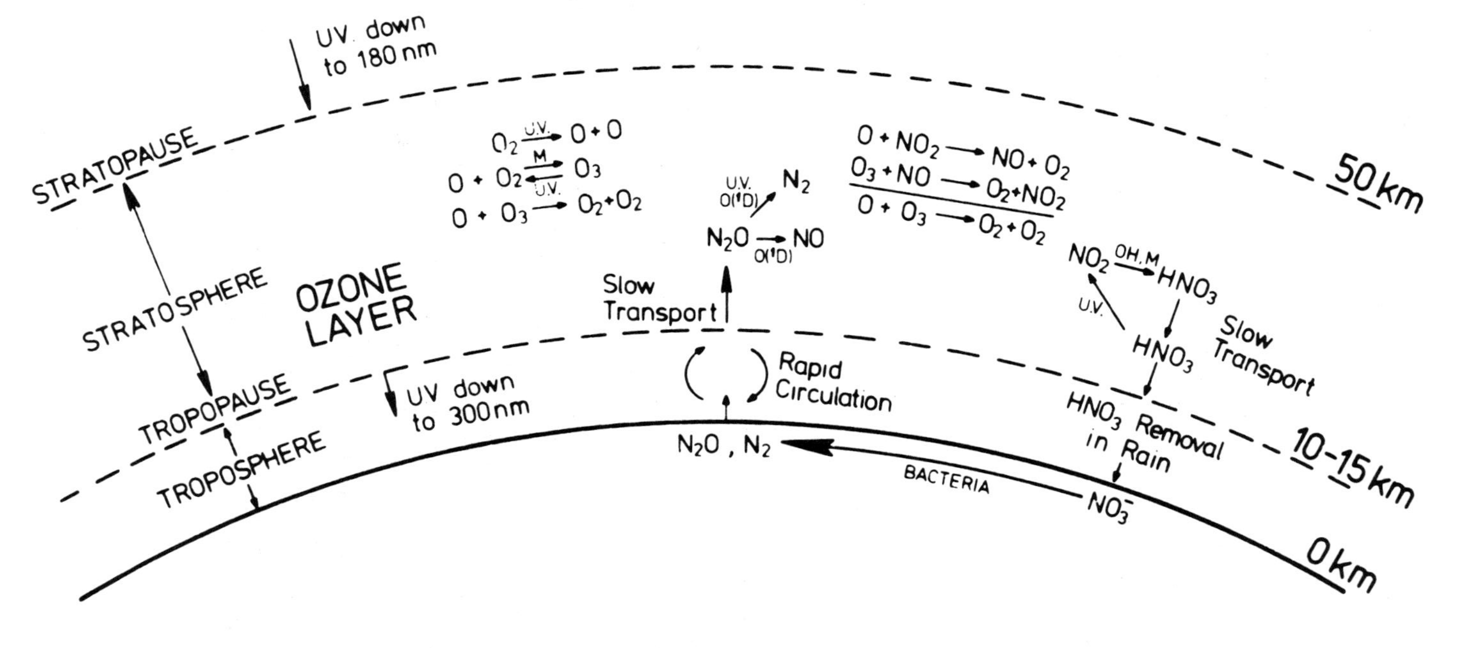

Fig. 2. *Ozone formation and removal processes under natural atmospheric conditions (13).*

Reactions of ClO_X species

It has been recognized for several years that the chlorofluoromethanes, (CFMs), $CFCl_3$ (F-11) and CF_2Cl_2 (F-12), which have been widely used as aerosol propellants, are accumulating in the atmosphere at rates approaching their rates of release (14,15). No significant sink reactions have been found for these species which are not attacked by hydroxyl radicals at any significant rate.

In 1974 Molina and Rowland (16,17) proposed that the CFMs could be transported to the stratosphere where at altitudes above 25 km they would be decomposed by ultraviolet light (wavelengths between 180 and 220 nm) or by reaction with $O(^1D)$ atoms to yield chlorine atoms which can attack O_3. It has subsequently been suggested that in addition to the man-made CFMs there is a natural source of chlorine atoms from methyl chloride.

The important active species in the ClO_X reactions are Cl and ClO:

$$Cl + O_3 \rightarrow ClO + O_2 \quad [37]$$

$$ClO + O \rightarrow Cl + O_2 \quad [38]$$

$$O + O_3 \rightarrow 2O_2 \quad \text{[net reaction]}$$

The major removal process for Cl atoms is by reaction with methane to form HCl, which is only slowly attacked by hydroxyl radicals, allowing transport of HCl to the troposphere where it is removed in rain:

$$Cl + CH_4 \rightarrow HCl + CH_3 \quad [39]$$

$$HO + HCl \rightarrow H_2O + Cl \quad [40]$$

A simplified reaction scheme illustrating the role of CFMs in ozone chemistry is shown in Figure 3, from which the similarity to the NO_X scheme is apparent.

Other compounds which are involved in stratospheric ClO_X chemistry are chlorine nitrate ($ClONO_2$) and hypochlorous acid (HOCl) formed in the following reactions:

$$ClO + NO_2 + M \rightarrow ClONO_2 + M \quad [41]$$

$$ClO + HO_2 \rightarrow HOCl + O_2 \quad [42]$$

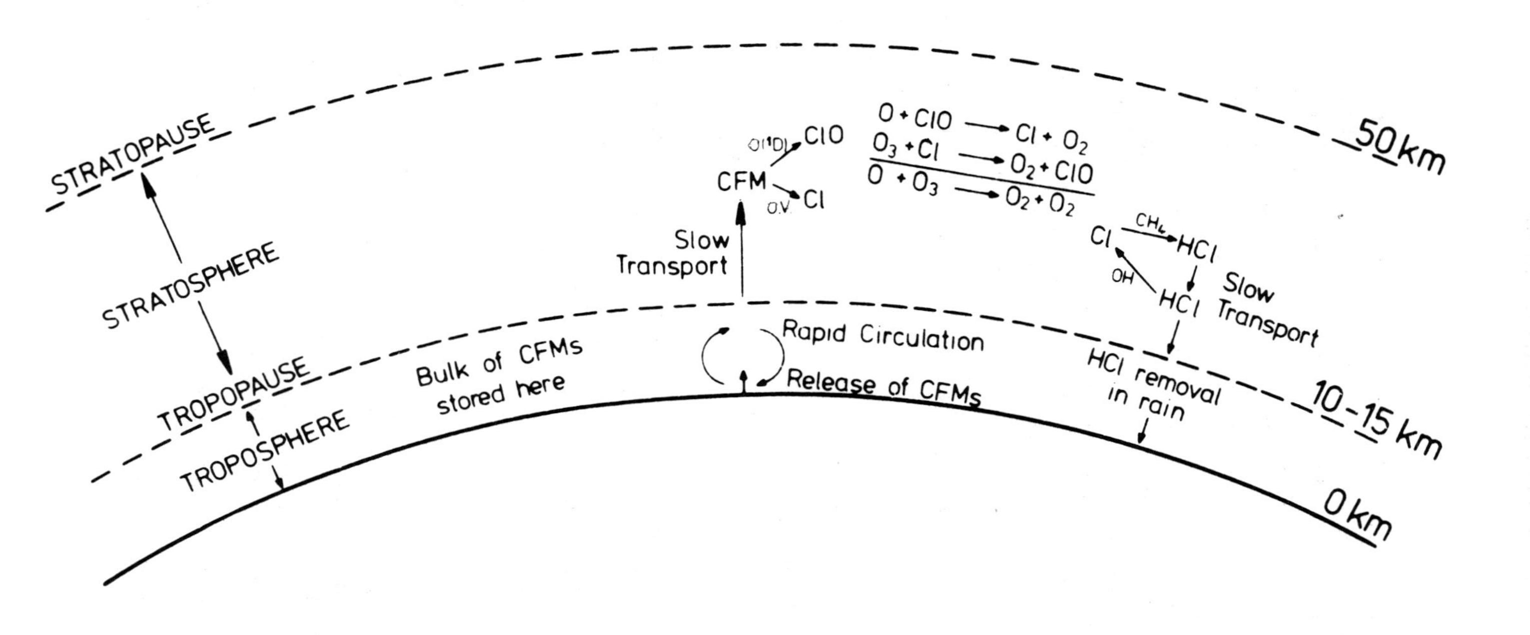

Fig. 3. Simplified reaction scheme for chlorofluoromethanes in the atmosphere (13).

These molecules are decomposed by ultraviolet light or by free radical attack. Both chlorine nitrate and hypochlorous acid reduce the effect of stratospheric chlorine, the former by 'tying up' the active species from both the NO_x and ClO_x catalytic cycles.

Another important reaction which links both the NO_x and ClO_x reaction cycles is the rapid process:

$$ClO + NO \rightarrow Cl + NO_2 \qquad [43]$$

Recent calculations indicate that the continued release of CFMs at the 1973 rate would eventually produce about a 7 per cent depletion in the ozone layer. During this time their tropospheric concentrations would rise from the present level of about 1 in 10^{10} to about 1 in 10^{9} (18).

Other potential catalytic cycles

Bromine atoms and BrO could play similar roles to those of Cl and ClO in an O_3 destruction cycle. At present at least two potential sources of stratospheric bromine have been identified, namely methyl bromide, used as a soil fumigant and also of natural origin, and trifluorobromomethane, used as a fire retardant.

Bromine is potentially more active than chlorine in the ozone chemistry of the stratosphere since HBr is less stable than HCl, which allows more bromine to remain in the active forms Br and BrO (19). The key reactions involving Br and BrO are included in the summary of reactions in this chapter.

While there are clearly large stratospheric sources of fluorine via CFMs, in contrast to Cl and Br, atomic fluorine reacts rapidly with all H-containing species, including water, to form hydrofluoric acid. HF is very stable towards photochemical and free radical breakdown and acts as a very effective sink for active FO_x species.

The thermochemistry of SO_x species, e.g. SO and SO_2, is unfavorable for their involvement in the 'odd-oxygen' chemistry of the stratosphere. Sulfur compounds are involved in stratospheric chemistry, however, via their participation in the formation of an aerosol layer in the lower atmosphere. Recent results indicate that the primary source of sulfur in the stratosphere is carbonyl sulfide, OCS, which is photodissociated in the middle stratosphere (20). The key reactions involving SO_x species are included in the summary of reactions.

Modelling studies of the atmosphere

Atmospheric modelling studies have largely been concerned with the calculation of the ozone column density and with the effects on this column density of perturbations of trace gas compositions, particularly involving NO_x and ClO_x species. At altitudes between 30 and 50 km the stratosphere can be treated in terms of a steady-state system of chemical reactions. At lower altitudes the rates of many of the reactions approach the rate of transport, and the computing in handling the combined equations for transport and chemical reactions is complex. A full three-dimensional calculation involving altitude, latitude and longitude as variables is a daunting task, and only a few two-dimensional calculations (omitting longitude) have so far been attempted on the basis of a simplified chemical mechanism. Most of the upper atmospheric modelling calculations have involved a one-dimensional approach with altitude as the variable.

CHEMISTRY OF THE TROPOSPHERE

Many of the trace gas constituents of the stratosphere are transported from the troposphere and originate initially at the earth's surface. The passage of these components through the atmosphere constitutes a cycle, consisting of the initial emission or injection of the gases, their transport and chemical transformation within the atmosphere, and finally their removal by rain-out or other sink processes. Clearly sink processes operating in the troposphere are important in determining the nature and amounts of trace components passing to and from the stratosphere and hence in potential stratospheric perturbations.

The principal agent involved in initiating these trace gas reaction cycles in the troposphere is the hydroxyl radical. As for stratospheric chemistry the main source of HO radicals is from the reaction of $O(^1D)$ with water vapor, reaction [8]. The $O(^1D)$ atoms are again generated from the photolysis of ozone, reaction [7]. In the sunlit lower atmosphere a steady-state concentration of HO radicals is maintained by reactions involving the oxidation of natural methane present to the extent of about 1.5 ppm. This photo-oxidation scheme proceeds via formaldehyde to CO and finally to CO_2, reactions [9-19]. It is also apparent from this mechanism that the oxides of nitrogen play a significant role in tropospheric HO_x chemistry through their ability to provide an additional source of 'odd-oxygen' via reactions such as:

$$HO_2 + NO \rightarrow HO + NO_2 \quad [18]$$

$$RO_2 + NO \rightarrow RO + NO_2 \quad [43]$$

followed by

$$NO_2 + h\nu \rightarrow NO + O \quad [28]$$

The photo-oxidation sequence of reactions involved in the tropospheric cycling of higher hydrocarbons is obviously more complex than for methane and must include at some stage radical fragmentation processes. Such schemes must, however, follow the same general pattern as for CH_4, i.e. initial attack by HO radicals followed by reaction with O_2, NO and radical decomposition (21). In addition, the lifetimes of higher hydrocarbons in the troposphere are much shorter than that for CH_4 since the initial rate of HO attack is always much faster than HO attack on CH_4.

SO_2 is also an important pollutant in the troposphere, arising from the burning of S-containing fossil fuels. Its oxidation can occur homogeneously, by reaction with HO, or heterogeneously in the precipitation elements. The oxidation products are sulfuric acid and sulfate aerosol.

EVALUATED KINETIC DATA FOR ATMOSPHERIC CHEMICAL REACTIONS

Table 1 contains a summary of the rate constants at 298 K for the key thermal reactions which participate in the chemistry of the stratosphere and the natural troposphere. These rate constants have been selected by the *CODATA Task Group on Chemical Kinetics* (2,3) on the basis of the existing literature data. Included in this Table are realistic estimates of the error limits to be assigned to these measurements. These are quoted as the term $\Delta\log k$, where $\Delta\log k = D$ and D is defined by the equation $\log_{10}k = C \pm D$, which is equivalent to the statement that k is uncertain to a factor of F where $D = \log_{10}F$.

It is apparent from the magnitude of these assigned error limits that many of the reaction rates are not known with great certainty. This arises from the considerable experimental difficulties involved in studying the kinetics of atomic and free radical reactions.

TABLE 1

Summary of key atmospheric thermal reactions and rate constants (2, 3)

Reaction	k_{298} / $cm^3 molecule^{-1} s^{-1}$	$\Delta \log k_{298}$
O_x REACTIONS		
$O + O_2 + M \rightarrow O_3 + M$	$3.6 \times 10^{-34}[Ar] (k_0)$	±0.1
	$5.6 \times 10^{-34}[N_2] (k_0)$	±0.1
	$2.8 \times 10^{-12} (k_\infty)$	±0.3
$O + O_3 \rightarrow 2O_2$	9.5×10^{-15}	±0.1
$O(^1D) + O_2 \rightarrow O(^3P) + O_2(^1\Sigma_g^+)$	3.7×10^{-11}	±0.15
$\rightarrow O(^3P) + O\ (^3\Sigma_g^-)$	0.9×10^{-11}	±0.15
HO_x REACTIONS		
$H + HO_2 \rightarrow H_2 + O_2$	1.4×10^{-11}	±0.4
$\rightarrow 2HO$	3.2×10^{-11}	±0.4
$\rightarrow H_2O + O$	$\leqslant 9.4 \times 10^{-13}$	+0 -0.5
$H + O_2 + M \rightarrow HO_2 + M$	$1.8 \times 10^{-32}[Ar] (k_0)$	±0.2
	$5.9 \times 10^{-32}[N_2] (k_0)$	±0.2
$H + O_3 \rightarrow HO + O_2$	2.8×10^{-11}	±0.2
$O + HO \rightarrow O_2 + H$	3.8×10^{-11}	±0.3
$O + HO_2 \rightarrow HO + O_2$	3.1×10^{-11}	±0.3
$O + H_2O_2 \rightarrow HO + HO_2$; $\rightarrow O_2 + H_2O$	2.1×10^{-15}	±0.3
$O(^1D) + H_2 \rightarrow HO + H$; $\rightarrow O(^3P) + H_2$	2.0×10^{-10}	±0.3
$O(^1D) + H_2O \rightarrow 2HO$; $\rightarrow O(^3P) + H_2O$	2.8×10^{-10}	±0.3
$HO + H_2 \rightarrow H_2O + H$	7.1×10^{-15}	±0.1
$HO + HO \rightarrow H_2O + O$	1.8×10^{-12}	±0.2
$HO + HO + M \rightarrow H_2O_2 + M$	$6.5 \times 10^{-31}[Ar] (k_0)$	±0.3
	$6.5 \times 10^{-31}[N_2] (k_0)$	±0.3
$HO + HO_2 \rightarrow H_2O + O_2$	3.5×10^{-11}	±0.5
$HO + O_3 \rightarrow HO_2 + O_2$	6.7×10^{-14}	±0.15

TABLE 1 (continued)

Reaction	$\frac{k_{298}}{cm^3 molecule^{-1} s^{-1}}$	$\Delta \log k_{298}$
$HO_2 + HO_2 \rightarrow H_2O_2 + O_2$	$2.3x10^{-12}$	±0.3
$HO_2 + O_3 \rightarrow HO + 2O_2$	$2.0x10^{-15}$	±0.2
NO_x REACTIONS		
$N + O_2 \rightarrow NO + O$	$8.9x10^{-17}$	±0.1
$N + O_3 \rightarrow NO + O_2$	$\leqslant 5x10^{-16}$	-
$N + NO \rightarrow N_2 + O$	$3.4x10^{-11}$	±0.15
$N + NO_2 \rightarrow N_2O + O$	$1.4x10^{-12}$	±0.2
$O + NO + M \rightarrow NO_2 + M$	$6.4x10^{-32}[Ar](k_0)$	±0.1
	$1.2x10^{-31}[N_2](k_0)$	±0.1
	$3.0x10^{-11}$ (k_∞)	±0.2
$O + NO_2 \rightarrow NO + O_2$	$9.3x10^{-12}$	±0.06
$O + NO_2 + M \rightarrow NO_3 + M$	$9x10^{-32}[N_2](k_0)$	±0.1
	$2.2x10^{-11}$ (k_∞)	±0.1
$O + NO_3 \rightarrow O_2 + NO_2$	$1x10^{-11}$	±0.5
$O(^1D) + N_2 \rightarrow O(^3P) + N_2$	$4.5x10^{-11}$	±0.15
$O(^1D) + N_2O \rightarrow N_2 + O_2$	$7.4x10^{-11}$	±0.15
$\rightarrow 2NO$	$8.6x10^{-11}$	±0.15
$\rightarrow O(^3P) + N_2O$	No recommendation	
$HO + NO + M \rightarrow HONO + M$	$6.5x10^{-31}[N_2](k_0)$	±0.1
	$1.0x10^{-11}$ (k_∞)	±0.2
$HO + NO_2 + M \rightarrow HONO_2 + M$	$1.0x10^{-30}[Ar](k_0)$	±0.1
	$2.6x10^{-30}[N_2](k_0)$	±0.1
	$1.6x10^{-11}$ (k_∞)	±0.2
$HO + HONO_2 \rightarrow H_2O + NO_3$	$8.5x10^{-14}$	±0.1
$HO + HO_2NO_2 \rightarrow$ products	$1x10^{-13}$	±1
$HO_2 + NO \rightarrow HO + NO_2$	$8.4x10^{-12}$	±0.08
$HO_2 + NO_2 + M \rightarrow HO_2NO_2 + M$	$2.1x10^{-31}[N_2](k_0)$	±0.1
	$5x10^{-12}$ (k_∞)	±0.4
$HO_2NO_2 + M \rightarrow HO_2 + NO_2 + M$	$1.2x10^{-20}[N_2](k_0/s^{-1})$	±0.1
	0.09 (k_∞/s^{-1})	±0.6

TABLE 1 (continued)

Reaction	k_{298} / $cm^3 molecule^{-1} s^{-1}$	$\Delta \log k_{298}$
$NO + O_3 \rightarrow NO_2 + O_2$	1.8×10^{-14}	±0.06
$NO + NO_3 \rightarrow 2NO_2$	2×10^{-11}	±0.5
$NO_2 + NO_3 + M \rightarrow N_2O_5 + M$	$1.5 \times 10^{-30}[N_2]$ (k_0)	±0.3
	5×10^{-12} (k_∞)	±0.3
$N_2O_5 + M \rightarrow NO_2 + NO_3 + M$	$6.4 \times 10^{-20}[N_2]$ (k_0/s^{-1})	±0.3
	0.20 (k_∞/s^{-1})	±0.3
$NO_2 + O_3 \rightarrow NO_3 + O_2$	3.2×10^{-17}	±0.06
<u>CH_4 REACTIONS</u>		
$O(^1D) + CH_4 \rightarrow HO + CH_3$	2.2×10^{-10}	±0.3
$\rightarrow HCHO + H_2$	2.4×10^{-11}	-
$HO + CH_4 \rightarrow H_2O + CH_3$	8.0×10^{-15}	±0.1
$HO + CO \rightarrow H + CO_2$	1.5×10^{-13} (≤100 Torr)	±0.05
$HO + CO \rightarrow$ products	2.8×10^{-13} (1 atm air)	±0.1
$HO + HCHO \rightarrow H_2O + HCO$ $\rightarrow H + HCOOH$	1.3×10^{-11}	±0.15
$HO_2 + CH_3O_2 \rightarrow O_2 + CH_3OOH$ $\rightarrow HO + O_2 + CH_3O$	6.5×10^{-12}	±0.3
$HCO + O_2 \rightarrow HO_2 + CO$	5.1×10^{-12}	±0.1
$HCO + O_2 + M \rightarrow HCO_3 + M$	No recommendation	
$CH_3 + O_2 \rightarrow HCHO + HO$	No recommendation	
$CH_3 + O_2 + M \rightarrow CH_3O_2 + M$	$2.6 \times 10^{-31}[N_2]$ (k_0)	±0.3
	2×10^{-12} (k_∞)	±0.3
$CH_3O + O_2 \rightarrow HCHO + HO_2$	6×10^{-16}	±0.6
$CH_3O_2 + CH_3O_2 \rightarrow CH_3OH + HCHO + O_2$ $\rightarrow 2CH_3O + O_2$ $\rightarrow CH_3OOCH_3 + O_2$	4.6×10^{-13}	±0.1
$CH_3O_2 + NO \rightarrow CH_3O + NO_2$	7.5×10^{-12}	±0.3
$CH_3O_2 + NO_2 + M \rightarrow CH_3O_2NO_2 + M$	1.6×10^{-12} (1 atm)	±0.5

TABLE 1 (continued)

Reaction	k_{298} / $cm^3 molecule^{-1} s^{-1}$	$\Delta \log k_{298}$
SO_x REACTIONS		
$O + H_2S \rightarrow HO + HS$	$2.7 x 10^{-14}$	±0.1
$O + CS \rightarrow CO + S$	$2.1 x 10^{-11}$	±0.1
$O + OCS \rightarrow SO + CO$	$1.4 x 10^{-14}$	±0.2
$O + CS_2 \rightarrow SO + CS$ $\rightarrow CO + S_2$ $\rightarrow OCS + S$	$5.5 x 10^{-12}$	±0.2
$O + SO_2 + M \rightarrow SO_3 + M$	$1.4 x 10^{-33}[N_2](k_0)$	±0.3
$HO + H_2S \rightarrow H_2O + HS$	$5.3 x 10^{-12}$	±0.1
$HO + OCS \rightarrow$ products	$\leqslant 6 x 10^{-14}$	+0.7 − ?
$HO + CS_2 \rightarrow$ products	$\leqslant 2 x 10^{-13}$	+0.7 − ?
$HO + SO_2 + M \rightarrow HOSO_2 + M$	$3 x 10^{-31}[N_2](k_0)$	±0.3
	$2 x 10^{-12}$ (k_∞)	±0.4
$S + O_2 \rightarrow SO + O$	$2.0 x 10^{-12}$	±0.15
$S + O_3 \rightarrow SO + O_2$	$1.2 x 10^{-11}$	±0.3
$SO + O_2 \rightarrow SO_2 + O$	$9 x 10^{-18}$	±0.5
$SO + O_3 \rightarrow SO_2 + O_2$	$6 x 10^{-14}$	±0.3
$SO + NO_2 \rightarrow SO_2 + NO$	$1.4 x 10^{-11}$	±0.3
$SO_3 + H_2O \rightarrow$ products	No recommendation	
$CH_3O_2 + SO_2 \rightarrow CH_3O + SO_3$ $\rightarrow CH_3O_2SO_2$	No recommendation	
FO_x REACTIONS		
$O + FO \rightarrow O_2 + F$	$5 x 10^{-11}$	±0.5
$O + FO_2 \rightarrow O_2 + FO$	$5 x 10^{-11}$	±0.7
$O(^1D) + HF \rightarrow HO + F$ $\rightarrow O(^3P) + HF$	$1 x 10^{-10}$	±0.5
$F + H_2 \rightarrow HF + H$	$2.5 x 10^{-11}$	±0.2

TABLE 1 (continued)

Reaction	k_{298} / $cm^3 molecule^{-1} s^{-1}$	$\Delta \log k_{298}$
$F + O_2 + M \rightarrow FO_2 + M$	$1.1 \times 10^{-32} [N_2] (k_0)$	±0.3
	3×10^{-11} (k_∞)	±0.5
$F + O_3 \rightarrow FO + O_2$	1.3×10^{-11}	±0.3
$F + H_2O \rightarrow HF + HO$	1.1×10^{-11}	±0.5
$F + CH_4 \rightarrow HF + CH_3$	8×10^{-11}	±0.2
$FO + O_3 \rightarrow F + 2O_2$ $\rightarrow FO_2 + O_2$	No recommendation	
$FO + NO \rightarrow F + NO_2$	2×10^{-11}	±0.5
$FO + NO_2 + M \rightarrow FONO_2 + M$	$1.7 \times 10^{-31} [N_2] (k_0)$	±0.4
	1.2×10^{-11} (k_∞)	±0.4
$FO + FO \rightarrow 2F + O_2$ $\rightarrow FO_2 + F$ $\rightarrow F_2 + O_2$	1.5×10^{-11}	±0.3
ClO_x REACTIONS		
$O + HCl \rightarrow HO + Cl$	1.4×10^{-16}	±0.3
$O + ClO \rightarrow O_2 + Cl$	5.0×10^{-11}	±0.1
$O + ClONO_2 \rightarrow ClO + NO_3$ $\rightarrow OClO + NO_2$ $\rightarrow O_2 + ClONO$	1.9×10^{-13}	±0.1
$O(^1D) + CF_2Cl_2 \rightarrow ClO + CF_2Cl$ $\rightarrow O(^3P) + CF_2Cl_2$	2.8×10^{-10}	±0.3
$O(^1D) + CFCl_3 \rightarrow ClO + CFCl_2$ $\rightarrow O(^3P) + CFCl_3$	3.5×10^{-10}	±0.2
$O(^1D) + CCl_4 \rightarrow ClO + CCl_3$ $\rightarrow O(^3P) + CCl_4$	4.8×10^{-10}	±0.2

TABLE 1 (continued)

Reaction	$\frac{k_{298}}{cm^3 molecule^{-1} s^{-1}}$	$\Delta \log k_{298}$
$Cl + H_2 \rightarrow HCl + H$	$1.8x10^{-14}$	±0.2
$Cl + HO_2 \rightarrow HCl + O_2$	$4.1x10^{-11}$	±0.3
$Cl + H_2O_2 \rightarrow HCl + HO_2$	$4.3x10^{-13}$	±0.2
$Cl + O_3 \rightarrow ClO + O_2$	$1.2x10^{-11}$	±0.06
$Cl + CH_4 \rightarrow HCl + CH_3$	$1.04x10^{-13}$	±0.06
$Cl + C_2H_6 \rightarrow HCl + C_2H_5$	$5.7x10^{-11}$	±0.06
$Cl + HCHO \rightarrow HCl + HCO$	$7.3x10^{-11}$	±0.06
$Cl + HONO_2 \rightarrow HCl + NO_3$	$\leqslant 7x10^{-15}$	+0.3 -2.5
$Cl + CH_3Cl \rightarrow HCl + CH_2Cl$	$4.9x10^{-13}$	±0.1
$Cl + ClONO_2 \rightarrow Cl_2 + NO_3$ $\rightarrow ClONO + ClO$	$2.2x10^{-13}$	±0.3
$HO + HCl \rightarrow H_2O + Cl$	$6.6x10^{-13}$	±0.06
$HO + ClO \rightarrow HO_2 + Cl$ $\rightarrow HCl + O_2$	$9.1x10^{-12}$	±0.3
$HO + ClONO_2 \rightarrow HOCl + NO_3$ $\rightarrow HO_2 + ClONO$ $\rightarrow HNO_3 + ClO$	$3.9x10^{-13}$	±0.2
$HO + CH_3Cl \rightarrow H_2O + CH_2Cl$	$4.1x10^{-14}$	±0.1
$HO + CHF_2Cl \rightarrow H_2O + CF_2Cl$	$4.4x10^{-15}$	±0.1
$HO + CHFCl_2 \rightarrow H_2O + CFCl_2$	$2.8x10^{-14}$	±0.1
$HO + CH_3CCl_3 \rightarrow H_2O + CH_2CCl_3$	$1.9x10^{-14}$	±0.15
$ClO + HO_2 \rightarrow HOCl + O_2$ $\rightarrow HCl + O_3$	$5.2x10^{-12}$	±0.2
$ClO + NO \rightarrow Cl + NO_2$	$1.8x10^{-11}$	±0.1

TABLE 1 (continued)

Reaction	k_{298} / $cm^3 molecule^{-1} s^{-1}$	$\Delta \log k_{298}$
$ClO + NO_2 + M \rightarrow ClONO_2 + M$	$1.7 \times 10^{-31} [N_2] (k_0)$	± 0.1
	1.2×10^{-11} (k_∞)	± 0.4
BrO_x REACTIONS		
$O + HBr \rightarrow HO + Br$	3.9×10^{-14}	± 0.2
$O + BrO \rightarrow O_2 + Br$	3×10^{-11}	± 0.5
$Br + HO_2 \rightarrow HBr + O_2$	1×10^{-11}	± 0.7
$Br + H_2O_2 \rightarrow HBr + HO_2$	$\leqslant 2 \times 10^{-14}$	$+0.3$ -1.7
$Br + O_3 \rightarrow BrO + O_2$	1.1×10^{-12}	± 0.1
$HO + HBr \rightarrow H_2O + Br$	8.5×10^{-12}	± 0.3
$HO + CH_3Br \rightarrow H_2O + CH_2Br$	3.8×10^{-14}	± 0.1
$BrO + HO_2 \rightarrow HOBr + O_2$; $\rightarrow HBr + O_3$	5×10^{-12}	± 0.5
$BrO + NO \rightarrow Br + NO_2$	2.1×10^{-11}	± 0.1
$BrO + NO_2 + M \rightarrow BrONO_2 + M$	$3 \times 10^{-31} [N_2] (k_0)$	± 0.4
	1.2×10^{-11} (k_∞)	± 0.4
$BrO + O_3 \rightarrow Br + 2O_2$	$\leqslant 5 \times 10^{-15}$	$+0.5$ $-$?
$BrO + ClO \rightarrow Br + OClO$	6.7×10^{-12}	± 0.3
$\rightarrow Br + Cl + O_2$; $\rightarrow BrCl + O_2$	6.7×10^{-12}	± 0.3
$BrO + BrO \rightarrow 2Br + O_2$; $\rightarrow Br_2 + O_2$	2.8×10^{-12}	± 0.1

Many of the rate constants in Table 1 have been obtained by the so-called 'direct measurement' techniques. These involve monitoring the concentration of the reactant atom or radical by a specific physical technique and following the decay kinetics in the presence of an added reactant. The techniques for detecting the atom or radical concentrations include emission spectrometry, ultraviolet spectrophotometry, mass spectrometry, resonance fluorescence and absorption, electron paramagnetic resonance spectrometry, infrared spectroscopy and, most recently, laser magnetic resonance spectroscopy. While these techniques ensure unique monitoring of the reactant in question, they yield no direct information on the nature of the reaction channels. Such information is obtained from careful analysis of the products, which is not always straightforward.

In addition to the thermal reactions listed in Table 1, there are also photochemical reactions which are important in atmospheric chemistry. Such processes for the stratosphere and natural troposphere are listed in Table 2, which lists the major primary processes where these are known.

TABLE 2

Summary of key atmospheric photochemical reactions (2,3)

$H_2O + h\nu \rightarrow HO + H$	$OCS + h\nu \rightarrow CO + S$
$H_2O_2 + h\nu \rightarrow 2HO$	$\rightarrow CO + S(^1D)$
$NO + h\nu \rightarrow N + O$	$CS_2 + h\nu \rightarrow CS + S$
$NO_2 + h\nu \rightarrow NO + O$	$\rightarrow CS + S(^1D)$
$NO_3 + h\nu \rightarrow NO_2 + O$	$HF + h\nu \rightarrow$ products
$\rightarrow NO + O_2$	$COF_2 + h\nu \rightarrow$ products
$N_2O + h\nu \rightarrow N_2 + O(^1S)$	$FONO_2 + h\nu \rightarrow$ products
$N_2O_5 + h\nu \rightarrow NO_2 + NO_3$	$HOCl + h\nu \rightarrow HO + Cl$
$HONO + h\nu \rightarrow HO + NO$	$COFCl + h\nu \rightarrow$ products
$HONO_2 + h\nu \rightarrow HO + NO_2$	$ClONO_2 + h\nu \rightarrow$ products
$HO_2NO_2 + h\nu \rightarrow$ products	$COCl_2 + h\nu \rightarrow$ products
$HCHO + h\nu \rightarrow H + HCO$	$CF_2Cl_2 + h\nu \rightarrow$ products
$\rightarrow H_2 + CO$	$CFCl_3 + h\nu \rightarrow$ products
$CH_3OOH + h\nu \rightarrow CH_3O + HO$	$CCl_4 + h\nu \rightarrow$ products
	$BrONO_2 + h\nu \rightarrow$ products

For many of the photochemical reactions the absorption cross-sections have been measured as well as the quantum yields of the primary processes, but it is not possible to present such data in a readily summarized form (2,3).

REFERENCES

1. 'Environmental Impact of Stratospheric Flight', National Academy of Sciences, Washington, D.C. 1975.

2. Kerr, J.A., Hampson, R.F., Baulch, D.L., Cox, R.A., Crutzen, P.J., Larin, I.K., Troe, J., and Watson, R.T., 'Evaluated Kinetic and Photochemical Data for Atmospheric Chemistry', CODATA Bulletin No.33, September 1979, International Council of Scientific Unions, I.S.S.N. 0366-757X.

3. Baulch, D.L., Cox, R.A., Hampson, R.F., Kerr, J.A., Troe, J., and Watson, R.T., *J. Phys. Chem. Ref. Data*, (1980), in press.

4. Chapman, S., *Mem. R. Meteorol. Soc.*, *3*, 103 (1930).

5. Johnston, H.S., *Ann. Rev. Phys. Chem.*, *26*, 315 (1975).

6. Hunt, B.G., *J. Geophys. Res.*, *71*, 1385 (1966).

7. Crutzen, P.J., *Quart. J. R. Meteorol. Soc.*, *96*, 320 (1970).

8. Stolarski, R.S., and Cicerone, R.J., *Can. J. Chem.*, *52*, 1610, (1974).

9. Wofsy, S.C., McElroy, M.B., and Yuk Ling Yung, *Geophys. Res. Lett.*, *2*, 215 (1975).

10. Bates, D.R., and Nicolet, M., *J. Geophys. Res.*, *55*, 30 (1950).

11. Nicolet, M., *Ann. Geophys.*, *26*, 531 (1970).

12. Hahn, J., and Junge, C., *Z. Naturforschung. A*, *32*, 190 (1977).

13. Thrush, B.A., *Endeavour, 1,* 3 (1977).

14. Lovelock, J.E., *Nature, 230,* 379 (1971).

15. Lovelock, J.E., Maggs, R.J., and Wade, R.J., *Nature, 241,* 194 (1973).

16. Molina, M.J., and Rowland, F.S., *Nature, 249,* 810 (1974).

17. Rowland, F.S., and Molina, M.J., *Rev. Geophys. Space Phys., 13,* 1 (1975).

18. 'Halocarbons: Their Effect on Stratospheric Ozone', National Academy of Sciences, Washington, D.C., 1976.

19. Derwent, R.G., and Eggleton, A.E.J., Atomic Energy Research Establishment Report, R9912, HMSO, London, 1978.

20. Crutzen, P.J., *Geophys. Res. Lett., 3,* 73 (1976).

21. Demerjian, K.L., Kerr, J.A., and Calvert, J.G., *Adv. Environ. Sci. Technol., 4,* 1 (1974).

FREE RADICAL REACTIONS RELATED TO FUEL RESEARCH[1]

ROBERT N. HAZLETT

Chemistry Division
Naval Research Laboratory
Washington, D. C.

Summary

Free radical reactions are important with respect to fuel properties and to fuel production. The property of jet fuel related to degradation in an aircraft fuel system, thermal oxidation stability, is related to autoxidation reactions. Fuel deposits are associated with hydroperoxide formation in some cases and to hydroperoxide decomposition in others.

Middle distillate fuels derived from shale oil contain large amounts of *n*-alkanes which adversely affect freezing and pour points. Carbon-13 n.m.r. measurements on and pyrolysis of fractions from shale oil indicate that large alkanes, olefins, and polar compounds may be the source of some of the *n*-alkanes found in fuels of interest to the U. S. Navy.

Coal can be fragmented to medium molecular weight (90-210) compounds by *controlled, limited, low temperature air oxidation*. The concept utilizes a non-reactive fluid to swell and suspend/dissolve the coal. The oxidation reactions appear to occur at bridges between aromatic groups.

INTRODUCTION

The U. S. Navy spent a billion dollars for liquid hydrocarbon products in 1977 (1) and this amount is expected to rise in the future. Further, the fleet's major demands for energy will continue to be served by liquid fuels for at least 25 years since weapons platforms in the inventory or in consturction require such fuels.

1 - The author thanks the Naval Air Systems Command, the Naval Material Command, and the Naval Air Propulsion Center for supporting portions of this research.

ISBN 0-12-566550-4

The Navy's primary fuels are middle distillates, JP-5 for aircraft and diesel fuel marine (DFM) for ships. The general character of these fuels are described by the physical properties shown in Table 1. The DFM has a higher end point and a higher specific gravity than JP-5. The low temperature requirement for the jet fuel is much more severe, however. JP-5 and DFM require high flash points since shipboard temperatures which fuels frequently encounter are extreme (2). The flash point is the most significant property difference for Navy jet fuel, JP-5, compared to commercial jet fuel, Jet A, and Air Force jet fuel, JP-4.

Fuel properties affected by free radical chemistry are storage stability, thermal oxidation stability, peroxidation (the formation of hydroperoxides), and combustion. The Naval Research Laboratory has done extensive research in thermal oxidation stability of jet fuels and a discussion of this topic will comprise a major part of this paper.

TABLE 1

Middle distillate fuel properties

	DFM	JP-5	Jet A	JP-4
Typical Distillation Range (°C)	200-345[a]	182-258	168-268	61-234
Pour Point, minimum (°C)	-7	-	-	-
Freezing Point, maximum (°C)	-	-47	-40	-58
Flash Point, minimum (°C)	60	60	38	no reqmt.
Specific Gravity, (60/60) minimum	b	0.788	0.775	0.751
Specific Gravity, (60/60) maximum		0.845	0.830	0.802

a. Estimated.
b. Average gravity value for DFM samples is 0.850.

Free radical chemistry is important for fuel production also. In this context we will discuss two topics: (a) the pyrolysis of shale oil and (b) the fragmentation and liquefaction of coal by controlled air oxidation.

JET FUEL THERMAL OXIDATION STABILITY

Many organic liquids react with oxygen in the first step of a sequence which forms solid material. Solids are desired in some cases but are definitely detrimental when formed in liquids which are used as fuels or lubricants. Certain fluids degrade in storage even at ambient temperatures (3-6), but other high-quality materials such as commerical and military aircraft turbine fuels form little, if any, deposit under storage conditions. At elevated temperatures, however, these fuels form varnishes on aircraft heat exchanger surfaces, degrade valve performance, and deposit solids in engine combustor nozzles (3,7-10). The higher speeds of advanced aircraft magnify solid formation since the heat load on the aircraft, engine, and fuel is increased (11,12). The solids formation is stimulated by oxygen and the amount of oxygen which dissolves in turbine fuel in equilibrium with air (50-80 mg/ℓ) is sufficient to cause deposits which seriously degrade heat transfer (13-15).

We are studying the sequence of reactions which lead to deposit formation during the thermal stress of jet fuels. In this study a normal alkane, *n*-dodecane, which is a significant component of JP-5 and Jet A fuels, has been stressed in a small flow apparatus, the Jet Fuel Thermal Oxidation Tester (JFTOT) (16). In this apparatus, the fuel flows at three ml/min over the outside of a metal specimen which is heated electrically. The residence time for fuel flow through the 5-inch heated section is 28 seconds.

Liquid phase chemistry

The liquid phase chemistry for *n*-dodecane stressed in the JFTOT has been described by Hazlett, Hall, and Matson (17). The chemistry of oxygenated species over the

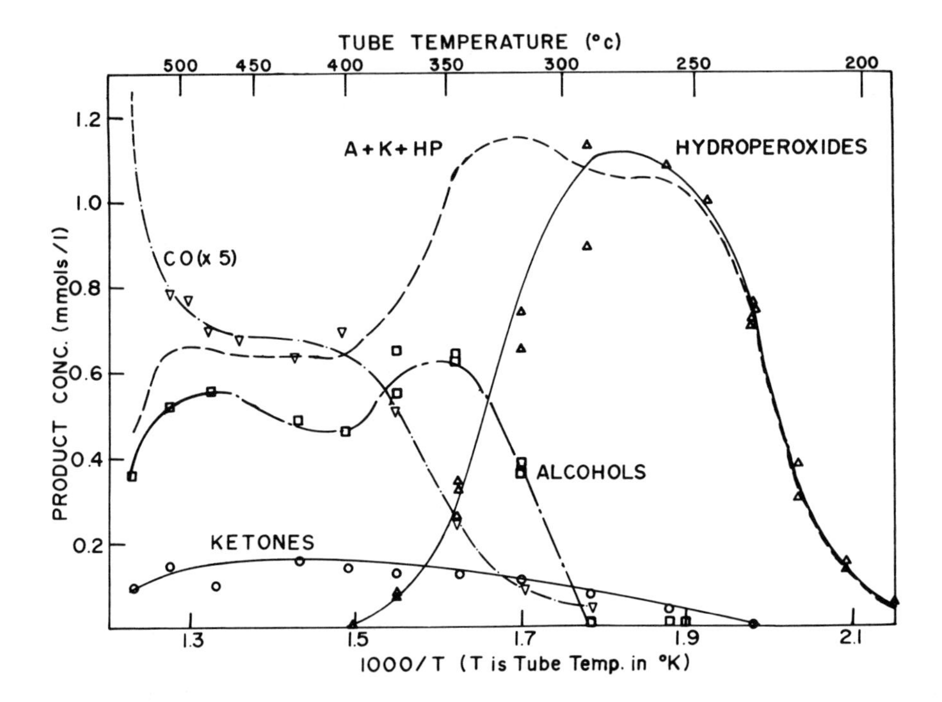

Fig. 1. Oxygenated species formed by reaction between n-dodecane and air; A + K + HP = alcohols + ketones + hydroperoxides.

temperature range 200-540°C is illustrated in Fig. 1. Hydroperoxides comprise the major initial products. These demonstrate thermal instability above 300°C, however, forming mainly alcohols. Some ketones and carbon monoxide also result from hydroperoxide decomposition.

In addition to oxygenated species, hydrocarbons smaller than *n*-dodecane appear as the dodecylhydroperoxides disappear. Both *n*-paraffins and 1-olefins are formed. The yield of four olefins is depicted in Fig. 2. The pattern for other alkenes in the series (C_4-C_8,C_{10}) are very similar to that for C_3 and C_9. The amount of C_2 and C_{11} olefins is much less than the others. *n*-Alkanes exhibit similar behavior to the olefins. Again the members at the end of the series - C_1, C_{10} and C_{11} - have lower yields than those in

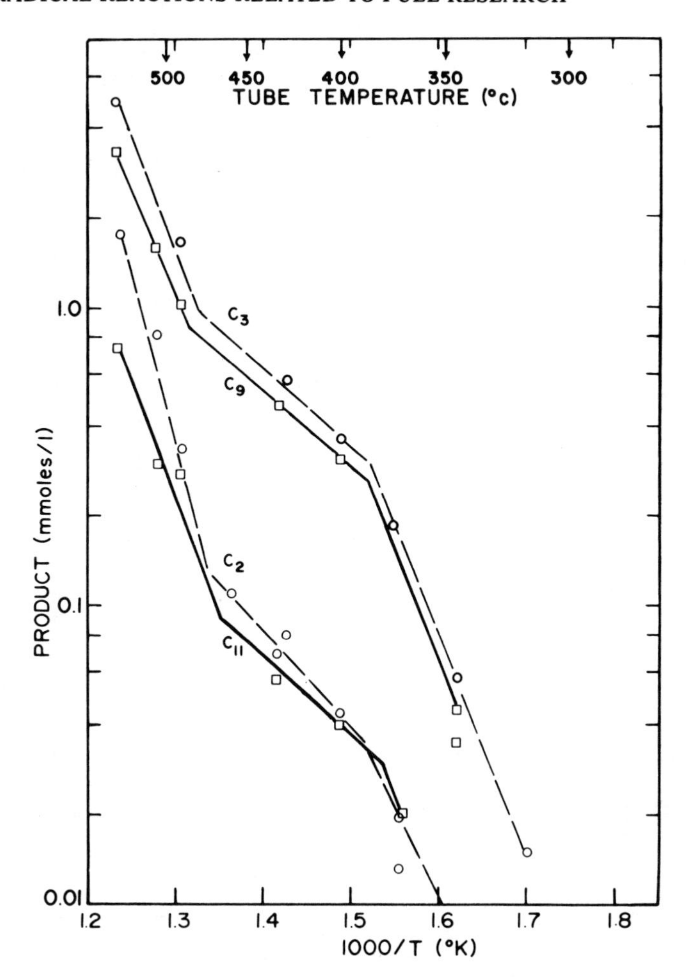

Fig. 2. Production of 1-olefins during decomposition of dodecylhydroperoxides; C_2 *= ethylene,* C_3 *= propene,* C_9 *- nonene,* C_{11} *= undecene.*

the center of the series. Fig. 3 indicates that the *n*-alkanes exceed the 1-olefins by a factor of two at 343°*C* but equal amounts are formed at temperatures above 430°C.

Most of the features of the reaction patterns observed in Figs. 1-3 can be interpreted on the basis of two schemes, i.e., autoxidation and pyrolysis. The latter controls the high temperature reactions (480°C and above), and autoxidation phenomena occur at the lower temperatures, 260°C and below.

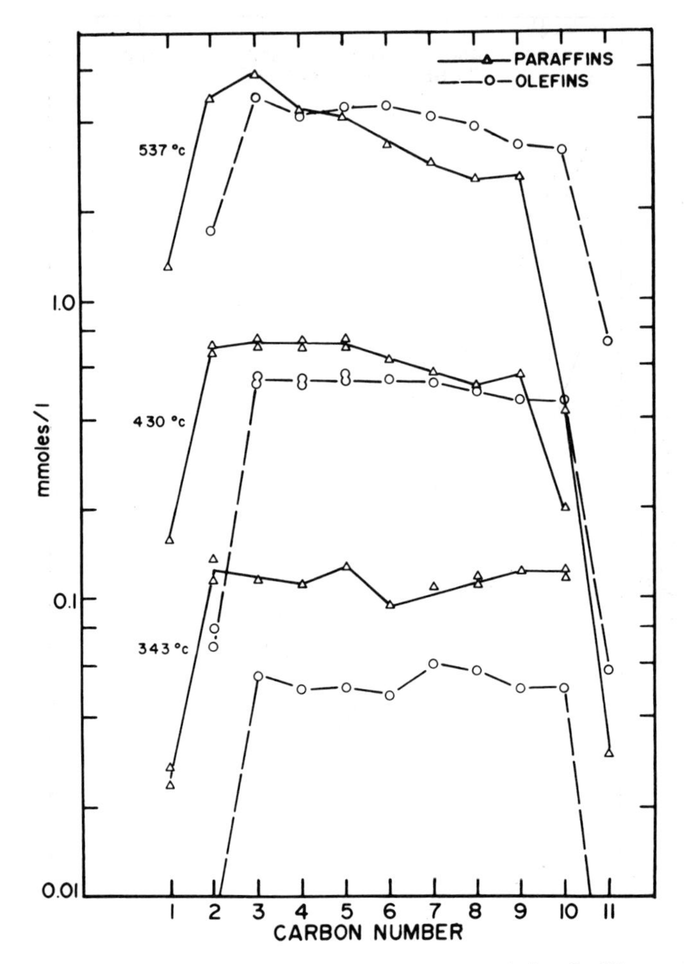

Fig. 3. Distribution of n-alkanes and 1-olefins at three temperatures.

In the intermediate regime, above the temperature at which oxygen is completely reacted but below pyrolysis temperatures, the reactions are more complex. The general features of these three regimes are summarized in Table 2.

TABLE 2

Temperature regimes in oxygen/dodecane reactions

Regime	Autoxidation	Intermediate	Pyrolysis
Temp (°C)	260 and below	290-480	480 and above
Reactions	O_2 with C_{12}	(1) ROOH decomposition (2) ROOH initiated pyrolysis	C_{12} cracking
Products	Hydro- peroxides	(1) alcohols, ketones, CO (2) *n*-alkanes, 1-alkenes, H_2	*n*-alkanes, 1-alkenes, H_2

The autoxidation reactions are those well established by many studies to explain liquid phase oxidation by elemental oxygen (18-22). In the flow apparatus used in our study, initiation occurs on the hot metal surface. The surfaces tested, aluminum and two stainless steels, exhibit similar behavior in the oxidation steps and do not appear to be significantly different in the rate of alkyl radical generation.

In the pyrolysis regime, Rice-Kossiakoff (23,24) reactions modified for high-pressure conditions can predict the pattern of hydrocarbon products. The Fabuss, Smith, and Satterfield (F-S-S) (25) one-step decomposition model for thermal cracking of *n*-hexadecane applies very well to the high pressure *n*-dodecane pyrolysis at the low conversions of this study. As in the Fabuss-Smith-Satterfield analysis, the first member of the *n*-alkane and 1-alkene series from *n*-dodecane reaction is formed in smaller amounts than the second member of these series. Likewise, the F-S-S model would predict the low yields of *n*-decane, *n*-undecane, and 1-undecene found in this work.

In the intermediate temperature regime, between autoxidation and hydrocarbon pyrolysis reactions, oxygenated products play an important role. The decomposition of hydroperoxides via alkoxy radicals can form alcohols, ketones, carbon monoxide, *n*-alkanes, and 1-olefins. The importance of this path is indicated by the fact that the yield patterns for CO, alkanes and alkenes exhibit a plateau above 385°C, the temperature at which ROOH is depleted.

The β-scission of alkoxy radicals contributes to the *n*-alkane production directly, as well as indirectly, via alkyl radical formation. The alkoxy radicals appear to give more *n*-alkanes than alkenes.

Effect of aromatics

The role of aromatic compounds in fuel oxidation and stability has been studied by several experimenters. Taylor (8) found that many aromatic compounds, particularly fluorene and diphenylmethane, inhibited deposit formation at low temperatures. On the other hand, Bol'shakov observed that tetralin or acenaphthene drastically increased sediment formation at 200°C (26). Bushueva and Bespolov found that alkyl-benzenes formed little sediment but naphthalenes and polycyclic naphthenic-aromatics stimulate deposit formation (27).

Some aromatics added to *n*-dodecane and stressed in the JFTOT flow apparatus definitely inhibit the rate of oxidation. This behavior is summarized in Table 3. The only compounds exhibiting inhibition are di- and tricyclic aromatics. Of the compounds studied, indane and cyclohexylbenzene are the only dicyclics showing no effect. Fluorene is by far the most effective inhibitor. We suggest that the active aromatics, which have bond strengths of 82 kcal/mol or less for their weakest C-H bond (28), behave as tradional fuel anti-oxidants. Thus the aromatic free radicals formed by hydrogen abstraction are relatively stable at the temperatures of these experiments. Reaction of oxygen with such radicals is significantly slower than with alkyl radicals. Consequently, the overall rate of oxidation is reduced.

The maximum quantities of hydroperoxides formed in the presence of the effective aromatics is less than that formed by 100% *n*-dodecane. Although the temperature/concentration pattern for decomposition of ROOH is the same in the presence or absence of the aromatics, the product yields from ROOH decomposition differ. The yields of hydrocarbon cleavage products are reduced, particularly in an intermediate temperature range, 370-430°C. Fluorene, for instance, decreases the *n*-alkane plus 1-olefin yield by 5-fold at 430°C. The inhibitory activity of the aromatics for small hydrocarbon formation differed from their oxidation activity although fluorene was most effective in both aspects. Fluorene also

TABLE 3

Reactivity with dissolved oxygen

Additive[a]	Concentration	Temperature[b] (°C)
None	---	225
Cumene	5.0 Mole%	223
Indane	"	224
n-Propylbenzene	"	224
Cyclohexylbenzene	"	225
1-Ethyl-2-Methylbenzene	"	226
1,2,4-Trimethylbenzene	"	226
2-Methylnaphthalene	"	236
Tetralin	"	236
Diphenylmethane	"	256
Triphenylmethane	2.5	256
Indene	5.0	276
Fluorene	1.0	277
Fluorene	2.5	297

a. Dissolved in *n*-dodecane.
b. Temperature at which 1/2 of dissolved oxygen (60 parts per million or 1.8 millimoles/liter) has reacted in the JFTOT; 5 inch 316 S.S. tube.

drastically alters the olefin to paraffin ratio of the cleavage products. The behavior is most evident at low temperatures and is not observed at 450°C and above.

Deposit formation chemistry

The deposits formed during thermal oxidative stress in the JFTOT were evaluated with a light reflectance tube rater. Although such ratings cannot be considered absolute, they are reliable for each specific fuel/metal combination. Quantitative estimation of deposit amount by direct weighing or by combustion to carbon dioxide is unreliable since only 5-50 μg is formed in a typical test and the background for unused tubes is high and erratic.

It is noteworthy that the tube deposit rating increases as the hydroperoxide decreases. In addition, Arrhenius plots for product formation and deposit rating exhibit the same

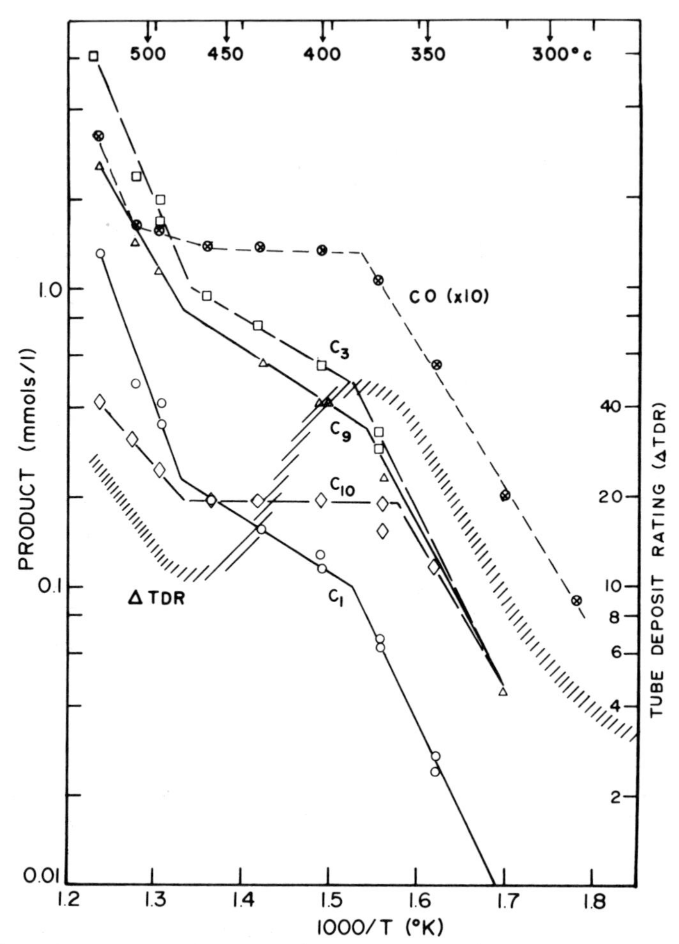

Fig. 4. Comparison of heater tube deposits with other reaction products; C_1 = methane, C_3 = propane, C_9 = n-nonane, C_{10} = n-decane.

slope for temperatures between 290 and 385°C (Fig. 4). This applies to CO and the *n*-alkanes. This implies that the solids on the heater tube are formed from the same precursor, hydroperoxide, as are the other products.

The deposit rating decreases above 385°C. Again, this behavior is similar to that of the other products - CO, alkanes, and alkenes - which exhibit a similar sharp change in slope at the same temperature. Since hydroperoxide is depleted at this temperature, the tie in between hydroperoxide decomposition and deposit formation is reinforced.

Although hydroperoxides are precursors to deposits, the reactions forming deposits are relatively minor in the overall scheme. For instance, the maximum hydroperoxide concentration found for *n*-dodecane is about 200 mg/ℓ whereas the typical deposit from a liter of fuel is 0.01 to 0.1 mg. Thus, the specific reactions that form deposits are definitely side branches to the primary chemical sequence. Nevertheless, it appears that dodecyloxy free radicals are key intermediates in deposit formation.

Since certain aromatics alter the oxidation of *n*-dodecane as well as modify product yields at high temperatures, their effect on deposit formation was examined in the JFTOT. No significant inhibition of deposit formation was observed for any aromatic studied over the temperature range 260-430°C. Several aromatics increased deposit formation, however. Indene, the greater promoter, yielded very heavy deposits at 315°C, greater than that observed for pure *n*-dodecane at 480°C. This bicyclic compound readily forms a polymer peroxide (29) which breaks down when stressed thermally.

Above 400°C, cumene, *n*-propylbenzene, triphenylmethane, and cyclohexylbenzene increase deposits. Indane, diphenylmethane, 2-methylnaphthalene, and 1-ethyl-2-methylbenzene exert no effect on deposit formation. Fluorene, the best oxidation inhibitor of the aromatics examined, was essentially innocuous with respect to deposit formation up to at least 480°C.

Several jet fuels have been examined for deposit formation as related to hydroperoxide content. For such fluids, comprised of 100 or more hydrocarbons plus sulfur and nitrogen compounds, the relationship is not as clearcut. However, deposit maxima on heater tubes are associated with temperature regimes typical of hydroperoxide formation as well as hydroperoxide decomposition. Thus, the deposits form during situations characterized by high free radical concentrations.

Conclusion

Deposit formation in fuels is triggered by autoxidation reactions. In some cases, the deposits are closely associated with hydroperoxide formation or decomposition. The deposit chemistry, however, does not appear to be in the main reaction sequence involving the oxgenated products. Rather it is in a closely related side branch.

PYROLYSIS OF SHALE OIL

Jet aircraft are exposed to low temperatures and the fuels must not interfere with flying operations by freezing and plugging filters. Commercial jet fuel (Jet A) has a specification requirement of -40°C maximum but that for military fuels is lower (Table 1). JP-5 must freeze below -47°C because the Navy jets operate world wide as well as at higher altitudes than commercial jets. U. S. Air Force bombers require an even lower freezing point, -58°C maximum, since long flights at high altitudes permit the fuel to reach lower temperatures. It has not been practical to make JP-5 from some petroleum crudes because the freezing point cannot be met along with the required flash point. Dimitroff et al (30) examined the influence of composition on freezing point of several types of fuels. They found the saturate fraction of a fuel usually exerted the greatest effect on freezing point but the aromatic fraction seemed to be important in some cases.

NRL has related the freezing point of JP-5 type fuels to the *n*-alkane content, specifically *n*-hexadecane (31). This relationship applies to jet fuels derived from alternate fossil fuel resources, such as shale oil, coal, and tar sands, as well as those derived from petroleum. The *n*-alkane content varies for jet fuels from different petroleum crudes but varies much more in jet fuels derived from alternate fossil energy sources such as shale oil, coal, and tar sands. Table 4 shows that, in general, jet fuels from shale oil have the highest and those from coal (COED's) have the lowest *n*-alkane content. The origin of these *n*-alkanes in the amounts observed, especially in shale derived fuels, is not readily explained on the basis of literature information. Studies of the processes, particularly the ones involving thermal stress, which produce these fuels are needed to define how the *n*-alkanes form from larger molecules which are cyclic or branched. The information developed will significantly contribute to the selection of processes and refining techniques for future production.

TABLE 4

n-Alkane content of jet fuels (weight percent)[a]

Source	C_{13}	C_{15}	C_{16}	ΣC_9 to C_{16}
Shale-Paraho	6.66	3.32	2.45	36.33
Shale-LERC	5.81	3.48	2.46	25.31
	5.36	3.17	2.58	23.73
Petroleum	3.49	0.81	0.14	16.00
	3.13	0.72	0.05	15.71
Tar Sands	2.43	0.73	0.15	9.50
COED-Utah Coal	1.51	1.08	0.66	7.14
	1.53	1.01	0.66	7.45
COED-W KY. Coal	0.66	0.46	0.20	3.90
	0.70	0.42	0.27	3.73
	0.76	0.11	0.02	5.48

a. Precision: $\pm$ 3%.

Pyrolysis mechanisms

A large *n*-alkane breaks apart by a free radical mechanism to yield smaller hydrocarbons, both *n*-alkanes and 1-olefins. This process, as originally diagnosed by Rice (23), ultimately yields mostly small olefins with 2 to 4 carbon atoms. This behavior is encouraged by high temperatures and low pressures. At conditions more typical of shale retorting and delayed coking, however, Fabuss-Smith-Satterfield (25) behavior occurs. In this situation, a single fragmentation step occurs and equal amounts of *n*-alkanes and olefins form. Further, the yield is about the same for hydrocarbons in the intermediate carbon range. Unbranched olefins formed in the pyrolysis reactions readily convert to *n*-alkanes by hydrogenation.

Thus, formation of *n*-alkanes in the jet fuel distillation range may come via pyrolysis of large *n*-alkanes present in the crude oil source. Other possible precursors to small straight chain molecules are substituted cyclic compounds.

Attack in the side chain obviously affords a path to an *n*-alkane. Esters, acids, amines, and ethers also have the potential to form *n*-alkanes if an unbranched alkyl chain is present in the molecule.

Since shale derived fuels have the highest concentration of *n*-alkanes, the main effort of this project has been devoted to jet fuels made from this fossil energy resource. Carbon-13 n.m.r. studies indicated that oil shale rock contains many long unbranched straight chain hydrocarbon groups (32). The shale oil derived from the rock also gives indications of considerable straight chain material.

Separation and G C analysis

In order to characterize the unbranched material, the Paraho crude shale oil was separated into 9 fractions. First the oil was distilled into an atmospheric distillate with an end point of 300°C, a vacuum distillate with an end point of 300°C (40 mm), and a residue. Each of these three cuts were separated on activated silicagel into a saturate fraction, an aromatic fraction, and a polar fraction. The saturate fraction was removed from the silica with n-pentane solvent, then the aromatic fraction was removed with a 25:75 benzene: n-pentane solvent. The polar fraction was desorbed with 25:75 benzene: methanol solvent. The nitrogen content of various eluates was determined as an indication of the separation efficiency. The pentane eluant for all distillation cuts contained no nitrogen. The polar fraction (benzene: methanol) contained 99.5% of the recovered nitrogen in the case of the atmospheric distillate, 96% in the case of the vacuum distillate, and 97% in the case of the residue. If 100% benzene was used to desorb the aromatic fraction, up to 20% of the nitrogen was found in this fraction.

The breakout of the three fractions and the nine subfractions is outlined in Fig. 5. The residue comprises almost one-half of the shale crude oil. On a chemical basis, the polar compounds comprise 53 percent of the total and the bulk of these are found in the residue. The saturate compounds are found primarily in the atmospheric and vacuum distillates and total 28.7% of the crude. The sums of all *n*-alkanes and 1-olefins found in the atmospheric and vacuum saturate fractions are also shown in Fig. 5.

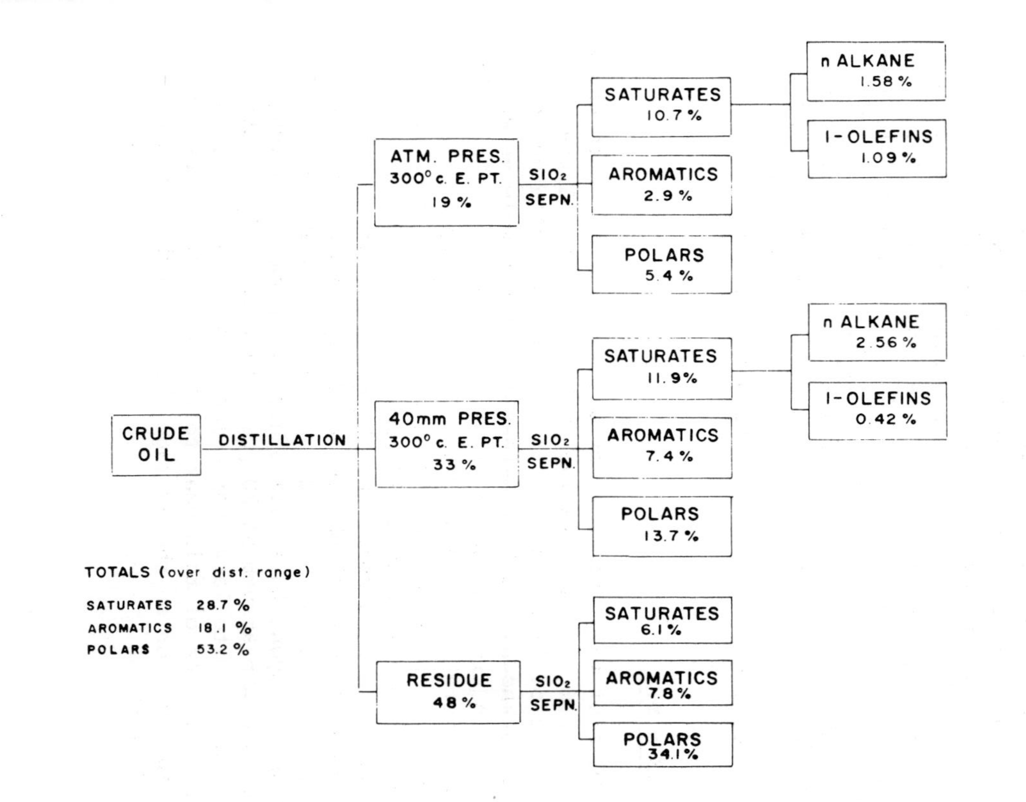

Fig. 5. Weight percent yields of fractions from physical and chemical separation of Paraho crude shale oil.

These data, obtained by glass capillary gas chromatography, indicate that the *n*-alkanes are a substantial portion of the two saturate fractions. The olefins represent a sizable fraction of the atmospheric distillate. A ^{13}C n.m.r. spectrum shows that the olefins in the atmospheric distillate saturate fraction are 1-olefins.

The bar graph in Fig. 6 gives quantitative information for each of the *n*-alkanes and 1-olefin hydrocarbons found in the saturate fraction of the atmospheric distillate. This distillate, with an end point of 300°C has a distillation range similar to JP-5 and the range of *n*-alkanes detected, C_9 to C_{16}, mirrors that of typical petroleum derived JP-5's. The *n*-alkanes + 1-olefins total only 14% of the atmospheric distillate cut (25% of the saturate fraction), however. This is far below the sums for *n*-alkanes listed in Table 4 for shale derived jet fuels (24 to 36%). Thus, a significant portion of the n-alkanes found in the jet fuels must be derived from larger molecules by thermal cracking and/or hydrotreatment.

Carbon-13 n.m.r.

Samples of the various fractions were submitted to analysis by ^{13}C n.m.r. Quantitative analysis of the aliphatic region, that of importance in this study, was attained by including a known amount of methanol in the sample as an internal standard. A long unbranched fragment will exhibit peaks at several positions in the aliphatic region of the spectrum. The peak corresponding to the methyl end group (α-carbon) appears at about 14 ppm with reference to tetramethylsilane at zero ppm. The CH_2 group adjacent to the methyl group (β-carbon) absorbs at 23 ppm and subsequent absorptions appear at about 32 and 29.4 ppm for the γ- and δ- carbons. Beyond this, all other CH_2 groups in a long unbranched chain absorb at 30 ppm. Therefore, this latter peak would be quite large for a long chain. In fact, the ratio of the area of this peak to the α- β- or γ-peak can afford information on the average chain length of the unbranched fragment.

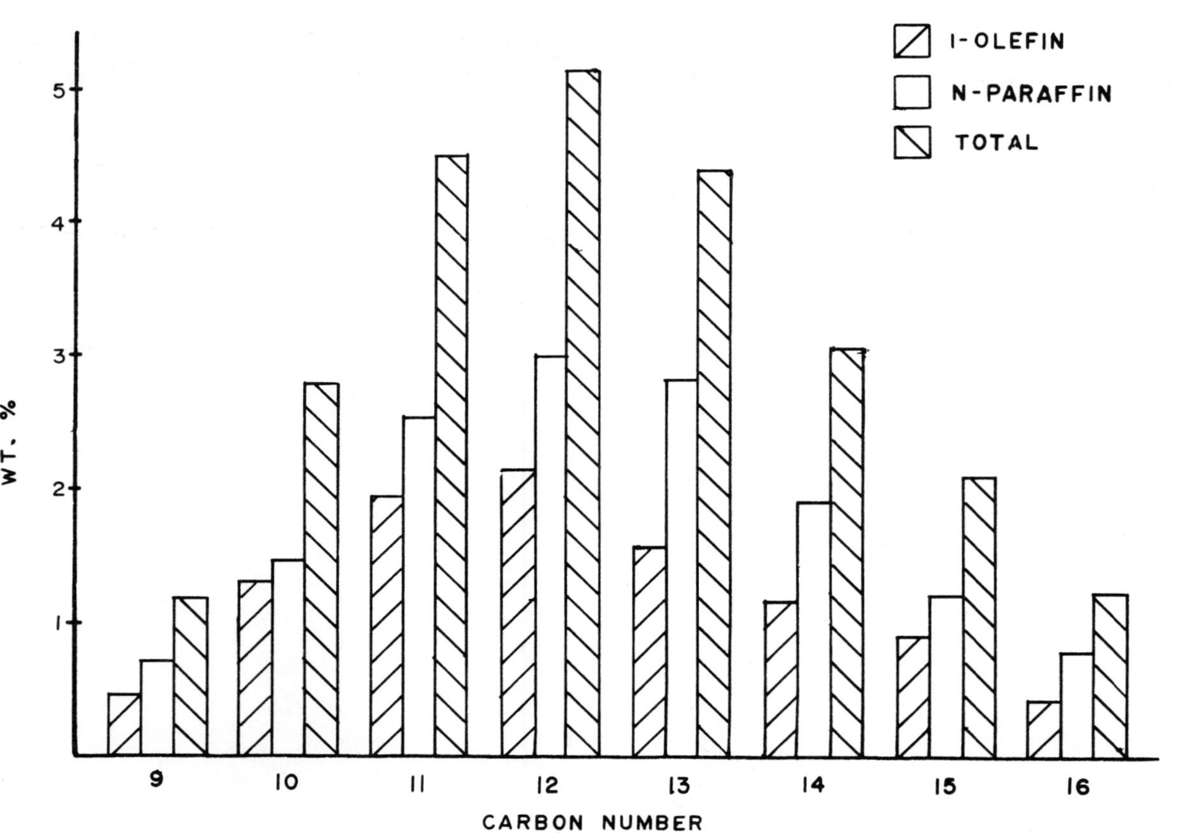

Fig. 6. Distribution of n-alkanes and 1-olefins in shale oil atmospheric distillate saturate fraction; data from glass capillary gas chromatography with internal standard quantitation.

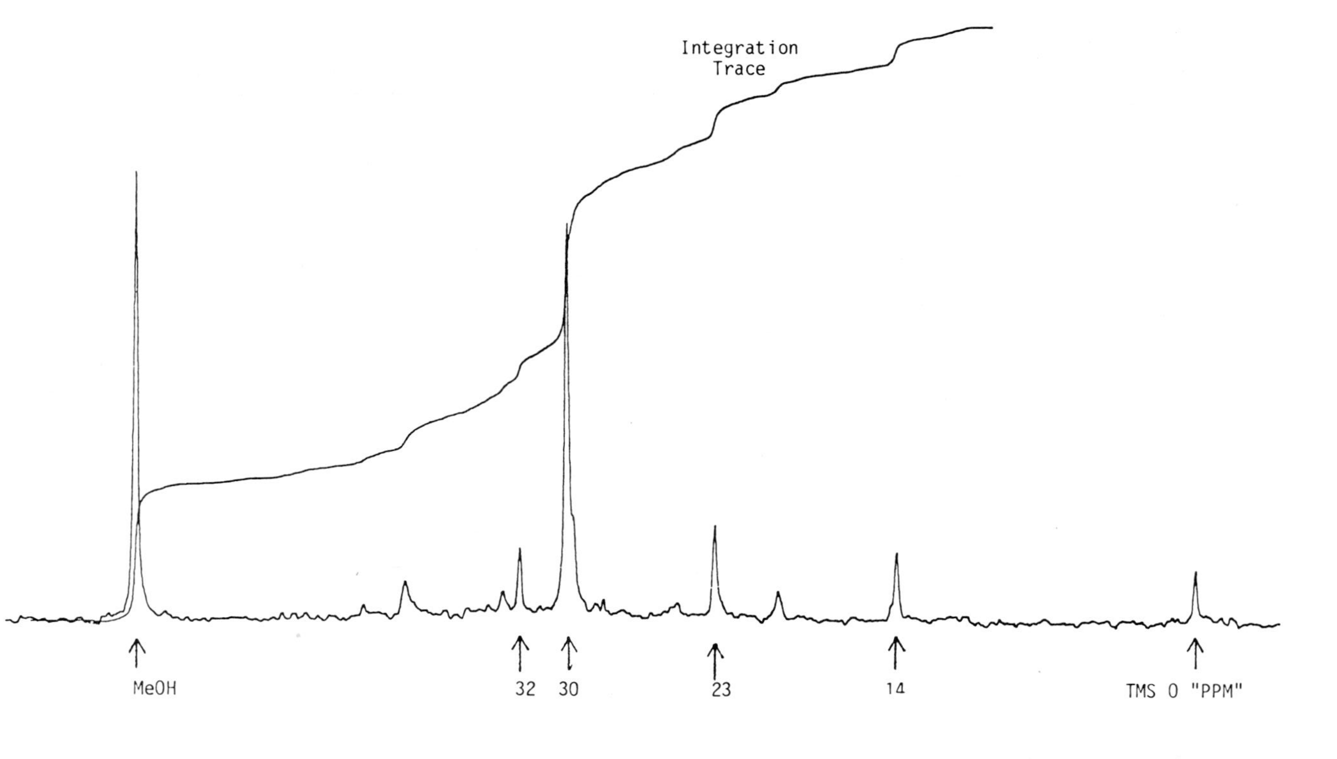

Fig. 7. Carbon-13 n.m.r. spectrum of shale oil vacuum distillate saturate fraction; MeOH peak is methanol internal standard.

TABLE 5

Carbon-13 n.m.r. examination of shale oil fractions

Fraction	Wt. % carbon in aliphatic region	Wt. % unbranched alkyl groups[a]	Average chain length[b]
Vacuum Distillate			
A. Saturate	100	58	19
B. Aromatic	48	15	13
C. Polar	55	36	13
Residue			
A. Saturate	100	40	43
B. Aromatic	60	21	14-22
C. Polar	56	30	20

Precision: $\pm$ 10%.

a. Sum of areas of absorption peaks at 14, 23, 30 and 32 ppm.

b. For unbranched alkyl groups: based on ratio of 30 ppm peak area to average of 14, 23, and 32 peak areas.

A spectrum for the aliphatic region of the saturate fraction from the vacuum distillate is shown in Fig. 7. The distinctive peaks at 14, 23, 32, and 30 ppm demonstrate the presence of significant amounts of long unbranched groups in this fuel fraction. The 29.5 peak appears as a shoulder on the 30 peak and these two peaks were integrated together.

Quantitation of the spectral information using the methanol internal standard gives the data listed in Table 5. As expected, the content of long unbranched alkyl groups is greatest for the saturate fractions. Further, the straight chain alkyl groups in the residue are longer on the average than those in the vacuum distillate. We conclude that there is a definite potential for making *n*-alkanes and 1-olefins in the jet fuel distillation range by cracking compounds found in the heavier shale oil cuts.

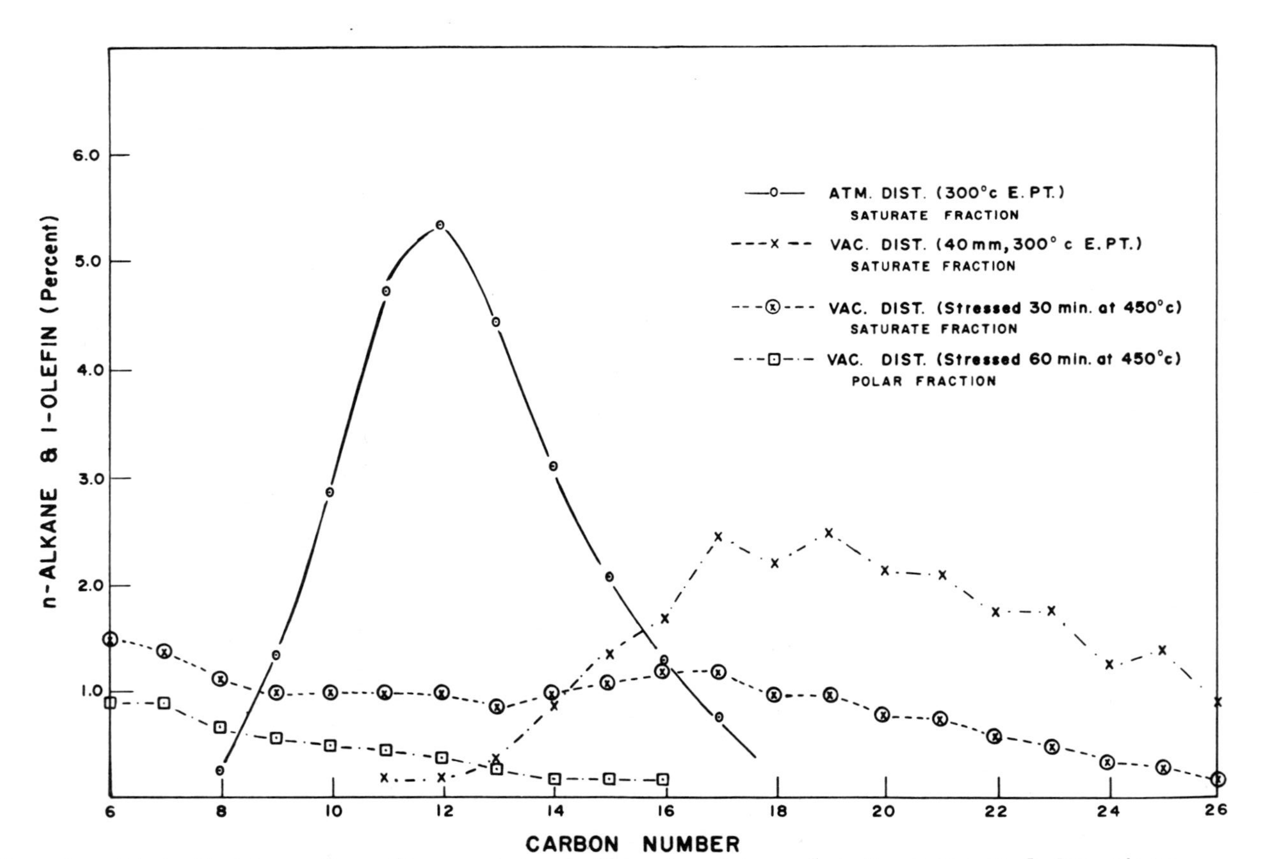

Fig. 8. Distribution of n-alkanes + 1-olefins in stressed and unstressed fractions of shale oil.

Pyrolysis

The vacuum distillate fractions have been stressed at conditions corresponding to the petroleum refining process known as delayed coking (33). These conditions are about 450°C and 90 psi pressure. The saturate fraction gave significant amounts of *n*-alkanes and 1-olefins after a 30 minute stress. The sum of the *n*-alkanes and 1-olefins for the unstressed sample are compared with those for the stressed material in Fig. 8. The larger alkanes have reacted more extensively in agreement with Fabuss et al's findings (25). Consequently, the net yield for carbon number 14 and below is positive. The nearly equal amounts of products for carbon numbers 8 through 12 indicate that the pyrolysis is taking place via the one-step F-S-S mechanism (25). The similar yield of *n*-alkanes and 1-olefins in this carbon number range also supports the one-step mechanism. The large olefins pyrolyze more rapidly than the large *n*-alkanes.

The polar fraction from the vacuum distillate gave straight chain compounds as indicated in Fig. 8. The yields shown, which were for a 60 minute stress, are distinctly less than for the saturate fraction at a 30 minute stress. The aromatic fraction of the vacuum distillate is a poor source of *n*-alkanes and 1-alkenes. The yields of *n*-alkanes varied little for 30 to 120 minute stress periods and the *n*-alkanes and 1-alkenes in the typical JP-5 carbon number range were less than 0.1%.

The ^{13}C n.m.r. data and the pyrolysis results for the vacuum distillate agree qualitatively. For instance, the n.m.r. data ranked the fractions in the proportion of unbranched alkyl groups in the order: saturate > polar > aromatic. This order prevailed also in the yield of *n*-alkanes and 1-olefins formed by thermal stress.

Conclusion

Shale oil made by the Paraho retort process was separated into nine fractions using combined physical and chemical techniques. The *n*-alkane plus 1-olefin content of the straight run jet fuel distillate is much lower than that found in jet fuels made from the total crude. Carbon-13 n.m.r. measurements indicate that potential for making unbranched material from heavier fractions is in the order: saturates > polars > aromatics. The pyrolysis of fractions from the vacuum distillate portion of the crude confirms the predictions made from the n.m.r. data.

COAL LIQUEFACTION BY CONTROLLED AIR OXIDATION

We have seen in the section on jet fuel thermal oxidation stability that air oxidation can cleave a hydrocarbon into smaller fragments. This goes through the sequence:

$$\text{hydroperoxide} \xrightarrow{\Delta} \text{alkoxy radical} \xrightarrow{\beta\text{-scission}} \text{small alkyl radical} \longrightarrow \text{alkanes} + \text{alkenes}$$

Autoxidative processes form the hydroperoxides. Each hydroperoxide decomposition event can yield many smaller molecules, the amount depending on the chain length. Therefore, a scission yield in excess of the oxidation stoichiometry is observed.

The application of the air oxidation concept to coal fragmentation has a decided appeal. The attraction of the air oxidation route to coal liquid is two fold: inexpensive reagent (air) and low-temperature and pressure conditions. However, the oxidation must be controlled and limited to prevent severe loss of C and H. It is worthwhile to note that the well known oxidations of coal in aqueous alkali produce high yields of benzene carboxylic acids but also give CO_2 amounting to approximately 50% of the available carbon (34,35).

Under mild conditions, initial stages of coal oxidation by air should proceed principally at aliphatic sites adjacent to aromatic groups. This results from the marked preference for autoxidation at benzylic positions (18) and the preponderence of aromatic structures in coal (36). Methylene and ethylene bridges between aromatic functions appear to be a common feature in coals (37).

We thus predict that air oxidation of coal should proceed with the initial formation of hydroperoxides at benzylic type sites. Subsequent thermolysis at relatively low temperatures in the absence of oxygen should result in 0-0 bond cleavage of the hydroperoxides followed by some β-scission to yield low molecular weight products.

The validity of the oxidation concept has been reported for solvent refined coal (SRC) (38). In that work, a solvent was utilized to encourage liquid phase autoxidation reactions to occur. The solvent, quinoline, was not oxidized at temperatures below 200°C but the SRC solutions reacted smoothly and readily with air at 105 to 180°C. When thermally stressed at about 400°C, the oxidized SRC solution afforded 30-40% yields of small molecules. The products, which were primarily aromatic hydrocarbons, were in a very useful molecular weight range, 90-210. The unoxidized, unstressed solution is compared in GC analysis with the oxidized, stressed solution in Fig. 9. Specific products formed are identified in the

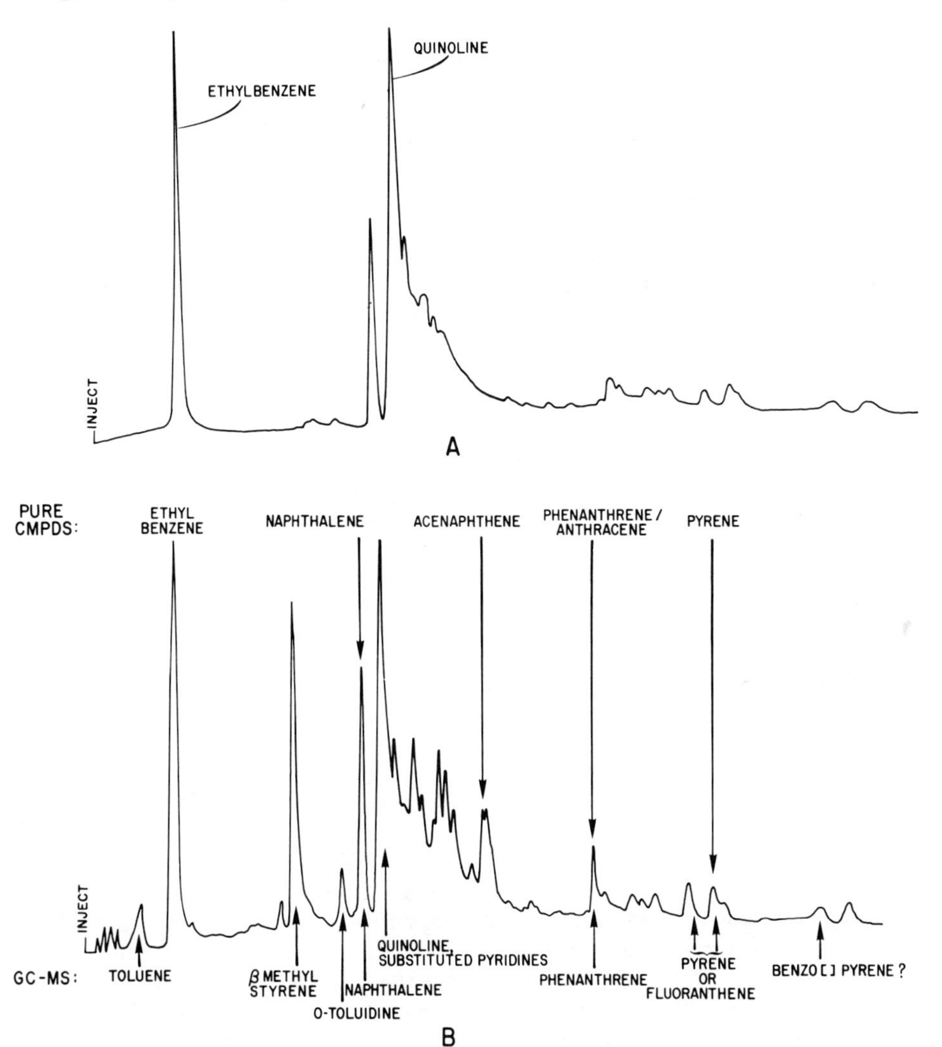

Fig. 9. GC analysis of 10% SRC dissolved in quinoline with added ethylbenzene as internal standard. Shown at the top of A and B are the results of retention time comparisons with authentic samples. Along the bottom are shown structural assignments for specific peaks based on GC-MS data; A-before oxidation and stress; B-after oxidation and stress.

figure. The ethylbenzene was added as a standard for quantitation. In the SRC - quinoline study, the extent of air oxidation was controlled and limited. Thus, the amount of coal reacted with oxygen was restricted to small conversions, equivalent to 3 to 6% of the carbon atoms.

The experimental work has been extended to "run-of-mine" (ROM) coals and to a more volatile solvent. Bituminous, subbitiminous and lignite coals have been subjected to the oxidation/thermolysis process in pyridine. As with SRC, the oxidation step occurs smoothly and a stress temperature of 400°C-415°C affords yields in the 10 to 32% range (39). Higher rank coals appear to give higher yields under the conditions tried thus far (Table 6). The products from ROM coal conversion were similar to those from SRC - aromatic with molecular weights between 90 and 210.

The role of the solvent is uncertain with ROM coals since only a small portion of the coal is soluble (6-24% at boiling point of pyridine, 115°C). Thus, complete autoxidation behavior, with multiple chain propagation steps and chain branching, seems unlikely. Solvents such as pyridine and quinoline probably effect swelling of the coals, however, and enhance the diffusion of reactants into and products out of the semi-rigid coal structure. The extent of coal solubility in pyridine seems to be important - the greater the solubility the better the yield of products.

The premise when this work began was that oxidation would form unstable hydroperoxides which would fragment during decomposition. The evidence favors an alternate process - formation and decomposition of carboxylic acids. This conclusion is based on indirect information. First, small molecules are observed only after stress above 350°C, a temperature 150°C above the stable region for hydroperoxides. Further, CO_2 is the main gas evolved during thermal stress rather than CO which would be expected from hydroperoxides. Campbell also found that carbon dioxide was the major gas evolved from heating sub-bituminous coal to temperatures above 400°C (40). Presumably, thermal decarboxylation also explains his data. Copper, a catalyst for decarboxylation of organic acids (41), increased the yield of gas during thermolysis of coal oxidate in quinoline solution, but yields of low molecular weight products were not increased.

TABLE 6

Liquefaction yields from SRC and various coals

Substrate	Solvent	Temperatures (°C) Oxidation	Temperatures (°C) Thermolysis	% Yield
SRC	Quinoline	120	415	16
SRC	Quinoline	120	435	32
SRC	Pyridine	No Oxid.	415	1
SRC	Pyridine	100	350	7
SRC	Pyridine	100	415	36
W. Ky. Bit.	Pyridine	100	400	13-20
Hi Vol A Bit.	Pyridine	No Oxid.	415	15
Hi Vol A Bit.	Pyridine	100	415	32
Hi Vol B Bit.	Pyridine	No Oxid.	400	5
Hi Vol B Bit.	Pyridine	100	400	13-20
Hi Vol C Bit.	Pyridine	No Oxid.	415	12
Hi Vol C Bit.	Pyridine	100	415	19
Sub-Bit. B	Pyridine	No Oxid.	415	4
Sub-Bit. B	Pyridine	100	415	10
Lignite	Pyridine	100	400	14-18

Oxidation: 3 hrs; thermolysis: 3 hrs; solutions (SRC) and slurries (ROM coals) were 10% by wt; yields determined by quantitative GLC for compounds with MW $\leq$ 210.

Free radical reactions occur during the thermal stress step. This statment is based on the finding that solvent dimers (e.g., bipyridine) appeared during pyrolysis of the oxidized ROM coals. We suggest that some fraction of the free radicals formed by decarboxylation of coal acids abstract a hydrogen atom from the solvent, the most readily available neighbor in the semi-rigid coal structure. Two solvent radicals then dimerize. This side reaction appears to limit the yield of desirable products. Thus, defining and understanding the chemistry of the two step oxidation/thermolysis process offers the potential of increasing the product yield. In addition, a significant reduction in the temperature required for thermolysis seems probable if we understand the decarboxylation mechanism.

Conclusion

The concept of coal fragmentation of medium molecular weight molecules by air oxidation has been proven. This process, which utilizes a controlled, limited oxidation step followed by a thermolysis step, has been applied with success to solvent refined coal and "run of mine" coals.

ACKNOWLEDGMENT

The author thanks the following NRL employees who participated in this work; James Hall, Jeffrey Solash, Jack Burnett, Martha Matson, Richard Sonntag, Peggy Climenson, and William Moniz.

REFERENCES

1. U. S. Navy Energy Plan and Program - 1978, *Document No. OPNAV 41P4, Chief of Naval Operations,* June 1978.

2. Affens, W. A., "Shipboard Safety: A Meaningful Flash Point Requirement for Navy Fuels", *NRL Rpt. No. 7999,* October 28, 1976.

3. Nixon, A. C. in *Autoxidation and Antioxidants,* Vol. II, W. O. Lundberg, ed., Interscience, N. Y., N. Y., 1962.

4. Chiantelli, A. J., and Johnson, J. E., *J. of Chem. & Eng. Data, 5,* 387 (1960).

5. Schwartz, F. G., Whisman, M. L. Allbright, C. S. Ward, C. C., *U. S. Bur. Mines Bull., No. 626,* "Storage Stability of Gasoline", 1964.

6. American Society for Testing Materials, "Symposium on Stability of Distillate Fuel Oils", *ASTM Special Tech. Pub. #244,* Philadelphia, Pa., June 25, 1958.

7. Schirmer, R. M. "Morphology of Deposits in Aircraft and Engine Fuel Systems", *SAE National Air Transportation Meeting, Paper 700258*, N. Y., N. Y., April 1970.

8. Taylor, W. F., *I & EC, Product Res. Develop., 8*, 375 (1969).

9. Taylor, W. F., *I & EC Product Res. Develop., 13*, 133 (1969).

10. Mayo, F. R., Richardson, H., and Mayorga, G. D., *Am. Chem. Soc., Div. of Petrol. Chem. Prepr., 20*, 38 (1975).

11. Dukek, W., J. *Inst. Petrol., 50*, 273 (1964).

12. Smith, J. D., *I & EC, Proc. Des. Develop., 8*, 299 (1969).

13. Watt, J. J., Evans, A., Hibbard, R. R., "Fouling Characteristics of ASTM Jet A Fuel When Heated to 700°F in a Simulated Heat Exchanger Tube", *NASA Technical Note D-4958*, December 1968.

14. Bradley, R. P., Bankhead, H. R., and Bucher, W. E., "High Temperature Hydrocarbon Fuels Research in an Advanced Fuel System Simulator on Fuel AFFB-14-70", *Air Force Aero Propulsion Lab., Rpt. No. AFAPL-TR-73-95*, 1974.

15. Astaf'ev, V. A., Borisov, B. D., Logvinyuk, V. P., Malyshev, V. V., and Popov, A. A., *Khim, Tekhnol. Topl. Masel, 11 (7)*, 41, (1975).

16. "Test for Thermal Oxidation Stability of Aviation Turbine Fuels", *ASTM Method D-3241-74*, Philadelphia, Pa., 1976.

17. Hazlett, R. N., Hall, J. M., and Matson, M., *I & EC Prod. R&D, 16*, 171 (1977).

18. Emanuel, N. M., Denisov, E. T., and Maizus, Z. K., *Liquid Phase Oxidation of Hydrocarbons*, Plenum Press, N. Y., N. Y., 1967.

19. Berezin, I. V., Denisov, E. T., and Emanuel, N. M., *The Oxidation of Cyclohexane*, The Pergamon Press, Ltd., Oxford, England, 1966.

20. Walling, C., *Free Radicals in Solution,* Ch. 9, Wiley and Sons, Inc., N. Y., N. Y., 1957.

21. Scott, G., *Atmospheric Oxidation and Antioxidants,* Elsevier Publishing Company, Amsterdam, 1965.

22. Brown, D. M., and Fish, A., *Proc. Roy. Soc. A., 308,* 547 (1969).

23. Rice, F. O., *J.A.C.S., 55,* 3035 (1933).

24. Kossiakoff, A., and Rice, F. O., *J.A.C.S., 65,* 590 (1943).

25. Fabuss, B. M., Smith, J. O., Satterfield, C. N., *Advances in Petroleum Chemistry and Refining,* Vol. IX, 157 (1964).

26. Bol'shakov, G. F., Bushueva, E. M., and Glebovskoya, E. A., *Khim. i Tekhnol. Topliv i Masel,* p. 40, April 1967.

27. Bushueva, E. M., and Bespolov, I. E., *Khim. i Teckhnol. Topliv i Masel,* p. 46, Sept. 1971.

28. Korcek, S., Chenier, J. H. B., Howard, J. A., and Ingold, K. U., *Can. J. Chem., 50,* 2285 (1972).

29. Russel, G. A., *J.A.C.S., 78,* 1035 (1956).

30. Dimitroff, E., Gray, J. T., Jr., Meckel, N. T., and Quillian, R. D., Jr., *Seventh World Petroleum Congress, Paper No. 47,* Mexico City, Mexico, Apr. 2-9, 1967.

31. Solash, J., Hazlett, R. N., Hall, J. M., and Nowack, C. J., *Fuel, 57,* 521 (1978).

32. Resing, H. A., Garroway, A. N., and Hazlett, R. N., *Fuel, 57,* 450 (1978).

33. Gary, J. H., and Handwerk, G. E., *Petroleum Refining,* Ch. 5, Marcel Dekker, Inc., N. Y., N. Y., 1975.

34. Smith, R. C. Tomarelli, R. C., and Howard, H. C., *J. A. C. S., 61,* 2398 (1939).

35. Montgomery, R. S., and Holly, E. D., *Fuel, 36,* 63 (1957).

36. Retcofsky, H. L., and Vanderhart, D. L., *Fuel, 57,* 421 (1978).

37. Deno, N. C., Greigger, B. A., and Stroud, S. A., *Fuel, 57,* 455 (1978).

38. Hazlett, R. N., Solash, J., Fielding, G. H., and Burnett, J. C., *Fuel, 57,* 631 (1978).

39. Solash, J., Hazlett, R. N., Burnett, J. C., Climenson, P. A., and Levine, J. R., *Am. Chem. Soc., Div. of Fuel Chem. Prepr. 24 (2),* 212 (1979).

40. Campbell, J. H., *Fuel, 57,* 217 (1978).

41. Cohen, T., and Schanbach, R. A., *J.A.C.S., 92,* 3189 (1970).

TWENTY YEARS OF PEROXIDE CHEMISTRY

RICHARD R. HIATT

Department of Chemistry
Brock University
St. Catharines, Ontario, Canada

INTRODUCTION

Our first hydroperoxide experiments, 20 years ago (1, 2) were intended as a minor footnote to an already vast, presumed accurate, literature. The idea was to construct a hydroperoxide that might reasonably give simultaneous 2-bond scission analogous to decompositions of some peroxyesters (3). Substantial yields of 1,1-dimethyl-2-phenylethanol from

$$PhCH_2CMe_2O_2H \xrightarrow{\Delta} PhCH_2\cdot + Me_2CO + \cdot OH \ ? \qquad (1)$$

thermal decomposition of the parent hydroperoxide disproved the thesis of reaction 1. Moreover, the rate of decomposition was not faster, but a good deal slower than the accepted "normal" rate for homolysis of hydroperoxides. Companion studies of t-BuO_2H (1) and t-$BuCMe_2O_2H$ (2) gave similar "slow" rates, t-BuO_2H proving remarkably stable; $k_d = 10^{15.8}\exp(-41600/RT)$. Clearly something was wrong.

Finding out what was right (or nearly right), has occupied us sporadically since then. The fairly complete picture of thermal peroxide reactions that obtains today is not attributable to our efforts alone, obviously. We have filled in a few details, and in trying to appreciate the whole canvas, have noted some areas that still seem rather murky. The details, mostly, will be the subject of this chapter.

ISBN 0-12-566550-4

FREE RADICAL REACTIONS OF HYDROPEROXIDES AND PEROXIDES

Unimolecular Homolysis

Thermal decompositions of organic peroxides almost invariably yield linear plots of concentration *vs* time. The resulting first-order rate constant, k_d, is frequently a good approximation to k_2, the homolysis constant (equation 2), and

$$RO_2R \xrightarrow[\Delta \text{ (solvent)}]{k_2} 2RO\cdot \qquad (2)$$

enough of such data exists as to throw suspicion on any k_d which differs by more than experimental error from $\log A = 15 - 15.5$ and $E_a = 37\text{-}38$ kcal/mole (4).

The experimentally observed behaviour of hydroperoxides parallels that of peroxides; the linearity of the first order plots, often extending to 90+% decomposition is remarkable, considering that it is an artifact bearing only a tenuous relation to k_3. The dependence of k_d on initial hydroperoxide concentration (ordinarily at $[RO_2H]_0 > 0.1$), shows this, but

$$RO_2H \xrightarrow[\Delta \text{ (solvent)}]{k_3} RO\cdot + \cdot OH \qquad (3)$$

doesn't solve the problem.

Isolation of the unimolecular homolysis of hydroperoxides from competing reactions by ordinary techniques is difficult, if not impossible. Following Benson's thermochemical prediction of the true, structure independent parameters for k_3 (5), we essayed a scrupulous reinvestigation of *t*-BuO_2H.(6) Using low initial concentrations (0.01-0.001M in toluene), to minimize higher order reactions, and high temperatures (180-210°C) to favor the 40-42 kcal/mole activation energy gave good kinetic results; $k_d = 2.6 \times 10^{-5} sec^{-1}$ at 182°C. The products, however, indicated that only some 50% of the overall decomposition was due to unimolecular homolysis. *Sec*-BuO_2H and *n*-BuO_2H behaved similarly.

The results suggest a probably value for k_3, in non-polar solvents, at least, on the order of $10^{15.8}\exp(-43000/RT)$. Any hydroperoxide decompositions that are significantly more rapid must in large part be due to other more interesting reactions.

Bimolecular Homolysis of Hydroperoxides

Several observations support a contention that two hydroperoxides are better than one. Rates of free radical production far exceeding those possible from unimolecular homolysis are not uncommon in hydroperoxide-containing mixtures (7, 8); the rate of initiation in autoxidizing olefins is approximately second order in the hydroperoxide concentration; hydroperoxides are partially dimerized in non-polar solvents; homolysis of the dimer equation (4) is

$$\begin{matrix} & \text{H-O-OR} \\ & \text{: :} \\ \text{R-O-O-H} & \end{matrix} \xrightarrow[\Delta]{} \text{RO}\cdot + H_2O + \cdot O_2R \qquad (4)$$

thermochemically attractive.

Two problems: 1, There is no evidence that t-BuO_2H, the single most thoroughly studied hydroperoxide, gives this reaction; 2, allylic hydroperoxides, for which the reaction is most frequently invoked, have received little careful attention until recently.(9, 10, 11).

Allylic hydroperoxides and peroxides have proved highly susceptible to free radical-induced decomposition (*vide infra*), the process effectively masking any possible contribution from a second-order term in the overall rate of thermal decomposition. However, comparison of t-Bu_2O_2-initiated decomposition with decomposition of the allylic hydroperoxide alone gives an indication of the homolysis rate required to spark the subsequent reactions. This rate turns out to be no faster than expected for normal unimolecular homolysis.

From the products, a sequence of reactions can be inferred.

$$C{=}C{-}C{-}O_2\cdot + C{=}C{-}CO_2H \longrightarrow C{=}C{-}C{-}O_2{-}C{-}\dot{C}{-}CO_2H \qquad (5)$$

$$C{=}C{-}C{-}O_2{-}C{-}\dot{C}{-}C{-}O_2H \longrightarrow C{=}C{-}C{-}O_2{-}C{-}\overset{O}{\widehat{C{-}C}} + \cdot OH \qquad (6)$$

$$C{=}C{-}CO_2{-}C{-}\overset{O}{\widehat{C{-}C}} \longrightarrow C{=}C{-}C{-}O\cdot \quad \cdot O{-}C{-}\overset{O}{\widehat{C{-}C}} \qquad (7)$$

The net result, not only produces three radicals for one in relatively rapid reactions, but yields a kinetic expression for rate of radical production which is approximately bimolecular in RO_2H.

Free Radical-Induced Decomposition

By far the major component in thermal decomposition of a hydroperoxide is its attack by free radicals in the system. Alkyl peroxides, on the other hand, while not immune, suffer much less. A detailed elucidation of the process is complicated, however, since several sites for attack present themselves, structural features may determine the course of subsequent reaction, and solvent may alter the nature or character of the attacking radicals.

H-abstraction from O-O-H. The chain process occurring at low temperatures with tertiary hydroperoxides in unreactive solvents is fairly simple. Initiated by a more thermally labile peroxyester (12), peroxide (13) or azo compound (14), it takes the form

$$RO_2H + RO\cdot \longrightarrow RO_2\cdot + ROH \qquad (8)$$

$$2RO_2\cdot \longrightarrow RO_4R \qquad (9)$$

$$RO_4R \longrightarrow O_2 + [2RO\cdot] \begin{cases} \longrightarrow 2RO\cdot \\ \longrightarrow RO_2R \end{cases} \qquad (10)$$

where [ ] represents the solvent cage. Under these conditions the rate of disappearance of RO_2H is independent of $[RO_2H]$ and first order in initiator concentration; the chain length is governed by the rate that the alkoxy radicals escape from the cage; *i.e.*, the viscosity of the medium.(15, 16)

If the alkyl group is non-tertiary, the decomposition becomes stoichiometric,(17) (or nearly so), due to a faster termination.(18)

$$\begin{array}{c} \text{cyclic tetroxide: } >C(H)-O-O-O-O-C(H)< \end{array} \longrightarrow O_2 + >C{=}O + -\overset{H}{\underset{|}{C}}-OH \qquad (11)$$

The latter reaction explains the perturbations on the tertiary system caused by the presence of primary or secondary alkyl radicals, which scavenge O_2 and participate in the rapid termination of equation (11). Obviously, alkyl radicals are produced if the solvent is subject to H-abstraction, or if the solvent promotes β-scission of the 3° alkoxy radical. The net result is to decrease the chain length and complicate the kinetics.

At high temperatures (>130°), where the hydroperoxide serves as its own initiator, chain lengths are much greater than the 1-15 found below 100°C and the kinetics change to 3/2 (4), (or 4/3 (5)), order. The reason is not hard to find; alkyl peroxides are so thermally labile that reaction (10) is no longer terminating.

This doesn't explain why primary and secondary hydroperoxides are just as subject to induced decomposition as tertiary ones under these conditions. One must suppose that for *p*- and *sec*-alkylperoxy radicals, reaction (10) competes effectively with reaction (11). Mill and Hendry (19) have recently estimated the activation energy difference to be 13 kcal/mole and the isokinetic temperature at 133°C. Thus at 170°C reaction (10) may be 4 times faster than reaction (11).

Homolytic substitution at O-O. For simple alkyl hydroperoxides, we found initial concentrations of 0.01 M (in toluene), low enough to effectively suppress the peroxy-type induced reaction. Thus unmasked, solvent-derived free radical attack on the peroxide bond can be inferred from the products.

$$PhCH_2\cdot + RO_2H \longrightarrow PhCH_2OH + RO\cdot \qquad (12)$$

At 182°C and 0.01 M *t*-BuO_2H, rates of homolysis and reaction (12) appear roughly equal,(6) giving $k_{12(182°)} \simeq 1 \times 10^4$.

H-Abstraction from $\overset{H}{\overset{|}{C}}$*-O-O-*. We had seen no clear cut evidence for this mode of attack, although suspecting it in the case of $PhCH(O_2H)CHMe_2$,(1) until a recent study of CH_2=CH-CH_2O_2-*t*-Bu,(20). Both pyrolysis (150°C) and low temperature decomposition (di-*t*-butyl peroxyoxalate) gave product compositions requiring contribution from reaction (13).

$$t\text{-}BuO\cdot + CH_2CHCH_2O_2\text{-}t\text{-}Bu \longrightarrow t\text{-}BuOH + CH_2CH\dot{C}HO_2\text{-}t\text{-}Bu \qquad (13)$$

Comparison with the rate constant for *t*-BuO• + $PhCH_3$ (the solvent) gives $k_{13} = 10^{8.7}\exp(-3910/RT)$.

This is slightly greater than the rate constant for *t*-BuO• abstraction from a sec-allyl position, but not as fast as attack on a secondary ether C-H (19, 20). The results is very much in line with the relative immunity of *p*- and *sec*-alkyl peroxides and strongly suggests that C-H α to O-O is not particularly activated towards free radical attack.

Alkyl radical addition to C=C-C-O_2•. Decomposition of Ac_2O_2 in the presence of CH_2=C(Me)C(Me)$_2O_2$H yields epoxide resulting from methyl addition to the carbon-carbon double bond.(9) Pyrolyses of allyl O_2-*t*-Bu in toluene yield the analogous benzyl addition product.(10) The addition is

$$PhCH_2\cdot + CH_2{=}CHCH_2O_2t\text{-Bu} \longrightarrow PhCH_2CH_2\dot{C}HCH_2O_2\text{-}t\text{-Bu} \quad (14)$$

$$PhCH_2CH_2\dot{C}HCH_2O_2\text{-}t\text{-Bu} \longrightarrow PhCH_2CH_2\overset{O}{\widehat{CHCH_2}} + t\text{-BuO}\cdot \quad (15)$$

fairly slow; modelling attempts suggest that $k_{14} \simeq 700\ M^{-1}sec^{-1}$ at 150°C. A similar pyrolysis of allyl O_2H yields no epoxide since the benzyl radicals are scavenged by O_2 much faster than they add to the hydroperoxide.(11)

NON-RADICAL REACTIONS

Electrocyclic Decompositions

The base-catalyzed production of H_2 from hydroxymethylene peroxide remained an historical curiosity until Mosher showed the thermal analog to be responsible for the autocatalytic decomposition of primary hydroperoxides (21).

$$n\text{-BuO}_2\text{H} + n\text{-PrCHO} \rightleftharpoons n\text{-BuO}_2\text{CH(OH)Pr} \quad (16)$$

$$\text{Pr(H)C(H)–O–O–C(H)(OH)Pr} \longrightarrow n\text{-PrCHO} + n\text{-PrCO}_2\text{H} + \text{H}_2 \quad (17)$$

It was not too surprising, then, to find that pyrolyses of *sec*-alkyl peroxides also generated H_2 in competition with homolytic decomposition, (22, 23). Yields of H_2 vary depending on structure and temperature; typically *s*-Bu_2O_2 gives 20-30% and $(Ph_2CHO)_2$, 90-95%.

The surprising feature is the limitation to pyrolses in solution. Vapor phase decomposition gives little or no H_2. The same limitation pertains to the electrocyclic decomposition of *p*- and *sec*-alkyl peroxyesters. In solution the reaction is virtually all non-radical, but in the vapor phase only unimolecular homolysis is observed.(24)

$$\text{-C}(=\text{O})\text{-O-O-C(H)}< \longrightarrow \text{-C}(=\text{O})\text{OH} + \text{O=C}< \qquad (17)$$

The explanation appears to be that the electrocyclic transition state is sufficiently product-like to register increase in polarity on going from peroxide to carbonyl compounds. Our recent study of i-Pr_2O_2 (25) in solvents ranging from toluene to H_2O showed a linear correlation of log k_H (equation (18)) with Kosower's Z-values,(26)

$$Me_2C(H)\text{-O-O-}C(H)Me_2 \; \begin{cases} \xrightarrow{k_H} H_2 + 2Me_2CO \\ \xrightarrow{k_R} 2Me_2CHO\cdot \end{cases} \qquad (18)$$

(6 + log k_H = 0.0419Z - 1.646). While we have not been able to measure k_H in the vapor, it can be no greater than 1% of k_R. Extrapolating the correlation places an upper limit of 36 on the Z-value for vacuum. This would be an interesting number to test, but unfortunately other comparisons of solution *vs* vapor phase for quasi-polar reactions are hard to find.

Addition of Hydroperoxides to Aldehydes and Ketones

The equilibrium between hydroperoxide + aldehyde and adduct (equation 16), is established rapidly and generally favors adduct, although values for the equilibrium constant, K, are solvent dependent. For example, Antonovskii's (27) infra red method gives $K_{20°}$ for t-BuO_2H + AcH as 54, 20.6 and 1.8 M^{-1} in CCl_4, $CHCl_3$ and i-Pr_2O, respectively, while Sauer and Edwards,(28) using NMR, obtain 4.2 in benzene. Our own preliminary work using the NMR technique indicates that in CCl_4, at least, t-BuO_2H dimerization introduces a sizeable correction factor, and that K for adduct formation with AcH may be as large as 125 M^{-1} at 37°C. There is a clear need for more accurate numbers if the course of aldehyde-catalyzed hydroperoxide decompositions is to be predicted with any degree of success.

Ketone adducts are not observed spectroscopically; the equilibrium predictably favors free hydroperoxide. But the adducts are implicated in thermal decompositions. With aromatic hydrocarbons as solvents the effect is small, the build-up of ketonic products leading to slight autoacceleration--and to the deceptive pseudo-linearity of first-order plots.

In alcoholic solvents the effect is more dramatic. Pyrolysis of *t*-BuO_2H and *s*-BuO_2H in MeOH are strongly autocatalytic, while addition of equimolar amounts of acetone, methyl ethyl ketone or diethyl ketone at the beginning increases the initial rate 5- to 6-fold.(11)

The products from *s*-BuO_2H decomposition include methyl acetate and formate as well as free acids, and show that MEK is both produced and consumed. The overall reaction scheme must be formidably complicated but a key feature seems to be adduct formation and decomposition, which leads to acetic

$$s\text{-}BuO_2H + MeEtC{=}O \rightleftharpoons s\text{-}BuO_2\text{-}C(OH)(Me)\text{-}Et \quad (19)$$

$$s\text{-}BuO_2C(OH)(Me)\text{-}Et \longrightarrow s\text{-}BuO\cdot + EtC(OH)(Me)\text{-}O\cdot \longrightarrow Et\cdot + MeCO_2H \quad (20)$$

acid. The formyl products arise from solvent since decompositions in ethanol yield only acyl substances.

Finding the acidic products answers a question that has puzzled us for some time: why heated ampoule-sealed dilute solutions of *t*-BuO_2H in *t*-BuOH invariably blow up.(6) Normal decomposition produces acetone, which with *t*-BuO_2H *via* adduct yields HCOH. Even catalytic amounts of acid in *t*-BuOH at 170° will eventually generate intolerable pressure of isobutylene. Ethanol dehydrates less readily, and we have as yet had no explosions using that solvent, while methanol avoids the problem altogether.

Oxidations with Hydroperoxides and Metal Ions

Epoxidation of olefins using a hydroperoxide plus catalytic amounts of vanadium or molybdenum, being commercially advantageous, occupies a vast literature which would be inappropriate to summarize here. But the reaction has turned out to be general for a number of nucleophiles; (29, 30, 31) (Table 1); the RO_2H + Mo (or V) combination mimicking a peroxyacid.

$$RO_2H + N{:} \xrightarrow{Mo} ROH + N{\rightarrow}O \quad (21)$$

Table 1

Metal-Ion Catalyzed Oxidations by t-BuO_2H[a]

Substrate	$k_{relative}$[b] $M^{-2}sec^{-1}$		*T°C*
	Vanadium	*Molybdenum*	
Ph_3Sb	1×10^6	1×10^7	25
Ph_3P	4×10^5	4×10^6	25
Ph_3As	2×10^5		35
Ph_2S		1.2	35
Ph_2SO	≃0.2	0.1	35
$PhCH{=}CH_2$[c]	0.6		60
$PhNH_2$[c]	1		

[a] *In ethanol unless otherwise noted.*
[b] *For $Ph_3P + MoO_2(Acac)_2$ + t-BuO_2H, $k = 1.0 \times 10^5\ M^{-1}sec^{-1}$.*
[c] *In benzene.*

Thus the metal ion appears to increase the electrophilicity of the hydroperoxide, perhaps by complexing. Certainly the

$$Mo \quad + RO_2H \quad \overset{K}{\rightleftharpoons} \quad \begin{matrix} \delta+ \\ RO-O-H \\ \downarrow \\ \delta- \\ Mo \end{matrix} \qquad (22)$$

combination *does not* lend itself to Baeyer-Williger oxidation of ketones, which in the first instance depend on the nucleophilic character of the peroxyacid.

An alternative hypothesis invokes formation of a peroxidized metal complex intermediate in the manner of peroxidases for which, then, RO_2H + Mo would serve as an inorganic model. This interpretation has been strengthened lately by isolation of a peroxomolybdate ($MoO(O_2)_2{\cdot}L$) from the reaction of MoO_3 with H_2O_2 (32,33), and its use as a very potent olefin epoxidizer.(34)

The formation of a reactive intermediate should display different kinetic behavior than reversible complex formation. In particular there should be circumstances where intermediate formation is rate determinint so that concentration or even character of the substrate would be irrelevant. No such kinetic behavior has been found as yet, but the possibility of more than one operative mechanism exists, and not all of the possible substrates have been examined.

With this in mind we have recently turned out attention to the oxidation of organic sulfides and sulfoxides. As expected, molybdenum is an excellent catalyst; at 0.001 M concentration, the rate of oxidation of Ph_2S to Ph_2SO is increased 50-fold, (relative to the uncatalyzed reaction, (EtOH, 65°C)). Sulfone is produced as well, and the kinetics (at constant [Mo]) fit a consecutive second-order scheme quite nicely, with k_{23} about 10 times as great as k_{24}. These

$$t\text{-BuO}_2\text{H} + \text{Ph}_2\text{S} \xrightarrow{\text{Mo}} t\text{-BuOH} + \text{Ph}_2\text{SO} \quad (23)$$

$$t\text{-BuO}_2\text{H} + \text{Ph}_2\text{SO} \xrightarrow{\text{Mo}} t\text{-BuOH} + \text{Ph}_2\text{SO}_2 \quad (24)$$

pseudo second-order constants are linearly dependent on [Mo], giving the third-order constants found in Table 1.

Clearly, sulfides and sulfoxides behave like the other substrates so far examined, and not in accordance with the peroxo-intermediate concept. A side-reaction of some interest has emerged, however.

When the Mo-catalyzed oxidation of Ph_2S is run in air, titratable hydroperoxide has a tendency to oscillate, from initial concentration to nearly zero and back again, for many cycles, and with a period of about 20 minutes. Unfortunately, the oscillation is not reproducible. Sometimes the reaction behaves exactly as it does in an inert atmosphere; at others there is oscillation, but severely damped. Until conditions yielding reproducibility are established, we can only speculate about the possibility of a parallel free radical oxidation of solvent, initiated by a side-reaction of t-BuO_2H and Mo producing free radicals.

CONCLUSION

Clearly, research in peroxide chemistry is not over, but a stage has been reached where moderately complicated experimental situations can be handled with some success. That is, a fair number of the operative reactions have been identified and rate constants assigned. Some others still require better definition, but that poses no great problem. What is needed, perhaps, is more widespread awareness that the data does exist, coupled with the competence to use it.

REFERENCES

1. Hiatt, R. and Strachan, W., *J. Org. Chem.*, *28*, 1893 (1963).
2. Hiatt, R. and Visser, T., *Can. J. Chem.* *42*, 1243 (1964).
3. Bartlett, P. D. and Hiatt, R., *J. Am. Chem. Soc.* *80*, 1398 (1958).
4. Hiatt, R., "*Organic Peroxides*", Vol. III, D. Swern ed., Wiley, New York, 1971, Ch. 1.
5. Benson, S. W., *J. Phys. Chem.*, *40*, 1007 (1964).
6. Hiatt, R. and Irwin, K. C., *J. Org. Chem.*, *33*, 1436 (1968).
7. Bateman, L. and Hughes, H., *J. Chem. Soc.*, 4594 (1952).
8. Howard, J. A., "*Free Radicals*", Vol. II, J. Kochi, ed., Wiley, New York, 1973, Ch. 1.
9. Hiatt, R. and McCarrick, T., *J. Am. Chem. Soc.*, *97*, 5234 (1975).
10. Hiatt, R. and Nair, V. K. G., *Can. J. Chem.*, (in press).
11. Unpublished work in these laboratories.
12. Hiatt, R., Clipsham, H. and Visser, T., *Can. J. Chem.*, *42*, 2754 (1964).
13. Factor, A., Russell, C. A. and Traylor, T. G., *J. Am. Chem. Soc.*, *87*, 3692 (1965).
14. Thomas, J. R., *J. Am. Chem. Soc.*, *89*, 4872 (1967).
15. Hiatt, R. and Traylor, T. G., *J. Am. Chem. Soc.*, *87*, 3766 (1965).
16. Hiatt, R., Mill, T., Irwin, K. C. and Castleman, J. K., *J. Org. Chem.*, *33*, 1421 (1968).
17. Hiatt, R., Mill, T., Irwin, K. C. and Castleman, J. K., *J. Org. Chem.*, *33*, 1428 (1968).
18. Russell, G. A., *J. Am. Chem. Soc.*, *79*, 3871 (1957).
19. Mill, T. and Hendry, D. G. "*Comprehensive Chemical Kinetics*". in press.
20. Hendry, D. G., Mill, T., Piszkiewicz, L., Howard, J. A. and Eigenmann, H. K., *J. Physical and Chemical Reference Data*, *3*, 937 (1974).
21. Durham, L. J. and Mosher, H. S., *J. Am. Chem. Soc.*, *84*, 2811 (1962).
22. Hiatt, R. and Szilagyi, S., *Can. J. Chem.*, *48*, 615 (1970).
23. Hiatt, R. and Thankachan, C., *Can. J. Chem.*, *52*, 4090 (1974).
24. Hiatt, R., Glover, L. C. and Mosher, H. S., *J. Am. Chem. Soc.*, *97*, 1556 (1975).
25. Hiatt, R. and Rahimi, P. M., *Int. J. Chemical Kinetics*, *10*, 185 (1978).

26. Kosower, E., *J. Am. Chem. Soc.*, *80*, 3253 (1968).
27. Antonovskii, V. L. and Terent'ev, V. A., *Zh. Org. Khim.*, *3*, 245 (1967).
28. Sauer, M. D. C. V. and Edwards, J. O., *J. Phys. Chem.*, *75*, 3377 (1971).
29. Howe, G. R. and Hiatt, R., *J. Org. Chem.*, *35*, 4007 (1970).
30. Howe, G. R. and Hiatt, R., *J. Org. Chem.*, *36*, 2493 (1971).
31. Hiatt, R., MeColeman, C. and Howe, G. R., *Can. J. Chem.*, *53*, 559 (1975).
32. Mimoun, H., Seree de Roch, I., and Sajus, L., *Tetrahedron*, *26*, 37 (1970).
33. Sharpless, K., Townsend, J. and Williams, C. R., *J. Am. Chem. Soc.*, *94*, 295 (1972).
34. Arakawa, K., Moro-oka, Y. and Ozaki, A., *Bull. Chem. Soc. Japan*, *47*, 2958 (1974).

INHIBITION OF HYDROCARBON AUTOXIDATION BY SOME SULPHUR CONTAINING TRANSITION METAL COMPLEXES (1)

J.A. HOWARD

Division of Chemistry
National Research Council of Canada
Ottawa, Ontario, Canada K1A OR9

INTRODUCTION

Liquid-phase autoxidation of most organic substances (RH) by atmospheric oxygen at temperatures below 150-200°C is a free-radical chain process which can be represented by the elementary reactions shown in scheme I(2).

Scheme I

Initiation:

Production of free-radicals [1]

Propagation:

$$R^{\cdot} + O_2 \xrightarrow{k_o} RO_2^{\cdot} \quad [2]$$

$$RO_2^{\cdot} + RH \xrightarrow{k_p} ROOH + R^{\cdot} \text{ (or } RO_2RH^{\cdot}) \quad [3]$$

Termination:

$$R^{\cdot} + R^{\cdot} \longrightarrow \quad [4]$$

$$RO_2^{\cdot} + R^{\cdot} \xrightarrow[2k_t]{} \quad \text{Non-radical products} \quad [5]$$

$$RO_2^{\cdot} + RO_2^{\cdot} \longrightarrow \quad [6]$$

where $R^{\cdot}$ is the alkyl radical derived from RH by abstraction of a labile hydrogen atom or radical addition to a double bond and $RO_2^{\cdot}$ and ROOH are the corresponding alkylperoxy radical and alkyl hydroperoxide, respectively.

At oxygen pressures above about 100 torr the steady-state concentration of alkylperoxy radicals is very much larger than the concentration of alkyl radicals because the rate constant for reaction of an alkyl radical with oxygen, k_o, is very much larger than the rate constant for reaction of $RO_2^{\cdot}$ with RH, k_p (2). The termination reaction [6] is,

ISBN 0-12-566550-4

therefore, much faster than reactions [4] and [5] and the steady-state rate of autoxidation of RH, $-d[O_2]/dt$, is given by the rate expression

$$\frac{-d[O_2]}{dt} = \frac{k_p[RH]R_i^{0.5}}{(2k_t)^{0.5}} \quad [7]$$

where k_p and $2k_t$ are the rate controlling propagation and termination rate constants and R_i is the rate of free-radical chain initiation.

Autoxidation usually has a detrimental effect on organic substances because of the loss of desirable properties of the original material. Thus the harmful effects of autoxidation are apparent in the rancidification of edible oils and fats, sludge formation in lubricating oils, and in the slow deterioration of rubber and plastics. It is, therefore, not surprising that inhibition of autoxidation has attracted the attention of research scientists for many years with the result that a wide spectrum of antioxidants are commercially available.

Probably the most important class of antioxidants and certainly the most thoroughly investigated are the hindered phenols and aromatic amines, AH(3-7). These compounds are known as chain breaking antioxidants because they reduce the rate of autoxidation by reacting with chain propagating alkylperoxy radicals. Reaction occurs by hydrogen atom transfer from the inhibitor to the peroxy radical to give a phenoxy or anilino radical, $A^{\cdot}$, which is either incapable of continuing the chain or does so at a very much reduced rate.

$$RO_2^{\cdot} + AH \longrightarrow ROOH + A^{\cdot} \quad [8]$$

$$RO_2^{\cdot} + A^{\cdot} \longrightarrow ROOA \quad [9]$$

These antioxidants do, however, suffer from several distinct disadvantages. First, they have no effect on the rate of chain initiation occurring by homolysis of peroxidic impurities and secondly, a hydroperoxide is formed in the slow rate controlling termination reaction [8] while a peroxide is formed in the rapid chain terminating reaction [9]. Clearly these inhibitors will lose their effectiveness under conditions where initiation by hydroperoxide and peroxide decomposition is important. In addition they rapidly lose their effectiveness at temperatures above about 100°C because of direct reaction with molecular oxygen at the labile hydrogen.

Because of these disadvantages a second class of antioxidants has been developed which are generally classified as preventive antioxidants (3). The principal function of these compounds is to destroy alkyl hydroperoxides by a non-radical mechanism and thus prevent chain initiation by homolytic hydroperoxide decomposition, although it should be noted that many of them are also alkylperoxy radical scavengers. Kinetics and mechanisms are, however, less firmly established for this class of antioxidants probably because of the experimental difficulties encountered in obtaining reliable kinetic data and proving reaction mechanisms.

The aim of this article is to critically review the kinetic and product studies that have been performed on sulphur containing transition metal complex preventive antioxidants with particular emphasis on zinc, nickel and cupric (M) dialkyldithiophosphates (I) and dialkyldithiocarbamates (II).

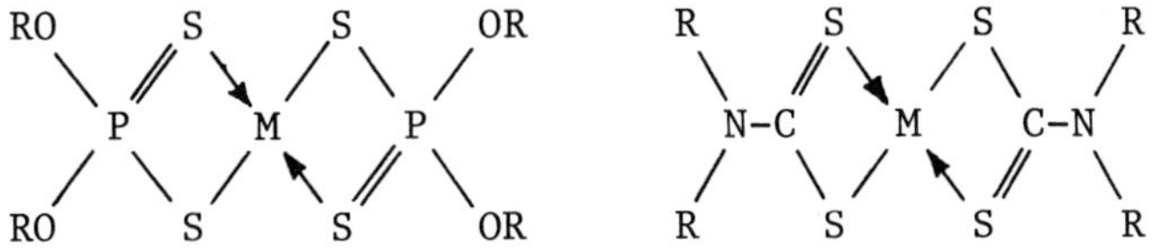

```
RO    S    S    OR        R    S    S    R
  \  //\  / \  /           \  //\  / \  /
   P    M    P              N-C    M    C-N
  /  \  / \\  \            /   \  / \\   \
RO    S    S    OR        R    S    S    R

         I                          II
```

The article has been divided into three main sections which deal respectively with the kinetics, products, and mechanisms that have been found and proposed for reactions of alkylperoxy radicals and alkyl hydroperoxides with dialkyldithiophosphates and dialkyldithiocarbamates. For brevity these complexes have been abbreviated to ML_x where L represents either $(RO)_2PS_2$ or R_2NCS_2 and x is the number of ligands.

INFLUENCE OF ML_x ON HYDROCARBON AUTOXIDATION

Larson (8) just over 20 years ago demonstrated that small concentrations of zinc dialkyldithiophosphates completely inhibited the autoxidation of a refined white mineral oil at 155°C, initiated by homolytic hydroperoxide decomposition , for lengths of time (induction periods) which depended on the nature of R in the ligand. Two years earlier Kennerly and Patterson (9) had shown that $Zn[(RO)_2PS_2]_2$ (R = 4-methyl-2-pentyl) accelerated the decomposition of cumene hydroperoxide, in mineral oil at 130-150°C, to phenol, the product characteristic of ionic decomposition. The antioxidant activity of zinc dialkyldithiophosphates was, therefore, attributed to heterolytic hydroperoxide decomposition. Since there was no obvious mechanism for acid catalysed decomposition by the original complex it was concluded that intermediates formed from the complex were responsible for ionic decomposition. These conclusions appeared to have been substantiated by Holdsworth Scott and Williams (10) who reported that $Zn[Et_2NCS_2]_2$ did not inhibit the α,α'-azo-bis-isobutyronitrile initiated autoxidation of tetralin at 50°C and that this complex was oxidized by hydroperoxide to SO_2, a potent catalyst for ionic hydroperoxide decomposition.

It was, however, soon realized that this mechanism presented an oversimplified picture of inhibition by ML_x. Thus several groups of workers (11-36) demonstrated that ML_x must have the ability to scavenge alkylperoxy radicals because they inhibited or retarded hydrocarbon autoxidation in the absence of hydroperoxide and at temperatures where initiation by hydroperoxide homolysis was unimportant (see e.g. Figures 1 and 2).

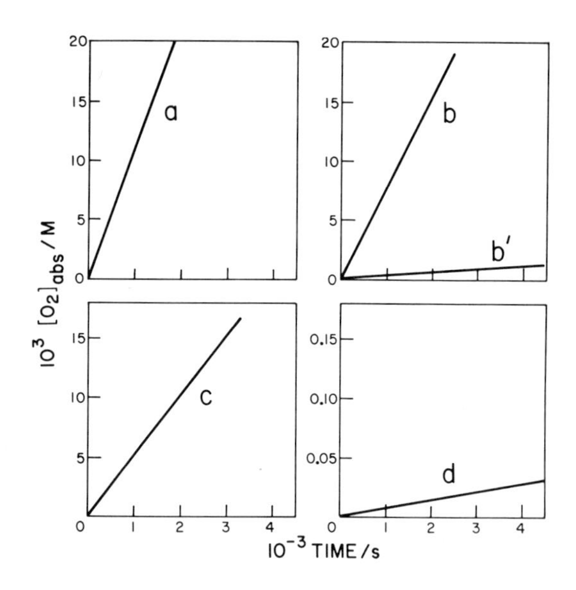

Fig. 1. Oxygen absorption as a function of time for the autoxidation of styrene (8.7M) containing 2,2,3,3-tetraphenylbutane (0.017 M) and $M[S_2P(OR)_2]_2$ at 30°C. (a) No inhibitor; (b) M = Zn(0.014M); (b') + t-BuOOH(0.02M); (c) M = Ni(0.013M); (d) M = Cu^{2+}(0.00125M).

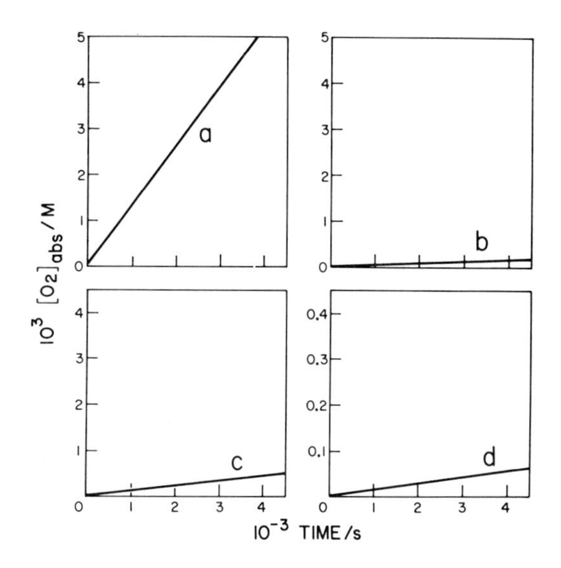

*Fig. 2. Oxygen absorption as a function of time for the autoxidation of cumene (7.17*M*) containing 2,2,3,3-tetraphenylbutane (0.0039*M*) and* $M[S_2P(OR)_2]_2$ *at 30°C. (a) No inhibitor; (b) M = Zn(0.015*M*); (c) M = Ni(0.015*M*); (d) M = Cu(0.000125*M*).*

If hydroperoxide is initially present in the hydrocarbon, initial inhibition by ML_x may not be observed because a radical generating reaction between the complex and hydroperoxide can overshadow inhibition. This *pro-oxidant* effect is particularly important for nickel and cupric complexes, see e.g., Figure 3. On the other hand, addition of *tert*-butyl hydroperoxide to the $Zn[(i\text{-PrO})_2PS_2]_2$ inhibited autoxidation of styrene at 30°C produces enhanced inhibition (Figure 1). In addition to these complications these systems may exhibit *autoinhibition* because of the accumulation of reaction products which are more effective peroxy radical scavengers than ML_x.

KINETICS OF THE REACTIONS OF ALKYLPEROXY RADICALS AND ALKYL HYDROPEROXIDES WITH ML_x

Peroxy Radicals

Indirect studies. The simplest form the reaction of

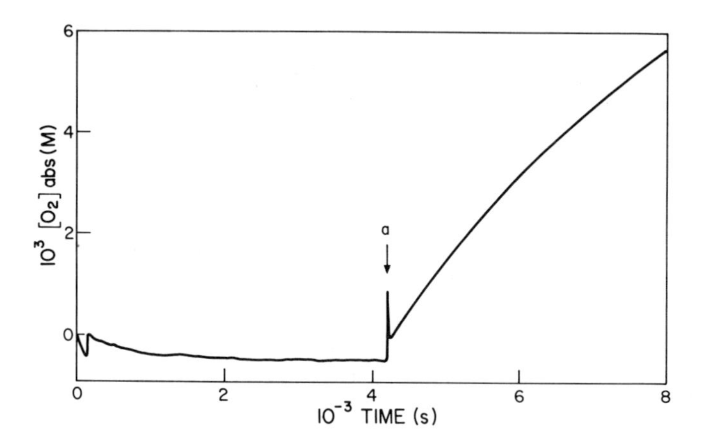

Fig. 3. Oxygen absorption as a function of time for autoxidation of styrene (1.74M) in chlorobenzene containing t-BuOOH(0.01M) and Ni[(i-PrO)$_2$PS$_2$]$_2$ (0.016M) at 30°C. a indicates the time when the hydroperoxide was added.

an alkylperoxy radical with a transition metal complex can take is

$$RO_2^{\cdot} + ML_x \xrightarrow{k_{inh}} X \qquad [10]$$

$$(n-1)\ RO_2^{\cdot} + X \xrightarrow{\text{fast}} \text{non-radical products} \qquad [11]$$

where k_{inh} is the inhibition rate constant, n is the number of peroxy radicals destroyed by each molecule of inhibitor and X is a transient intermediate (19, 25). If reaction [10] is the rate controlling termination reaction initial rates of oxygen absorption by RH in the presence of ML_x will be given by the rate law,

$$\left\{\frac{-d[O_2]}{dt}\right\}_o = \frac{k_p[RH]_o R_i}{nk_{inh}[ML_x]_o} \qquad [12]$$

i.e., the initial rate will be linearly dependent on the rate of chain initiation and the substrate concentration and inversely proportional to the first power of the inhibitor concentration.

Several workers have demonstrated a linear dependence on R_i for Zn[(RO)$_2$PS$_2$]$_2$ inhibited autoxidation of cumene and tetralin whereas concentration exponents for RH and ML_x were found to be less than one (14, 17, 19). More recently we have reported that zinc, nickel, and cupric dialkyldithiocarbamate and dialkyldithiophosphate inhibited autoxidation of styrene, cyclohexene, tetralin, benzaldehyde, cyclohexa-1,3-diene, cyclohexa-1,4-diene obey equation [12] (25, 34, 36).

It should, however, be emphasized that precise reactant exponents are difficult to obtain for these systems because oxygen absorption as a function of time is often not linear and shows quite pronounced downward curvature even at low extents of oxidation. This means that initial oxidation rates have to be measured which is not particularly easy for systems that may still be approaching thermal equilibrium. Difficulties associated with thermal equilibrium can be overcome using photochemical initiation. This method of initiation has, however, only been successfully used for the zinc complexes (25) because most other transition metals give highly coloured dialkyldithiophosphates and dialkyldithiocarbamates (34, 36). Despite the limitations posed by these systems, rate constants have been estimated for several transition metal complexes from inhibited rates of autoxidation. It would appear from the inhibition rate constants given in Table 1 that the nickel and zinc complexes have values in the range $1 \times 10^3 - 2 \times 10^4$ $M^{-1}s^{-1}$ at 50° (25, 34) while cupric complexes are at least two orders of magnitude more reactive (36). Unfortunately the agreement between rate constants obtained by different groups of workers is generally not that good. For instance, Gervitis, Zolotova and Denisov (32) have reported a value of 2.3×10^3 $M^{-1}s^{-1}$ for reaction of the cumylperoxy radical with $Ni[n\text{-}Bu_2NCS_2]_2$ at 60°C while Howard and Chenier (34) found a value of 1.4×10^4 $M^{-1}s^{-1}$ for reaction of the tetralylperoxy radical with this complex.

Direct studies. Reaction of the *tert*-butylperoxy radical with zinc and nickel dialkyldithiophosphates and dialkyldithiocarbamates has been studied directly at low temperatures (<273 K) by kinetic electron spin resonance spectroscopy (25, 34). This method involves photochemically generating tertiary alkylperoxy radicals directly in the cavity of an ESR spectrometer in the presence of excess complex and monitoring the radical concentration after the light has been masked.

In the case of the zinc complexes it was originally reported (25) that radical decays obey the kinetic expression

$$\frac{-d[t\text{-}BuO_2\cdot]}{dt} = nk_{inh}[t\text{-}BuO_2\cdot][ZnL_2] \qquad [13]$$

TABLE 1

Rate constants for reaction of alkylperoxy radicals with sulphur containing transition metal complexes from inhibited rates of autoxidation

Peroxy Radical	Solvent	Temp. (°C)	ML_x	$10^{-3}(nk_{inh})$ ($M^{-1}s^{-1}$)	Reference
Poly(peroxystyryl)	Styrene	50	$Zn[(i\text{-}PrO)_2PS_2]_2$	16	(25)
1-Phenylethyl	Ethylbenzene	80	$Zn[(c\text{-}C_6H_{11}O)_2PS_2]_2$	21	(14)
Poly(peroxystyryl)	Styrene	50	$Ni[(i\text{-}PrO)_2PS_2]_2$	0.9	(34)
α-Tetralyl	Tetralin	50	$Ni[n\text{-}Bu_2NCS_2]_2$	14	(34)
α-Cumyl	Cumene	60	$Ni[n\text{-}Bu_2NCS_2]_2$	2.3	(32)
9,10-Dihydro-9-anthracyl	o-Dichlorobenzene	50	$Cu[(i\text{-}PrO)_2PS_2]_2$	∿100	(36)

and that the activation parameters for dithiophosphates are log $(A_{inh}/M^{-1}s^{-1}) \sim 6.0$ and $E_{inh} \sim 4$ kcal mol^{-1}. It was, however, subsequently discovered (34) that although radical decays are exponential in the presence of excess complex, pseudo-first order rate constants are almost independent of the complex concentration (Table 2). A first-order radical decay and an almost zero-order dependence on the complex concentration was also observed for $Ni[(i\text{-}PrO)_2PS_2]_2$ (34). The insensitivity of the rate to changes in the complex concentration may be due to self-association of the complex although these complexes are believed to exist as monomers in hydrocarbon solvents at ambient temperature.

An alternative explanation for the apparent zero-order dependence on complex concentration involves reaction of the peroxy radical with the complex to give an associated species which can either dissociate back to the original reactants or decompose to give non-radical products

$$RO_2^{\bullet} + ZnL_2 \rightleftharpoons RO_2.....ZnL_2 \quad [14]$$

$$RO_2^{\bullet}.....ZnL_2 \xrightarrow{k_{15}} \text{non-radical products} \quad [15]$$

TABLE 2

Kinetic data for reaction of t-$BuO_2^{\bullet}$ with $Zn[(i\text{-}PrO)_2PS_2]_2$ at 173K

$Zn[(i\text{-}PrO)_2PS_2]_2 \times 10^3$ (M)	$k_\psi (s^{-1})$ [a]
0.0025	0.1
0.025	0.055
0.25	0.024
2.5	0.09

[a] k_ψ is the pseudo-first order rate constant

If these reactions are involved in the overall process, the rate of radical decay is given by

$$\frac{-d[RO_2^\cdot]}{dt} = \frac{k_{15}K_{14}[RO_2^\cdot][ZnL_2]}{1 + K_{14}[ZnL_2]} \quad [16]$$

where K_{14} is the equilibrium constant for [14]. Consequently if $K_{14}[ZnL_2] > 1$

$$\frac{-d[RO_2^\cdot]}{dt} = k_{15}[RO_2^\cdot] \quad [17]$$

This explanation does, however, require an exceedingly large value for K_{14}.

The unusual kinetic behaviour exhibited by the t-$BuO_2^\cdot$-$Zn[(i\text{-}PrO)_2PS_2]_2$ system at low temperatures has prompted us to reinvestigate very carefully the kinetics for inhibition of hydrocarbon autoxidation by this complex (37). It would now appear that inhibited rates of autoxidation of styrene at 30°C are best described by the rate law

$$\left\{\frac{-d[O_2]}{dt}\right\}_o \propto \frac{[RH]^{1.4}R_i^{0.79}}{ML_x^{0.54}} \quad [18]$$

A rate expression of this form has an important bearing on the reaction mechanism although it offers no explanation for the apparent difference between the direct and indirect kinetics for reaction of $RO_2^\cdot$ with $Zn[(RO)_2PS_2]_2$. It should, however, be emphasized that this rate expression has only been demonstrated for one complex and one substrate, and we do not know yet whether it is applicable to other complexes and other oxidizable substrates.

In contrast to the behaviour of the dialkyldithiophosphates reaction of t-$BuO_2^\cdot$ with $Ni[n\text{-}Bu_2NCS_2]_2$ in toluene and CCl_3F obeys the rate expression

$$\frac{-d[t\text{-}BuO_2^\cdot]}{dt} = nk_{inh}[t\text{-}BuO_2^\cdot][Ni(n\text{-}Bu_2NCS_2)_2] \quad [19]$$

and $nk_{inh} \sim 2 \times 10^4\ M^{-1}s^{-1}$ over the temperature range 179 to 258K (34), implying that this reaction has a very small activation energy and a rather low pre-exponential factor (34). Rates of disappearance of t-$BuO_2^\cdot$ are much slower in isopentane and it has been suggested (34) that reaction of t-$BuO_2^\cdot$ with $Ni[n\text{-}Bu_2NCS_2]_2$ gives a transient capable of abstracting a tertiary hydrogen atom from the solvent to regenerate a tertiary alkylperoxy radical.

Stoichiometric factors. The total number of peroxy radicals destroyed by each molecule of scavenger, n, is an important parameter in inhibition studies because it determines the length of time the antioxidant can be expected to be effective. This is because

$$R_i = \frac{n[ML_x]}{\tau} \qquad [20]$$

where τ is the induction period.

Values of n determined by the induction period method for several ML_x are listed in Table 3 and it would seem that n is close to 1 for dialkyldithiophosphates while dialkyldithiocarbamates give significantly larger values. There would, however, appear to be experimental difficulties in obtaining reliable values of this parameter since values between 0.7 and 3 have been reported for $Zn[(i\text{-}PrO)_2PS_2]_2$ using thermal initiation (13, 18, 25, 38). Furthermore, a value as large as 4.5 has been obtained for this complex from photochemically initiated autoxidations (25).

n-Factors can also be determined from the initial rate of complex disappearance at a constant rate of initiation since

$$n = \frac{R_i}{(-d[ML_x]/dt)} \qquad [21]$$

This method generally gives n-values close to 1. The larger values obtained by the induction period method must, therefore, arise because oxidation products of the complex and in some instances of the hydrocarbon scavenge peroxy radicals with rate constants equal to or less than the original complex.

The ability of oxidation products to scavenge peroxy radicals is demonstrated for the $Cu[Et_2NCS_2]_2$ inhibited autoxidation of 9,10-dihydroanthracene in Figure 4.

Hydroperoxides

Burn, Cecil and Young (39) have reported that initial rates of disappearance of $Zn[(i\text{-}PrO)_2PS_2]_2$ in PhCl containing excess $PhCMe_2OOH$ are first-order with respect to the complex concentration and that the second-order rate constant for this reaction is 0.015 $M^{-1}s^{-1}$ at 70°C and that the activation energy is 17±3 kcal mol^{-1}. These workers also reported that initial rates of hydroperoxide decomposition are equal to initial rates of complex disappearance and that neither rates are influenced by radical scavengers.

TABLE 3

Overall stoichiometric factors for some metal complexes [a]

Metal Complex	Initiation	Hydrocarbon	Temperature (°C)	n	Reference
$Zn[(i\text{-}PrO)_2PS_2]_2$	Thermal[b]	Cumene	60	2	(18)
$Zn[(i\text{-}PrO)_2PS_2]_2$	Thermal[b]	Various[c]	50	0.7–1.5	(25)
$Zn[(i\text{-}PrO)_2PS_2]_2$	Photochemical[d]	Various[e]	30	3.1–4.5	(25)
$Zn[n\text{-}Am_2NCS_2]_2$	Thermal[b]	Ethylbenzene	60	2.8	(26)
$Ni[(i\text{-}PrO)_2PS_2]_2$	Thermal[b]	Cumene	50	1	(34)
$Ni[n\text{-}Bu_2NCS_2]_2$	Thermal[b]	Cumene	50	4	(34)
$Cu[(i\text{-}PrO)_2PS_2]_2$	Thermal[b]	DHA[f]	50	0.8	(36)
$Cu[Et_2NCS_2]_2$	Thermal[b]	DHA[f]	50	1.6–3.6	(36)

[a] calculated from [20]

[b] AIBN; [c] Tetralin, cyclohexene, and 9,10-dihydroanthracene; [d] α,α'-azo-bis-cyclohexanecarbonitrile; [e] Tetralin, cumene and styrene; [f] 9,10-dihydroanthracene (0.5*M*) in o-dichlorobenzene

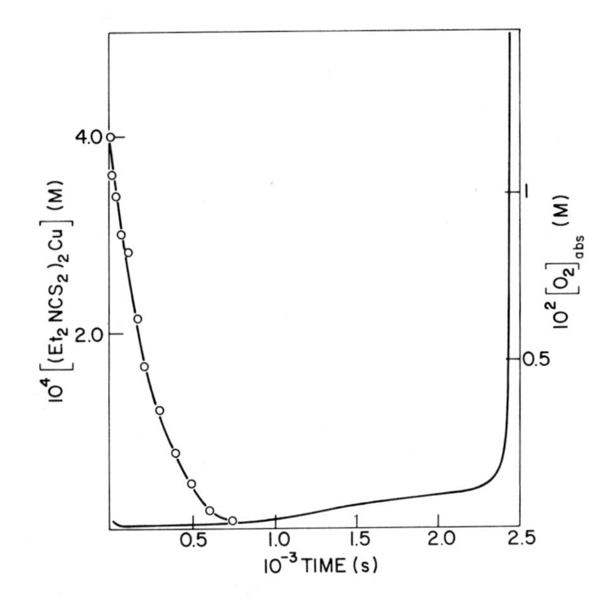

Fig. 4. Concentration-time profiles for $Cu[Et_2NCS_2]_2$ (0.0004M) inhibited autoxidation of 9,10-dihydroanthracene (0.5M) in o-dichlorobenzene at 50°C.

More recently Okatsu, Kikkawa and Osa (40) have found that initial rates of decomposition of $PhCMe_2OOH$ by $Zn[(i\text{-}BuO)_2PS_2]_2$ in PhCl obey the rate expression

$$\frac{-d[PhCMe_2OOH]}{dt} = k[ZnL_2][ROOH]$$

with k = 0.0207 $M^{-1}s^{-1}$ at 70°C. This rate constant is close to the value reported by Burn and it is claimed to be much larger than the rate constant for the initial reaction between $PhCMe_2OOH$ and $Zn[i\text{-}Bu_2NCS_2]_2$.

The rate of disappearance of both reactants does, however, slow down after the initial bimolecular reaction and an induction period occurs before the onset of rapid first-order hydroperoxide decomposition. Consequently the overall reaction consists of three distinct stages with the length of each stage depending on the initial hydroperoxide to complex ratio (see, e.g., Figure 5). Hydrogen atom donating radical scavengers increase the induction period but have no effect on the third stage. Oxygen also has a marked effect on the

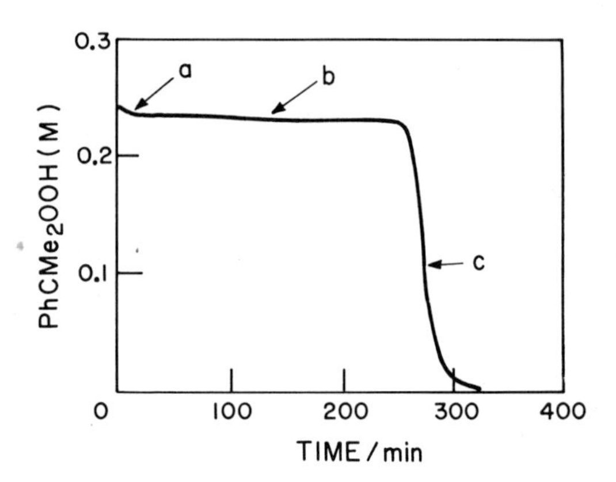

Fig. 5. Decomposition of $PhCMe_2OOH$(0.25M) in PhCl at 70°C by $Zn[(i\text{-}PrO)_2PS_2]_2$ (0.0025M). (a) First stage, (b) Induction period, (c) Third stage.

reaction, increasing the induction period, and slowing down the third stage.

Ivanov and coworkers (41-49) have made several studies of the decomposition of ROOH (R = *t*-Bu, $PhCMe_2$, α-tetralyl) by $Zn[(i\text{-}PrO)_2PS_2]_2$ using a variety of experimental techniques (chemiluminescence, radical scavenging, and hydroperoxide consumption) and have concluded that the hydroperoxide and zinc complex exist in equilibrium with a co-ordination complex which can decompose to give free-radicals or molecular products with a low efficiency, *f*, of "free" radical formation.

$$y\text{ROOH} + \text{ZnL}_2 \rightleftharpoons y\text{ROOH}\ldots\ldots\text{ZnL}_2$$

$$y\text{ROOH}\ldots\ldots\text{ZnL}_2 \begin{cases} \xrightarrow{f} \text{Free radicals} \\ \xrightarrow{1-f} \text{Molecular products} \end{cases}$$

These workers reported that *y*=2 for *t*-BuOOH, $PhCMe_2OOH$, and α-tetralin hydroperoxide (above 90°C) and 1 for α-$C_{10}H_{11}OOH$ below 90°C and that the equilibrium constant for the α-$C_{10}H_{11}OOH$ - $Zn[(i\text{-}PrO)_2PS_2]_2$ system is $1.6 \times 10^6\ M^{-1}$ at 70°C (44). It was also found that decomposition of *t*-BuOOH and $PhCMe_2OOH$ follows three stages whereas decomposition of α-$C_{10}H_{11}OOH$ exhibits only one stage. Activation energies and pre-exponential factors were reported for hydroperoxide decomposition and consumption of the radical scavenger N-phenyl-β-naphthylamine (48).

Initial rates of disappearance of $Ni[(RO)_2PS_2]_2$, $Cu[(RO)_2PS_2]_2$ and $Ni[R_2NCS_2]_2$ in the presence of excess alkyl hydroperoxide and the disappearance of the hydroperoxide in the presence of excess complex obey the rate law

$$\frac{-d[ML_x]}{dt} = \frac{-1}{m}\frac{d[ROOH]}{dt} = k_{22}[ROOH][ML_x] \qquad [22]$$

where the number of moles of hydroperoxide decomposed by each mole of complex m is 5 - 15 (50, 51).

Values of the bimolecular rate constant k_{22}, given in Table 4, indicate that the rate of disappearance of $Cu[(RO)_2PS_2]_2$ is ∿600 times faster than the rate of disappearance of $Ni[(RO)_2PS_2]_2$ and that the rate of disappearance of $Ni[(RO)_2PS_2]_2$ is about 40 times slower than the rate of disappearance of $Ni[R_2NCS_2]_2$.

TABLE 4

Rate constants for the initial reaction between transition metal complexes and alkyl hydroperoxides at 30°C

Complex	Hydroperoxide	k_{22} ($M^{-1}s^{-1}$)	Reference
$Ni[(i\text{-}PrO)_2PS_2]_2$	Cumene	0.004	(50)
$Ni[Et_2NCS_2]_2$	Cumene	0.15	(50)
$Cu[(i\text{-}PrO)_2PS_2]_2$	*t*-Butyl	2.4	(51)

The initial reaction of *t*-BuOOH with $Cu[(i\text{-}PrO)_2PS_2]_2$ has been studied over a temperature range and the Arrhenius equation $\log (k_{22}/M^{-1}s^{-1}) = 6.7 - (8.5 \text{ kcal mol}^{-1})/RT$ was obtained (51).

The magnitude of k_{22} for a particular complex is independent of the nature of R in the ligand, the structure of the hydroperoxide and the presence of oxygen and *t*-butyl alcohol (50, 51).

Hydrocarbon autoxidation is initiated by mixtures of nickel complex and alkyl hydroperoxide (50) indicating that free-radicals are produced by this reaction. These autoxidations are completely inhibited by 2,6-di-*t*-butyl-4-methoxyphenol (Figure 6) and values of R_i calculated from the induction periods produced by this inhibitor obey the expression

$$R_i = k_{23}[NiL_2][ROOH] \qquad [23]$$

The rate constant $k_{23} = 0.0026\ M^{-1}s^{-1}$ for $Ni[(i\text{-}PrO)_2PS_2]_2$ and $0.16\ M^{-1}s^{-1}$ for $Ni[i\text{-}Pr_2NCS_2]_2$ at 30°C and these rate constants are close to the values of k_{22} obtained from rates of complex and hydroperoxide disappearance. This indicates a high efficiency of "free" radical production for the initial homolytic reaction between these reactants.

Although nickel complexes disappear completely by a second-order reaction (50) concentration-time profiles for $Cu[(i\text{-}PrO)_2PS_2]_2$ exhibit three stages (see e.g., Figure 7) (51). The initial second-order reaction is followed by an induction period which commences when 70-90% of the complex has been destroyed. The length of this induction period is approximately linearly dependent on the complex concentration and inversely proportional to the first power of the hydroperoxide concentration. After the induction period the complex disappears relatively rapidly.

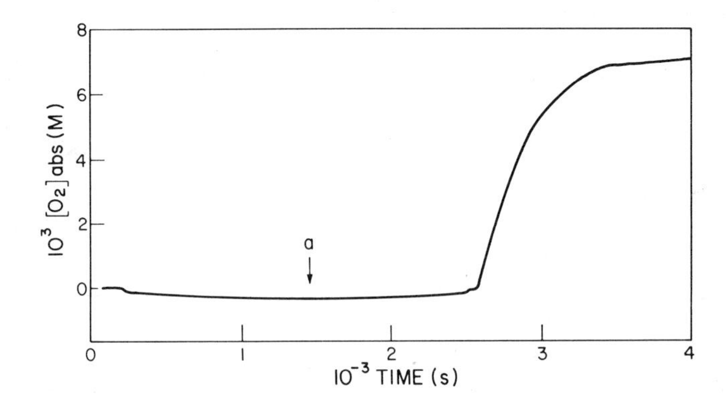

Fig. 6. Oxygen absorption as a function of time for autoxidation of styrene (1.74M) in chlorobenzene at 30°C initiated by a mixture of $Ni[n\text{-}Bu_2NCS_2]_2$ *(0.0016M) and t-BuOOH (0.005M) and inhibited by 2,6-di-t-butyl-4-methoxyphenol (0.007M).* a *indicates the time when the hydroperoxide was added.*

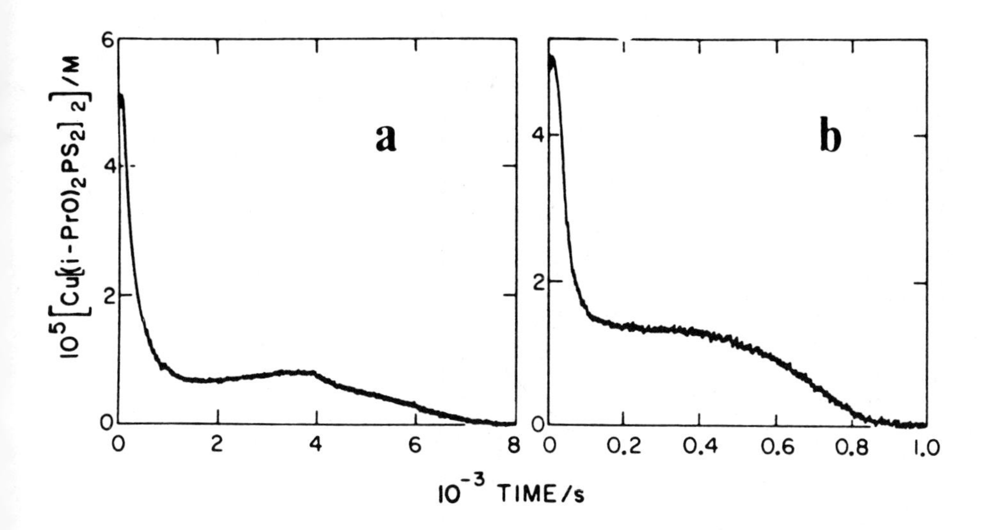

Fig. 7. The change in the concentration of $Cu[(i\text{-}PrO)_2PS_2]_2$ *with respect to time after the addition of 0.001*M *(a) and 0.01*M *(b)* t-*BuOOH at 30°C.*

Cupric dialkyldithiocarbamates behave quite differently to cupric dialkyldithiophosphates (51) as can be seen from Figure 8. For these complexes $-d[ML_x]/dt$ is generally zero-order with respect to the complex concentration and second-order with respect to the hydroperoxide concentration, i.e.,

$$\frac{-d[Cu[R_2NCS_2]_2]}{dt} = k_{24}[ROOH]^2 \qquad [24]$$

Initial rates of disappearance of nickel and cupric complexes in the presence of excess ROOH are slower if 2,6-di-t-butyl-4-methylphenol (BMP) is added to the system

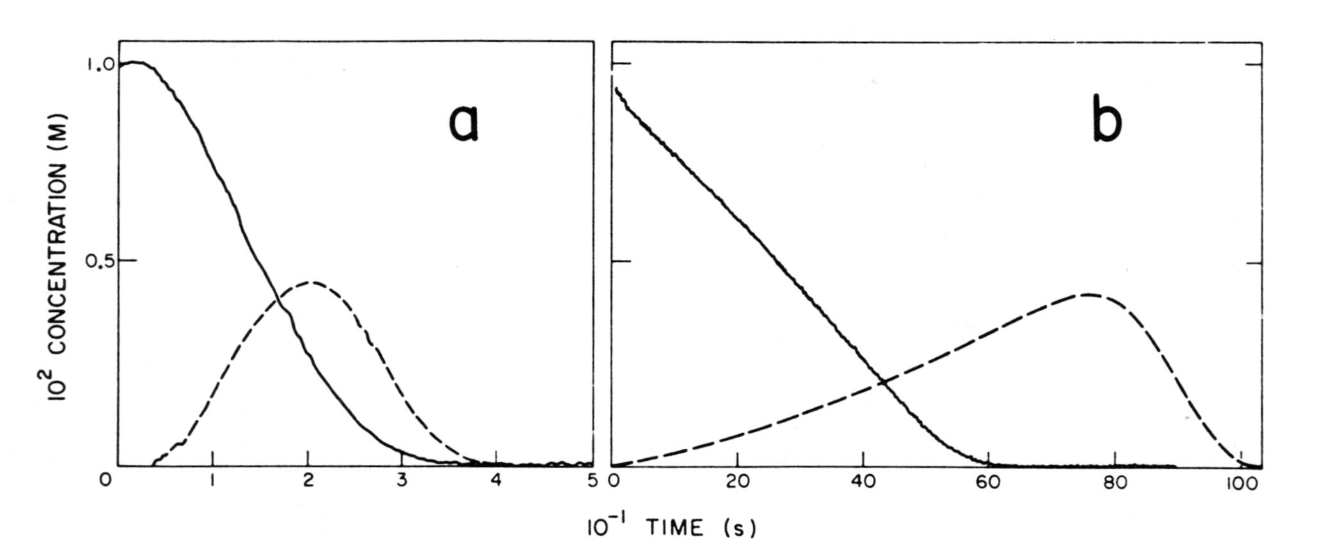

Fig. 8. The change in the concentration of $Cu[Et_2NCS_2]_2$ *with respect to time after the addition of 0.11M (a) and 0.011M (b)* t-*BuOOH at 30°. The broken line shows the concentration-time profile for* **1**.

(50, 51) and for example, -d[Ni[(*i*-PrO)$_2$PS$_2$]$_2$/d*t* and -d[Cu[(*i*-PrO)$_2$PS$_2$]$_2$/d*t* are reduced by a factor of 100 and 4, respectively. Reactant orders are the same in the absence and presence of BMP and this additive appears to function solely by protecting the complex from free radical attack. The number of moles of hydroperoxide decomposed homolytically by ML_x are much larger in the presence of BMP. For instance, each mole of Ni[*n*-Bu$_2$NCS$_2$]$_2$ decomposes 40-60 moles of *tert*-butyl hydroperoxide (50).

Initial rates of disappearance of Cu[(*i*-PrO)$_2$PS$_2$]$_2$ are also reduced by (*i*-PrO)$_2$PS$_2$H and [(*i*-PrO)$_2$PS$_2$]$_2$ (51) probably because of regeneration of the cupric complex by reaction of the additive with copper ions. In addition compounds which can co-ordinate with ML_x, e.g., pyridine, reduce the rate of reaction with hydroperoxide (50).

REACTION PRODUCTS

Peroxy radicals

Cumylperoxy radicals produced by decomposition of azocumene in oxygen saturated chlorobenzene undergo self-reaction to give acetophenone (50%) and α-cumyl alcohol (27%) based on the number of radicals generated. When these radicals are generated in the presence of Zn[(*i*-PrO)$_2$PS$_2$]$_2$ a lower yield of acetophenone (33-43%) is obtained and no α-cumyl alcohol is produced (37).

The yields of the major radical derived products from reaction of cumylperoxy radicals with nickel and cupric dialkyldithiophosphates and dialkyldithiocarbamates (34, 36) are given in Table 5. In these cases much lower yields of acetophenone are produced along with significant yields of α-cumyl alcohol and α-methylstyrene and these products account for 60-90% of the radicals generated from the initiator. α-Methylstyrene is almost certainly produced by an acid catalyzed dehydration of α-cumyl alcohol. The ratio of α-cumyl alcohol to acetophenone in these systems depends on the nature of ML_x. Thus the nickel complexes give a value of 3.4 while this ratio is 89 for Cu[*i*-Pr$_2$NCS$_2$]$_2$.

The formation of α-cumyl alcohol and acetophenone suggests the intermediacy of cumyloxy radicals and this has been substantiated by the observation that thermally generated cumyloxy radicals react with Ni[*n*-Bu$_2$NCS$_2$]$_2$ to give an α-cumyl alcohol and α-methylstyrene to acetophenone ratio identical to the ratio produced by reaction of the complex with cumylperoxy radicals (34).

Radical derived products have also been determined for the 1,1-diphenylethylperoxy radical (34). This radical in the absence of ML_x undergoes self-reaction to give aceto-

TABLE 5

Radical derived products from reaction of cumylperoxy radicals with ML_x at 50°C

ML_x	Acetophenone	α-Methylstyrene	α-Cumyl alcohol	Reference
-	0.5	-	0.27	(50)
$Zn[(i\text{-}PrO)_2PS_2]_2$	0.35	-	-	(37)
$Ni[(i\text{-}PrO)_2PS_2]_2$	0.16	0.41	0.13	(50)
$Ni[n\text{-}Bu_2NCS_2]_2$	0.14	0.30	0.17	(50)
$Cu[(i\text{-}PrO)_2PS_2]_2$	0.11	0.65	0.04	(51)
$Cu[i\text{-}Pr_2NCS_2]_2$	0.01	0.62	0.27	(51)

phenone (14%), benzophenone (2%) and 1,1-diphenylethanol (3%) as the products that have so far been identified (34, 52). Reaction of $Ph_2CMeO_2^{\bullet}$ with Ni[n-Bu_2NCS_2]$_2$ gives high yields of 1,1-diphenylethylene (50%) and acetophenone (35%) products which account for 85% of the peroxy radicals produced from tetraphenylbutane (34). This olefin must be formed by efficient dehydration of 1,1-diphenylethanol and it has been concluded (34) that formation of this alcohol and acetophenone are consistent with removal of an oxygen atom from $Ph_2CMeO_2^{\bullet}$ by Ni[n-Bu_2NCS_2]$_2$ to give $Ph_2CMeO^{\bullet}$ which is efficiently reduced by the complex. This alkoxy radical must undergo some rearrangement because acetophenone can only be formed according to reaction scheme II.

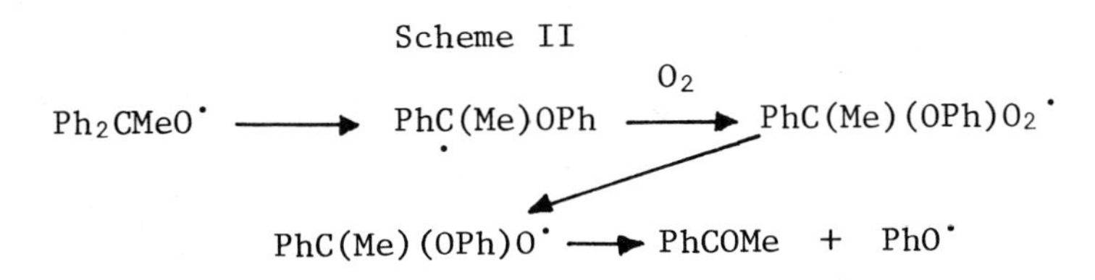

There have been fewer successful attempts to identify reaction products derived from the complex although Burn (13) found that 2-cyano-2-propylperoxy radicals react with Zn[(i-PrO)$_2$PS$_2$]$_2$ to give low yields of di-isopropyldithiophosphoryl disulphide, (i-PrO)$_2$P(S)SSP(S)(OPr-i)$_2$, while Fe[(i-PrO)$_2$PS$_2$]$_3$ in cyclohexane at 80°C absorbs oxygen to give high yields of the disulphide. We have found that cumylperoxy radicals react with Zn[(i-PrO)$_2$PS$_2$]$_2$ to give the disulphide in ∿10% yield along with at least seven (by t.l.c.) involatile products which have so far evaded identification (37).

The most successful study of complex derived products has been achieved with Cu[R_2NCS_2]$_2$ (36, 53). This complex is rapidly destroyed by photochemically generated t-butylperoxy radicals to give at least three transient cupric complexes **1**, **2**, and **3** before the copper ions are quantitatively precipitated as $CuSO_4$. These transients and Cu[R_2NCS_2]$_2$ are readily distinguished from each other by electron spin resonance spectroscopy (36, 53) because $g(\mathrm{Cu[R_2NCS_2]_2}) < g(\mathbf{1}) < g(\mathbf{2}) < g(\mathbf{3})$ (Table 6).

The concentration-time profiles for the disappearance of Cu[Et_2NCS_2]$_2$ and the appearance and disappearance of **1**, **2**, and **3** and the appearance of t-$BuO_2^{\bullet}$ are shown in Figure 9. The rate of disappearance of Cu[Et_2NCS_2]$_2$ is ∿5-6 times slower than d[t-$BuO_2^{\bullet}$]/dt while d[**1**]/dt is ∿-d[Cu[Et_2NCS_2]$_2$/dt.

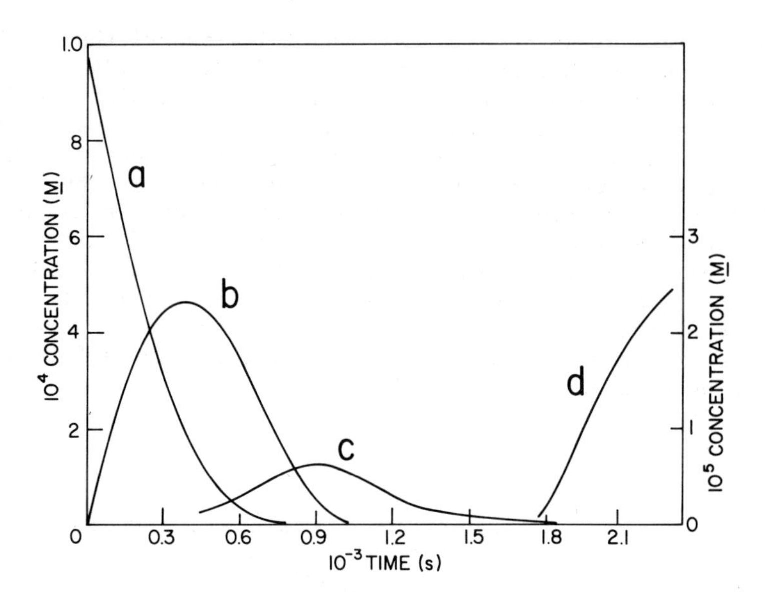

*Fig. 9. Concentration-time profiles for the products of the photolyses of di-*t*-butyl ketone (0.5M) in air-saturated toluene-chlorobenzene (5:1) containing $Cu[Et_2NCS_2]_2$ (10^{-4}M) at -40°C (a) $Cu[Et_2NCS_2]_2$, (b) 1, (c) 2 and 3, (d)* t-*BuO_2.*

The complex 1 reaches a maximum concentration of approximately one half the initial concentration of $Cu[Et_2NCS_2]_2$ and it is completely destroyed just after complete destruction of the original complex. The maximum concentrations of 2 and 3 occurs soon after 1 has disappeared and t-$BuO_2{}^{\cdot}$ are not detected in the system until all the cupric complexes have disappeared. If the light is switched off when 1 has reached its maximum concentration its concentration grows slowly by a further 20% while the concentration of $Cu[Et_2NCS_2]_2$ continues to decrease slowly, implying the formation of a fairly stable intermediate between $RO_2{}^{\cdot}$ and $Cu[Et_2NCS_2]_2$ which decomposes to give 1.

A transient with the same g-factor as 1 has been detected by several groups of Russian workers (23, 27-31) during the $Cu[R_2NCS_2]_2$ inhibited autoxidation of ethylbenzene while we have observed the formation of 1, 2, and 3 during the inhibited autoxidation of cumene (36) (see Figure 10). The Russians originally concluded (23) that 1 is a cupric

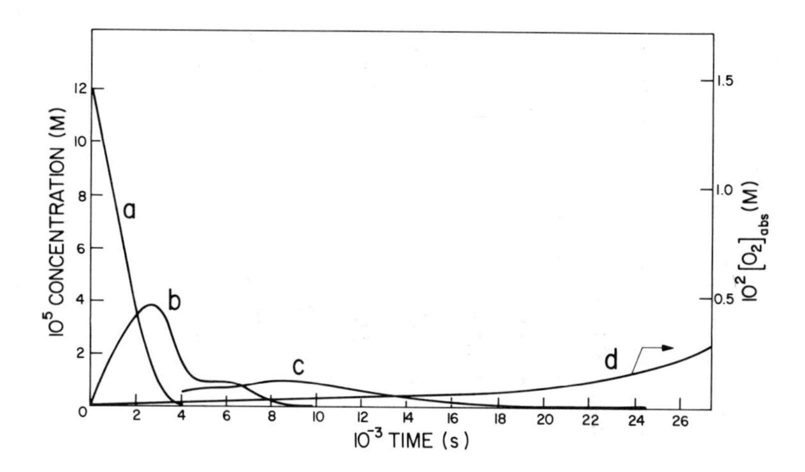

Fig. 10. Concentration-time profiles for $Cu[Et_2NCS_2]_2$ *(*1.25×10^{-4}M*) inhibited autoxidation of cumene (7.17*M*) at 30°C. (a)* $Cu[Et_2NCS_2]_2$*, (b) 1, (c) 2 and 3, (d)* O_2*.*

complex with an oxidized ligand but more recently they have attributed it to an alkylperoxo dialkyldithiocarbamato copper complex (30).

The structures of the transients 1, 2, and 3 have recently been established (53) by reacting $Cu[Me_2NCS_2]_2$ with *t*-butylperoxy radicals enriched to 50% with O-17. The isotropic e.s.r. spectrum obtained from a solution containing mainly 1 is shown in Figure 11. Hyperfine structure due to the interaction of the unpaired electron with one ^{17}O nucleus can be observed on the high field side of the isotropic (-3/2) line of the $Cu[Me_2NCS_2]_2$. This spectrum is consistent with species 1 containing *one* ^{17}O nucleus ($A_{iso}(^{17}O)$=19.2G) since there is no evidence for a second equivalent oxygen or an oxygen with a hyperfine splitting constant ≥ 1.5G. The e.s.r. spectrum of a frozen solution containing 1 is shown in Figure 12 and hyperfine structure can be observed in both $g_{\parallel}$ and $g_{\perp}$ regions and it is again consistent with a species which contains only one ^{17}O nucleus.

The e.s.r. spectrum of a solution containing principally 2 prepared from ^{17}O enriched *t*-butylperoxy radicals has been observed and it appears that this species contains *two*

TABLE 6

Esr parameters for $Cu[Me_2NCS_2]_2$ *and its oxidation products (53)*

Parameter	$Cu[Me_2NCS_2]_2$	$Cu[Me_2NCS_2][Me_2NCS_2O]$	$Cu[Me_2NCS_2O]_2$	**3**
$g_{\parallel}$	2.0881	2.1288	2.1827	-
$g_{\perp}$	2.0215	2.0345	2.0385	-
g_{iso}	2.0448	2.0641	2.0908	2.1012
$A_{\parallel}(Cu)/G$	163.5	165.5	169	-
$A_{\perp}(Cu)/G$	45	43	32	-
$A_{iso}(Cu)/G$	79.4	77.6	72.9	~72.9
$A_{\parallel}(O)/G$	-	-17.5	-	-
$A_{\perp}(O)/G$	-	-20	-	-
$A_{iso}(O)/G$	-	-19.2	-	-

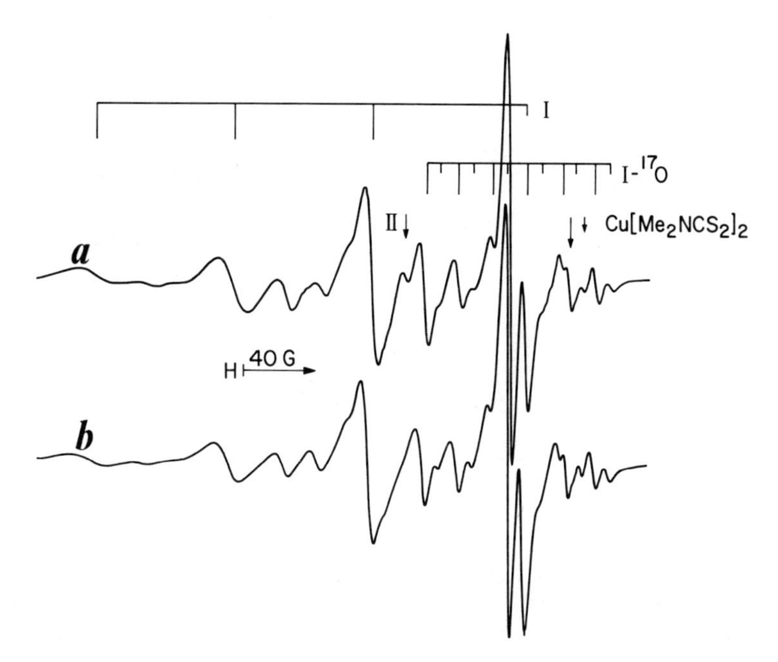

Fig. 11. (a) The epr spectrum at 243K of 1 in 60% toluene -40% $CHCl_3$ prepared from t-$BuO_2^{\cdot}$ enriched with ^{17}O and Cu[Me_2NCS_2]$_2$. (b) Computer simulation using the parameters in Table 6.

equivalent oxygen nuclei with $A_{iso}(^{17}O)$ = 19.2G. Unfortunately the number of oxygens associated with 3 could not be unequivocally established because of loss of spectral resolution as Cu[Me_2NCS_2]$_2$ was successively oxidized.

The e.s.r. parameters for Cu[Me_2NCS_2]$_2$ and its oxidized forms 1, 2, and 3 are gathered in Table 6.

The g-factor for 1 is very similar to g-factors for other Cu(II) species with square planar co-ordination indicating that the unpaired electron in 1 must occupy the $d_{x^2-y^2}$ or d_{xy} orbital on Cu. The ^{17}O hyperfine tensor of 1 is nearly isotropic and the largest component of the dipolar tensor is associated with $g_{\parallel}$ suggesting that the bonding orbital is directed parallel to the z axis of the Cu complex. It would, therefore, appear that 1 contains one oxygen bonded to sulphur and that it takes a position in which the S-O bond is directed out of the plane of the complex, i.e.,

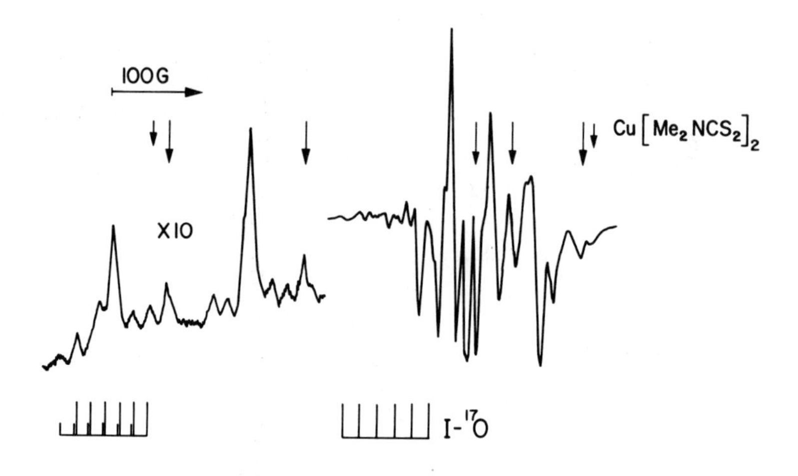

Fig. 12. The epr spectrum at 135K of a frozen solution of 1. The stick diagram shows ^{17}O *hyperfine lines for 1 in the parallel and perpendicular region.*

O
S S
Me_2NC Cu $CNMe_2$
S S

1

The more highly oxidized species 2 and 3 almost certainly have the following structures.

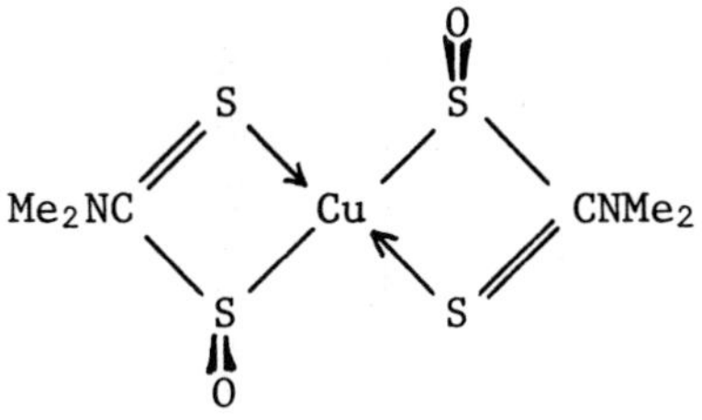

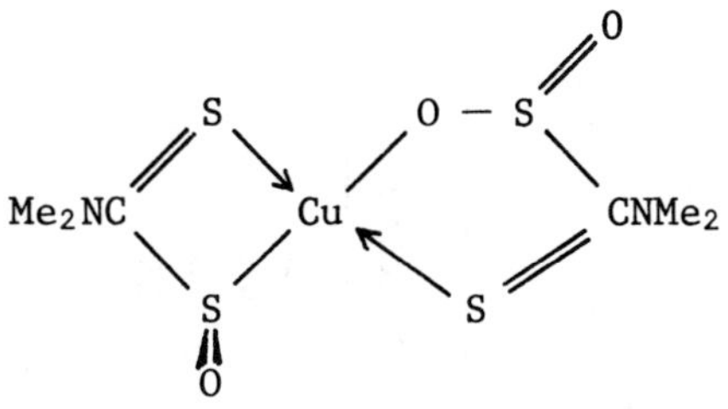

$$\mathrm{Me_2NC}\langle\mathrm{S(=O)},\mathrm{S}\rangle\mathrm{Cu}\langle\mathrm{S(=O)},\mathrm{S}\rangle\mathrm{CNMe_2}$$

2

$$\mathrm{Me_2NC}\langle\mathrm{S(=O)},\mathrm{S}\rangle\mathrm{Cu}\langle\mathrm{O{-}S(=O)},\mathrm{S}\rangle\mathrm{CNMe_2}$$

3

Hydroperoxides

Decomposition of alkyl hydroperoxides by ML_x gives products which depend on the initial ratio of hydroperoxide to complex (39, 40). At high ratios products associated with heterolytic decomposition are formed while if the reaction is performed under stoichiometric conditions and high initial concentrations of ROOH products associated with homolytic decomposition predominate (39, 40). Thus catalytic decomposition of cumene hydroperoxide gives phenol and acetone while homolytic decomposition gives acetophenone, α-methylstyrene, and α-cumyl alcohol. It has, however, recently been noted (40) that even at $ROOH/Zn[(RO)_2PS_2]_2 = 1$ in PhCl at 70°C the products depend on the initial concentration of ROOH. Thus, if $PhCMe_2OOH = 0.001$ *M* phenol and acetone are formed in 86-87% yield.

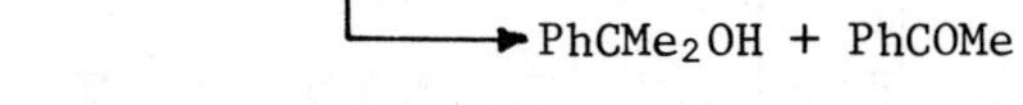

$$PhCMe_2OOH \begin{cases} \xrightarrow{H^+} PhOH + Me_2CO \\ \xrightarrow{R^\cdot} PhCMe_2OH + PhCOMe \end{cases}$$

The major reaction products derived from cumene hydroperoxide for a variety of ML_x are summarized in Table 7. Under stoichiometric conditions zinc complexes give almost quantitative reduction to α-cumyl alcohol whereas nickel and cupric complexes give substantial yields of acetophenone. Although phenol and acetone are the major products at $ROOH/ML_x = 10^4$, detectable concentrations of α-cumyl alcohol, α-methylstyrene, and acetophenone are produced and it would seem from these results that up to several hundred moles of ROOH are homolytically decomposed by each mole of ML_x.

TABLE 7

Products of the decomposition of cumene hydroperoxide by transition metal complexes at 30°C

Complex	$\frac{[ROOH]}{[ML_x]}$ [a]	Acetophenone	α-Methylstyrene	α-Cumyl alcohol	Phenol	Acetone
$Zn[(i\text{-}PrO)_2PS_2]_2$	1	-	5	100	-	-
$Zn[(i\text{-}PrO)_2PS_2]_2$	10^4	16	6.5	7	80	80
$Zn[Et_2NCS_2]_2$	1	5	-	95	-	-
$Zn[Et_2NCS_2]_2$	10^4	24	10	20	44	44
$Ni[(i\text{-}PrO)_2PS_2]_2$	1	23	27	58	-	-
$Ni[(i\text{-}PrO)_2PS_2]_2$	10^4	13	9	6	74	75
$Ni[i\text{-}Pr_2NCS_2]_2$	1	8	22	64	-	-
$Ni[i\text{-}Pr_2NCS_2]_2$	10^4	5	21	6	71	71
$Cu[(i\text{-}PrO)_2PS_2]_2$	1	6	4	24	-	-
$Cu[(i\text{-}PrO)_2PS_2]_2$	10^4	7	5.5	6	76	80
$Cu[Et_2NCS_2]_2$	1.2	7	18	66	-	-
$Cu[Et_2NCS_2]_2$	10^4	8	7.5	7	89	-

a $[ROOH]_o$ = 0.1 M for catalytic decompositions and 0.02 M for stoichiometric decompositions

b % based on hydroperoxide decomposed

It has been concluded (51) from concentration-time profiles for phenol, α-methylstyrene, and complex during the decomposition of $PhCMe_2OOH$ by $Cu[Et_2NCS_2]_2$ that α-methylstyrene is an initial reaction product and that phenol is only formed after the complex has disappeared. An insoluble precipitate is formed after destruction of the complex and it is this precipitate which brings about heterolytic decomposition of cumene hydroperoxide. Interestingly this precipitate does not have the ability to heterolytically decompose *t*-butyl hydroperoxide (51).

Decomposition of cumene hydroperoxide by $Ni[n\text{-}Bu_2NCS_2]_2$ ($ROOH/ML_x$=33) in the presence of 2,6-di-*t*-butyl-4-methylphenol gives α-methylstyrene (38%), acetophenone (6%), α-cumyl alcohol (28%) and 1-methyl-1-α-cumylperoxy-3,5-di-*t*-butylcyclohexadiene-2,5-dione-4 (50). Thus at this ratio of hydroperoxide to complex decomposition in the presence of BMP is principally homolytic.

The disappearance of reactants and the appearance of products for the decomposition of *t*-butyl hydroperoxide by $Ni[n\text{-}Bu_2NCS_2]_2$ ($ROOH/ML_x$=100) in the presence of BMP are shown in Figure 13 (50). In this system *t*-butyl alcohol (50%) and 1-methyl-1-*t*-butylperoxy-3,5-di-*t*-butylcyclohexadiene-2,5-dione-4 (50%) are produced which is indicative of the formation of *t*-butoxy and *t*-butylperoxy radicals.

tert-Butyl hydroperoxide is reduced quantitatively to *tert*-butyl alcohol by $Cu[Et_2NCS_2]_2$ while the complex is oxidized to $Cu[Et_2NCS_2][Et_2NCS_2O]$ (51). Subsequent oxidation gives $Cu[Et_2NCS_2O]_2$ and $Cu[Et_2NCS_2O][Et_2NCS_2O_2]$ and in the presence of excess hydroperoxide $Cu[Et_2NCS_2]_2$ is eventually quantiatively oxidized to copper sulphate.

A concentration-time profile of the reaction of $Cu[Et_2NCS_2]_2$ with *tert*-BuOOH ($ROOH/ML_x$=5.6) is shown in Figure 14. An interesting feature of this reaction is that although $Cu[Et_2NCS_2]_2$ is destroyed in about 800 sec and $Cu[Et_2NCS_2][Et_2NCS_2O]$ is produced and destroyed in about the same length of time, both these complexes are regenerated after an induction period. In fact the concentration of $Cu[Et_2NCS_2]_2$ reaches almost 75% of its original concentration while $Cu[Et_2NCS_2]_2[Et_2NCS_2O]$ is regenerated to about 50% of its maximum concentration and then slowly disappears. Regeneration of these complexes is almost certainly associated with reaction of cupric ions with the disulphides $[Et_2NCS_2]_2$ and $Et_2NC(S)SS(O)C(S)NEt_2$.

Free radicals must be produced by the initial reaction between ZnL_x and ROOH because mixtures of these reactants initiate hydrocarbon autoxidation (10, 12) and destroy 2,2,6,6-tetramethylpiperidin-1-oxyl (43, 44) and N-phenyl-β-naphthylamine (45, 46). The rate of radical production is about two orders of magnitude slower than the rate of hydro-

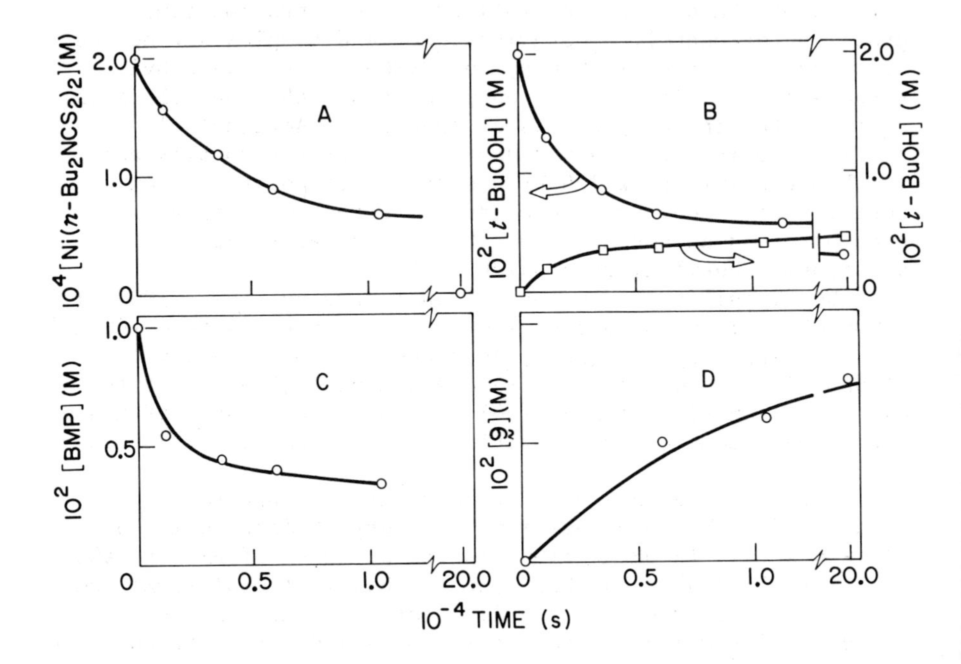

Fig. 13. The concentration of reactants and products as a function of time from reaction of t-*BuOOH (0.02*M*) with Ni*[n-*Bu*$_2$*NCS*$_2$]$_2$ *(0.0002*M*) in the presence of 2,6-di-*t-*butyl-4-methylphenol. (A) Ni*[n-*Bu*$_2$*NCS*$_2$]$_2$*; (B)* t-*BuOOH and* t-*BuOH; (C) BMP; (D) 1-methyl-1-*t-*butylperoxy-3,5-di-*t-*butylcyclohexadiene-2,5-dione* 9.

peroxide decomposition, a difference that has been attributed to a low efficiency of radical escape from the solvent cage (43). This interpretation does, however, seem unlikely in view of the fact that nickel and cupric complexes interact with ROOH to give radicals with almost 100% efficiency (50, 51). It seems much more likely that zinc complexes react initially with ROOH by a non-radical molecular reaction.

Although attempts to detect directly the radicals produced by reaction of ROOH with Zn[(i-PrO)$_2$PS$_2$]$_2$ have failed (54) nitroxides were detected when N-methylene-t-butylamine N-oxide and phenyl t-butyl nitrone were added to a mixture of t-butyl hydroperoxide and the complex (54). Interpretation of these results is, however, complicated by a direct reaction

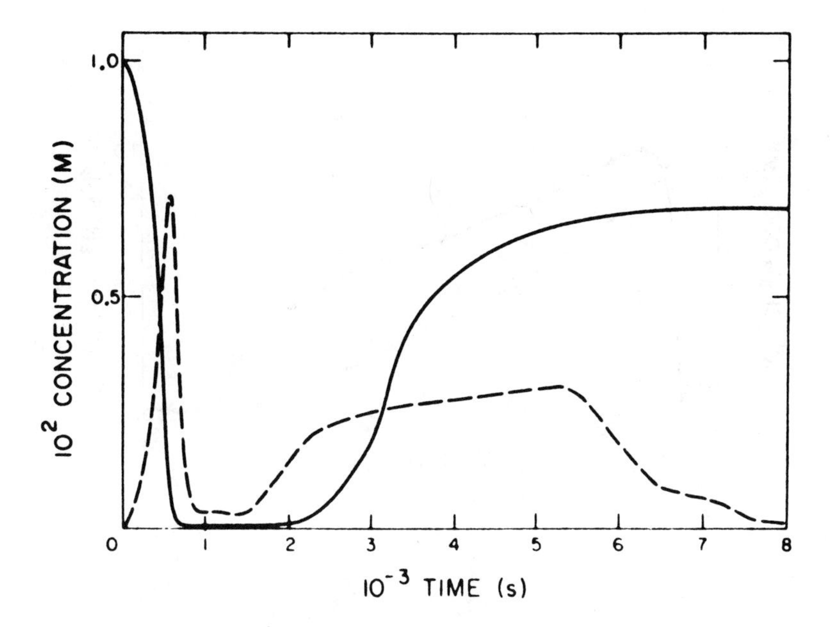

Fig. 14. The variation in the concentrations of $Cu[Et_2NCS_2]_2$ and 1 (broken line) during reaction of $Cu[Et_2NCS_2]_2$ (0.01M) with t-BuOOH (0.056M) in argon saturated chlorobenzene at 30°C.

between the hydroperoxide and the spin traps to give nitroxides.

t-Butylperoxy radicals are readily detected by e.s.r. spectroscopy during the decomposition of *t*-BuOOH by a trace of Ni[*n*-$Bu_2NCS_2]_2$ (50). The variation of the radical and complex concentrations with time at ROOH/ML_x=150 is shown in Figure 15 and it can be seen that the concentration of *t*-$BuO_2^{\cdot}$ increases to a maximum value and then slowly decreases as the complex is consumed. The maximum concentration of *t*-$BuO_2^{\cdot}$ generated is approximately linearly dependent on the complex concentration and the hydroperoxide concentration (50). Detection of alkylperoxy radicals in these systems has not been confined to *t*-$BuO_2^{\cdot}$ since α-tetralylperoxy radicals have been detected during the decomposition of α-tetralin hydroperoxide by Ni[*n*-$Bu_2NCS_2]_2$ (50).

Decomposition of *t*-BuOOH by Ni[(*i*-PrO)$_2PS_2]_2$ also produces *t*-$BuO_2^{\cdot}$ (50). The maximum radical yields for this complex are, however, about 10 times smaller than for Ni[*n*-$Bu_2NCS_2]_2$ with radical generation lasting very much longer because the complex is destroyed more slowly.

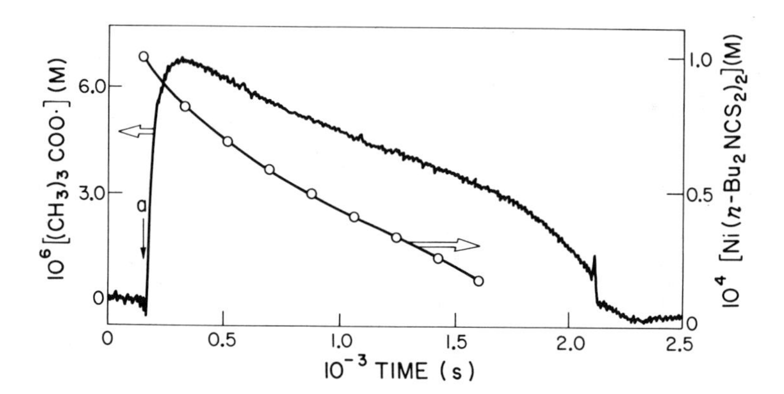

Fig. 15. The variation in the concentration of t-*BuO*$_2$$^\cdot$ *and Ni*[n-*Bu*$_2$*NCS*$_2$]$_2$ *with respect to time during reaction of* t-*BuOOH (0.015*M*) with Ni*[n-*Bu*$_2$*NCS*$_2$]$_2$ *in PhCl at 30°C.*

Rossi and Imparato (55) have demonstrated that α-cumyl alcohol, 0,0'-di-isopropyldithiophosphoryl disulphide, and a basic zinc salt are formed by reaction of cumene hydroperoxide with a four fold excess of $Zn[(i\text{-}PrO)_2PS_2]_2$ in CCl_4 at room temperature.

$$PhCMe_2OOH + 4Zn[(i\text{-}PrO)_2PS_2]_2 \longrightarrow PhCMe_2OH + [(i\text{-}PrO)_2PS_2]_2 + [(i\text{-}PrO)_2PS_2]_6Zn_4O \quad [25]$$

This basic salt then reacts with excess hydroperoxide according to reaction [26]

$$[(i\text{-}PrO)_2PS_2]_6Zn_4O + 15PhCMe_2OOH \longrightarrow 1.5[(i\text{-}PrO)_2PS_2]_2 + 1.5[(i\text{-}PrO)_2P(O)S]_2 + 3ZnSO_4 + ZnO + 15PhCMe_2OH \quad [26]$$

Interestingly this basic zinc salt is the only transition metal complex that has so far been isolated from reaction of

ML_x with ROOH and appears to have a structure with an oxygen atom surrounded tetrahedrally by four zinc atoms and the dialkyldithiophosphate ligands attached symmetrically to the six edges of the tetrahedron (56). Neither nickel nor cupric dialkyldithiophosphates give analogous salts and oxidized forms of $Cu[R_2NCS_2]_2$ are too unstable and amorphous to be positively identified. It should perhaps be noted here that neither the basic salt nor O,O'-dialkyldithiophosphoryl disulphides have the ability to scavenge alkylperoxy radicals whereas O,O'-dialkyldithiophosphoric acids, $(RO)_2PS_2H$, do so quite efficiently (25).

Interestingly cupric di-ethyldiselenocarbamate, $Cu[Et_2NCSe_2]_2$, reacts with *t*-BuOOH at about the same rate and with the same kinetics as $Cu[Et_2NCS_2]_2$; the selenium analogues of **1**, **2**, and **3** have, however, not been detected (57).

REACTION MECHANISMS

Peroxy radicals

Colclough and Cunneen (11) suggested that zinc complexes react with alkylperoxy radicals by an electron transfer mechanism in which an electron is abstracted from an electron rich sulphur atom in the ligand by the radical.

$$RO_2^{\cdot} + Zn[(RO)_2PS_2]_2 \longrightarrow RO_2^{-} + \underset{\underset{\mathbf{4}}{S^{+\cdot}}}{(RO)_2\overset{}{P}(\!=\!)SZnS_2P(OR)_2} \longrightarrow (RO)_2PS_2Zn^{+} + (RO_2)_2PS_2^{\cdot} \qquad [27]$$

The radical cation **4** would be expected to decompose rapidly to give dialkyldithiophosphoryl radicals which could undergo mutual combination to give $[(RO)_2PS_2]_2$, a minor reaction product.

Burn has argued that "free" dialkyldithiophosphoryl radicals are unlikely to be formed in the termination reaction because $Zn[(RO)_2PS_2]_2$ inhibit olefin autoxidation (13). Thus in these substrates the zinc complex might be expected to function simply as a chain transfer agent because of the known ability of $(RO)_2PS_2^{\cdot}$ to add to olefins (58, 59). As a way around this objection, Burn suggested that the intermediate **4** could perhaps exist long enough to react with a second peroxy radical.

$$RO_2^{\cdot} + (RO)_2\overset{S^{+\cdot}}{\overset{\|}{P}}SZnS_2P(OR)_2 \longrightarrow \text{non-radical products} \qquad [28]$$

An alternative mechanism proposed by Burn (13) involves a stabilized peroxy radical - zinc complex intermediate which, on attack by a second $RO_2^\cdot$, leads to intramolecular dimerization of the incipient $(RO)_2PS_2^\cdot$ before addition to a double bond can occur.

$$RO_2^\cdot + Zn\left[(RO)_2PS_2\right]_2 \longrightarrow \underset{\underset{\displaystyle 5}{}}{\overset{ROOS}{\underset{(OR)_2}{\overset{|}{\underset{|}{P^\cdot}}}}} - S\text{-}ZnS_2P(OR)_2 \xrightarrow{RO_2^\cdot}$$

$$\left[(RO)_2PS_2\right]_2 + 2RO_2^- + Zn^{2+} \qquad [29]$$

At one time it did appear that there was e.s.r. evidence for either 4 or 5 because Liston, Ingersoll and Adams (21) had reported that photochemically generated *t*-butoxy radicals react with $Zn[(RO)_2PS_2]_2$ and $Et_2NH_2S_2P(OR)_2$ but not $(RO)_2PS_2H$ to give a doublet e.s.r. spectrum with a_P = 24.8 - 26.7 G and g=2.0203. It was, however, later shown that photolysis of $Zn[(RO)_2PS_2]_2$ (and also $(RO)_2PS_2H$) in the absence of di-*t*-butyl peroxide gives a doublet e.s.r. spectrum (a_P = 24.7 - 24.9 G), with about the same intensity as the spectrum obtained in the presence of the peroxide (25). Thus, although this doublet can almost certainly be assigned to $(RO)_2PS_2^\cdot$ it does mean that there is no direct evidence that alkoxy and by analogy alkylperoxy radicals react with $Zn[(RO)_2PS_2]_2$ to give $(RO)_2PS_2^\cdot$.

Burn more recently concluded (19) that peroxy radicals abstract a hydrogen atom from the alkyl moiety of the ligand to give, after reaction with oxygen, $LZnLO_2^\cdot$ which was assumed to react with a second $RO_2^\cdot$ to give inactive products. There is, however, apart from some very unusual reaction kinetics, no experimental evidence for this mechanism and it does require a remarkable reactivity for the hydrogen atom.

It has been argued by Howard, Ohkatsu, Chenier and Ingold (25) that alkylperoxy radicals react with $Zn[(RO)_2PS_2]_2$ and related complexes at the metal either by the electron transfer mechanism [27] or by an S_H2 process [30] which involves a zinc (III) species as a transition state or transient intermediate.

$$RO_2^\cdot + ZnL_2 \longrightarrow ROOZnL_2 \longrightarrow ROOZnL + L^\cdot \qquad [30]$$

This mechanism does, of course, lead to the formation of $L^\cdot$. The arguments advanced by Burn (13) to discount the intermediacy of $L^\cdot$ are, however, not entirely valid because dialkyldithiophosphoric acids have been found to inhibit autoxidation of styrene and cyclohexene and to react rapidly with t-$BuO_2^\cdot$ at 173K (25). The most probable mechanism for inhibition by the free acid involves abstraction of the hydrogen attached to sulphur by $RO_2^\cdot$, i.e.,

$$(RO)_2P(S)SH + RO_2^\cdot \longrightarrow (RO)_2PS_2^\cdot + ROOH \qquad [31]$$

with the $(RO)_2PS_2^\cdot$ radical being less reactive than $RO_2^\cdot$ towards propagation.

Our recent finding that inhibited rates of oxidation of styrene containing $Zn[(i\text{-}PrO)_2PS_2]_2$ are proportional to the three-halves power of the substrate concentration can only be rationalized in terms of a mechanism in which there is extensive chain transfer from substrate derived peroxy radicals to radicals derived from the inhibitor (60). This mechanism is further supported by the finding that poly-(peroxystyryl)peroxide prepared in the presence of $Zn[(i\text{-}PrO)_2PS_2]_2$ contains Zn, P, and S in the expected proportions (37).

Zverev, Vinogradova and Maizus (30) have concluded that the transient cupric complex formed during the $Cu[Et_2NCS_2]_2$ inhibited autoxidation of ethylbenzene is $PhCH(CH_3)OOCuS_2CNEt_2$ formed by a substitution reaction analogous to [30]. Furthermore it was suggested that this complex is destroyed by reaction with a second alkylperoxy radical to give bis(1-phenylethylperoxo) copper co-ordinated to a diethyldithiocarbamate radical which subsequently dimerizes to give an insoluble species.

$$Cu[Et_2NCS_2]_2 + RO_2^\cdot \longrightarrow \{Cu[Et_2NCS_2]_2 \ldots RO_2\} \longrightarrow$$

$$\{Cu[Et_2NCS_2]_2^+ \ldots RO_2^-\} \longrightarrow \qquad [32]$$

$$ROOCuS_2CNEt_2 + Et_2NCS_2^\cdot$$

$\underset{\sim}{6}$

$$Et_2NC(=S\rightarrow)(-S-)CuOOR + RO_2^\cdot \longrightarrow Et_2NC(=S\rightarrow Cu(OOR)_2)(-S^\cdot)$$

$\underset{\sim}{7}$

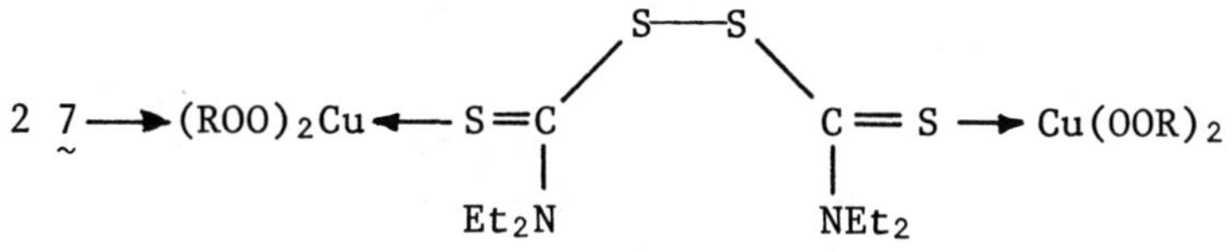

$$2\ \underset{\sim}{7} \longrightarrow (ROO)_2Cu \leftarrow S{=}C(NEt_2){-}S{-}S{-}C(NEt_2){=}S \rightarrow Cu(OOR)_2 \quad [33]$$

$\underset{\sim}{8}$

We have, however, shown (53) that $\underset{\sim}{6}$ is in fact $[R_2NCS_2][R_2NCS_2O]Cu$ and we have concluded that alkylperoxy radicals react with cupric dialkyldithiocarbamates according to the mechanism shown in scheme III.

Scheme III

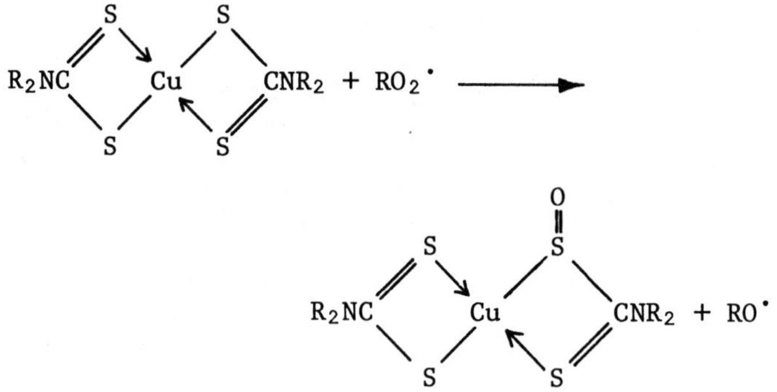

$$[R_2NCS_2]_2Cu + RO_2^{\bullet} \longrightarrow [R_2NCS_2][R_2NCS_2O]Cu + RO^{\bullet} \quad [34]$$

$\underset{\sim}{1}$

$$[R_2NCS_2][R_2NCS_2O]Cu + RO_2^{\bullet} \longrightarrow [R_2NCS_2O]_2Cu + RO^{\bullet} \quad [35]$$

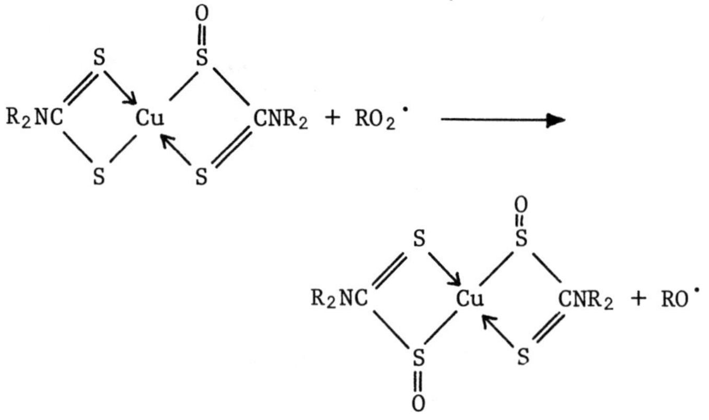

$\underset{\sim}{2}$

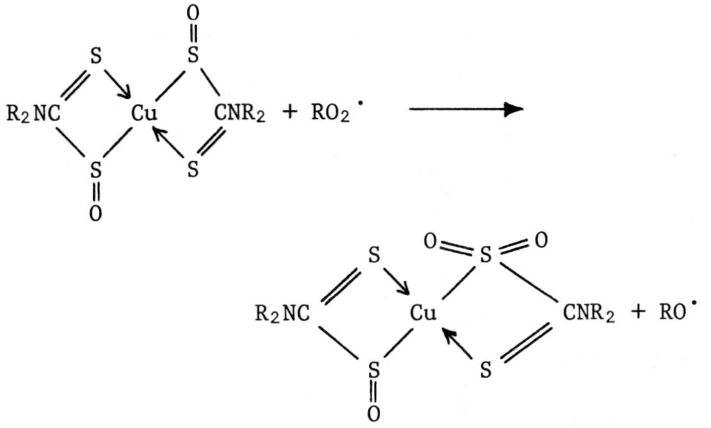

O
S S
R_2NC Cu CNR_2 + $RO_2^{\bullet}$ ⟶
S S
O

O S O
S
R_2NC Cu CNR_2 + $RO^{\bullet}$ [36]
S S
O

3

In conclusion alkylperoxy radicals appear to react at the metal centre of dialkyldithiophosphates to displace a ligand radical and to react with an electron rich sulphur atom of dialkyldithiocarbamates. These conclusions are, of course, purely speculative because there is no evidence that zinc and nickel dialkyldithiocarbamates give complexes analogous to 1, 2, and 3 and that transition metal dialkyldithiophosphates other than zinc react at the metal centre.

It is interesting to note here that Chien and Boss (61) have proposed that the oxidation of $Ni[R_2NCS_2]_2$ by ROOH involves the intermediacy of nickel complexes analogous to 1 and 2. These complexes are, however, more likely to be formed by reaction of $Ni[R_2NCS_2]_2$ with $RO_2^{\bullet}$. On the other hand it could be argued that the alkylperoxy anion formed by electron transfer picks up a proton to give a hydroperoxide and that the reaction products observed from reaction of $RO_2^{\bullet}$ with ML_x result from decomposition of ROOH by ML_x.

Hydroperoxides

Burn (39) has suggested that there are at least four mechanisms, shown in Schemes IV, V, and VI, which can account for the initial reaction of $Zn[(i\text{-}PrO)_2PS_2]_2$ with cumene hydroperoxide.

Scheme IV

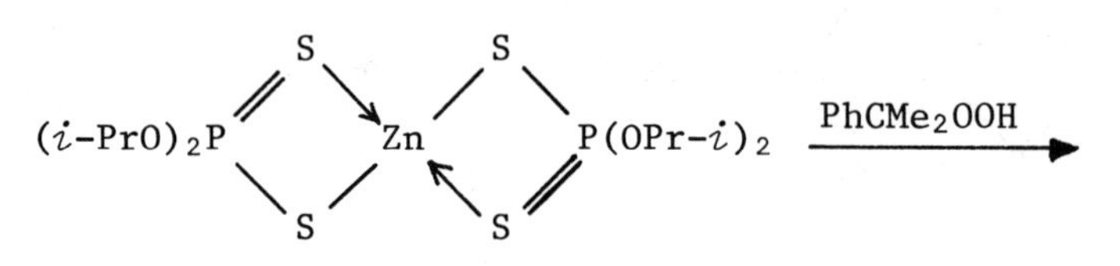

$$(i\text{-PrO})_2P(S^{+\cdot})(S)Zn(S)(S)P(OPr\text{-}i)_2 + PhCMe_2O^{\cdot} + OH^{-}$$

$\downarrow PhCMe_2O^{\cdot}$

$$\longleftarrow (i\text{-PrO})_2P(S^{+\cdot})(S)Zn(S)(S^{\cdot +})P(OPr\text{-}i)_2$$

$$[(i\text{-PrO})_2PS_2]_2 + Zn(OH)(OCMe_2Ph)$$

$$\downarrow$$

$$ZnO + PhCMe_2OH$$

Scheme V

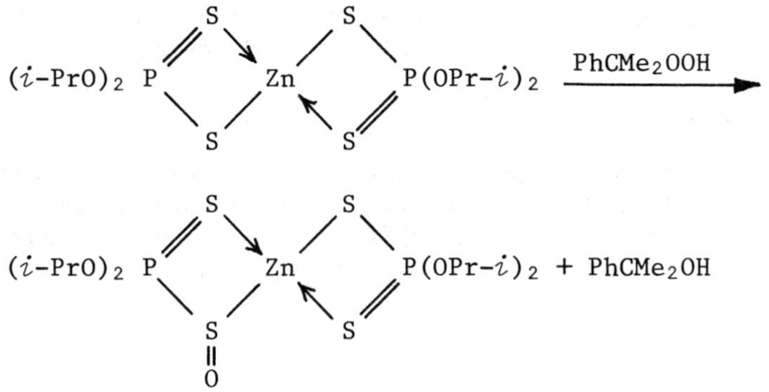

$$(i\text{-PrO})_2P(S)(S)Zn(S)(S)P(OPr\text{-}i)_2 \xrightarrow{PhCMe_2OOH}$$

$$(i\text{-PrO})_2P(S)(S{=}O)Zn(S)(S)P(OPr\text{-}i)_2 + PhCMe_2OH \qquad [37]$$

Scheme VI

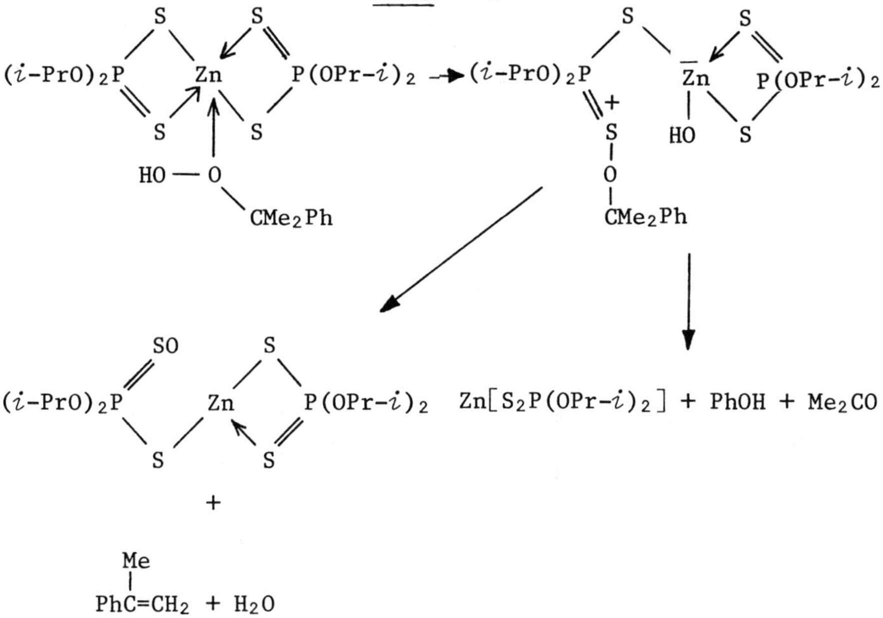

$$(i\text{-PrO})_2P(S)(S)Zn(S)(S)P(OPr\text{-}i)_2 \cdot HO{-}O{-}CMe_2Ph \rightarrow (i\text{-PrO})_2P^{+}(=S{-}O{-}CMe_2Ph)(S)\overset{-}{Zn}(HO)(S)(S)P(OPr\text{-}i)_2$$

$$\rightarrow (i\text{-PrO})_2P(=SO)(S)Zn(S)(S)P(OPr\text{-}i)_2 + PhC(Me){=}CH_2 + H_2O$$

$$\rightarrow Zn[S_2P(OPr\text{-}i)_2] + PhOH + Me_2CO$$

The mechanism shown in Scheme IV was originally favoured because it accounts for the concurrent formation of disulphide and $PhCMe_2OH$. It does, however, require rapid reduction of cumyloxy radicals because low yields of acetophenone are produced and the rate of free radical production is very much less than the rate of hydroperoxide decomposition. Scheme V was included because it is well known that hydroperoxides are reduced to alcohols by organosulphur compounds. The reactions shown in Scheme VI were included to rationalize decomposition of more than one mole of $PhCMe_2OOH$ by each mole of complex and the production of α-methylstyrene and phenol. The mechanisms shown in this scheme are, however, not necessary because phenol is not formed in the initial reaction and α-methylstyrene is formed by acid catalyzed dehydration of α-cumyl alcohol.

Ivanov and co-workers (41-49) have concluded from their work that most co-ordination complexes between $Zn[(RO)_2PS_2]_2$ and ROOH contain two moles of ROOH and that these complexes decompose to give radical and molecular products.

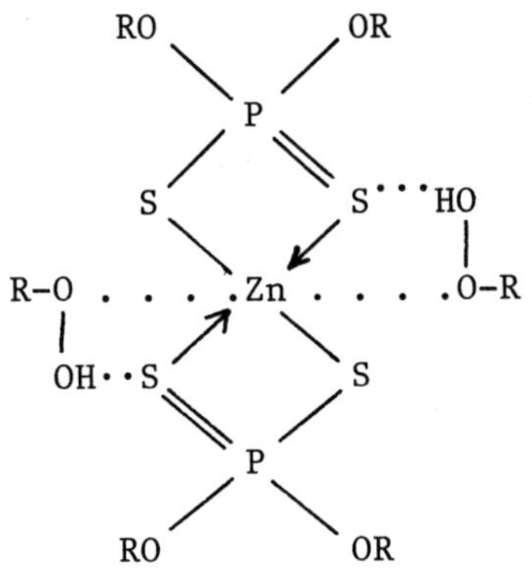

$$
\begin{array}{ccccc}
 & RO & & OR & \\
 & & P & & \\
 & S & & S\cdots HO & \\
R\text{-}O & \cdots & Zn & \cdots & O\text{-}R \\
OH\cdot\cdot S & & & S & \\
 & & P & & \\
 & RO & & OR &
\end{array}
$$

The initial reaction between nickel and cupric complexes and ROOH is almost entirely homolytic and probably involves the redox cycle shown by reactions 38 and 39

$$ML_x + ROOH \longrightarrow ML_x^{+\cdot} + OH^- + RO^\cdot \qquad [38]$$

$$ML_x^{+\cdot} + ROOH \longrightarrow ML_x + RO_2^\cdot + H^+ \qquad [39]$$

The alkoxy radicals produced by reaction [38] either abstract a H-atom to give alcohol or undergo β-scission. Thus cumyloxy radicals give α-cumyl alcohol and acetophenone by these two reactions. This mechanism is quite consistent with the reaction kinetics (first-order in each reactant, the m-values of 5-10, and the hydroperoxide derived products). The complex must be destroyed principally by reaction with $RO_2^\cdot$ because no oxygen is evolved. This reaction could lead to the formation of L_2. In addition there could be a small fraction of a molecular reaction analogous to reaction [37].

The unusual kinetics exhibited by the initial reaction of ROOH with $Cu[R_2NCS_2]_2$, zero-order with respect to the complex and second-order with respect to the hydroperoxide, are difficult to rationalize on the basis of a radical generating reaction between the complex and hydroperoxide. A second-order dependence on the hydroperoxide concentration would be obtained if the hydroperoxide and the complex are in equilibrium with a co-ordination complex and this co-ordination complex reacts with a second molecule of hydroperoxide or if two molecules of ROOH are co-ordinated by $Cu[R_2NCS_2]_2$ and $2ROOH\ldots Cu[R_2NCS_2]_2$ decomposes irreversibly. We could, however, find no e.s.r. evidence for species such as

$ROOH\cdots Cu[R_2NCS_2]_2$ or $2ROOH\cdots Cu[R_2NCS_2]_2$ (53).

The second stage or induction period that is observed in hydroperoxide decompositions at high ratios of $ROOH/ML_x$ must be the time that is necessary to form the ionic catalyst from oxidation products of ML_x. This induction period is lengthened by free-radical scavengers implying that production of the ionic catalyst is a homolytic process. The length of the induction period also depends to some extent on the nature of ML_x and ROOH. Thus $PhCMe_2OOH$, $Zn[(RO)_2PS_2]_2$, $Cu[(RO)_2PS_2]_2$, and $Cu[R_2NCS_2]_2$ give the shortest induction periods while *t*-BuOOH, $Zn[R_2NCS_2]_2$, $Ni[(RO)_2PS_2]_2$, and $Ni[R_2NCS_2]_2$ give very long ones.

Ohkatsu, Kikkawa and Osa (40) have recently made the very interesting suggestion that the induction period occurs because H_2SO_4, produced by oxidation of $[(RO)_2PS_2]_2$ by ROOH in the initial reaction, is rendered inactive by reaction with the metal ions. Ionic decomposition is believed to only occur when all the metal ions have been precipitated as sulphate. This mechanism does, however, not explain the apparent heterogeneous nature of the ionic decomposition or the marked difference in the stability of *t*-BuOOH and $PhCMe_2OOH$.

In one instance, $Cu[(RO)_2PS_2]_2$, a three stage concentration time profile has been observed for ML_x which is, of course, over very much sooner than the three stage hydroperoxide decomposition. The induction period in this system has been attributed to regeneration of the complex and a period when the rate of regeneration equals the rate of destruction (51).

The third stage in the decomposition of ROOH by ML_x is the rapid catalytic decomposition of the hydroperoxide to give products characteristic of ionic decomposition. The general mechanism by which these products are formed is well established as a rearrangement catalyzed by an electrophilic species (X)

$$PhCMe_2OOH \xrightarrow{X} PhCMe_2O^+ + XOH^-$$

$$PhCMe_2O^+ \longrightarrow Ph\overset{+}{O}CMe_2$$

$$Ph\overset{+}{O}CMe_2 + PhCMe_2OOH \longrightarrow PhCMe_2\overset{+}{O} + PhOH + Me_2CO$$

$$Ph\overset{+}{O}CMe_2 + XOH \longrightarrow PhOH + Me_2CO + X$$

There have been several suggestions as to the nature of the ionic catalyst (X). Larson (8) proposed that the catalytic agent is formed by thermal decomposition of the complex since the more readily decomposed zinc complexes are the more efficient antioxidants. Burn (19) has, however, discounted this mechanism because ionic decomposition occurs at temperatures where the complexes are thermally stable. Scott (10, 62) favours sulphur dioxide formation with sulphur containing antioxidants and quantitative production of $NiSO_4$ and $CuSO_4$ from nickel and cupric complexes appears to support this mechanism (50, 51). In most cases, however, heterolytic decomposition of cumene hydroperoxide is achieved by a heterogeneous catalyst and the catalyst does not have the ability to decompose *t*-butyl hydroperoxide. These observations are not entirely consistent with decomposition by SO_2 or a related inorganic acid.

It has been suggested that the radical cation $ML_x^{+\cdot}$ is the catalyst responsible for heterolytic decomposition

$$ML_x^{+\cdot} + ROOH \longrightarrow ML_x + RO^+ + OH^\cdot$$

This mechanism receives some support from the fact that free-radical scavenging antioxidants inhibit heterolytic decomposition perhaps by interfering with the reaction of ML_x with $RO_2^\cdot$ to give $ML_x^{+\cdot}$. It is also conceivable that this radical cation could react with the hydroxylic oxygen of cumene hydroperoxide but not cause C-O heterolysis of *t*-butyl hydroperoxide because of steric hindrance.

The mechanism for heterolytic hydroperoxide decomposition shown in Scheme VI could perhaps account for the difference between cumene and *t*-butyl hydroperoxides. This mechanism is, however, not consistent with the heterogeneous nature of the catalyst or the observation that the complex is destroyed before the appearance of heterolytic products.

The observation that $RO_2^\cdot$ reacts with $Cu[R_2NCS_2]_2$ to give ligands with the SO_2 function is an obvious route to a heterogeneous catalyst and warrants further investigation.

EPILOGUE

Although many facets of the chemistry of inhibition of hydrocarbon autoxidation by transition metal dialkyldithiophosphates and dialkyldithiocarbamates have been established, it is clear that a great deal of experimental work is still needed before our understanding of this class of antioxidants is as complete as our understanding of inhibition by H-atom donating alkylperoxy radical scavengers.

ACKNOWLEDGMENT

I wish to express my gratitude to Dr. K.U. Ingold, Dr. S. Korcek, Dr. G. Brunton, and Dr. R.F. Bridger for the helpful comments that they made on a preliminary version of this manuscript.

REFERENCES

1. Issued as NRCC No. 17737

2. Howard, J.A. "Free Radicals" Vol. II, J.K. Kochi, Ed., Wiley, New York, Chap. 12, 1973.

3. Ingold, K.U. *Chem. Revs. 61,* 563 (1961).

4. Ingold, K.U. *Advan. Chem. Ser. 75,* 296 (1968).

5. Mahoney, L.R. *Angew. Int. Ed. Engl. 8,* 547 (1969).

6. Adamic, K., Bowman, D.F. and Ingold, K.U. *J. Amer. Oil Chem. Soc. 47,* 109 (1970).

7. Ingold, K.U. *Chem. Soc. Spec. Publ. No. 24,* 285 (1970).

8. Larson, R. *Scientific Lubrication 10,* 12 (1958).

9. Kennerley, G.W. and Patterson, W.L. *Ind. Eng. Chem. 48,* 1917 (1956).

10. Holdsworth, J.D., Scott, G. and Williams, D. *J. Chem. Soc.* 4692 (1964).

11. Colcough, T. and Cunneen, J.I. *J. Chem. Soc.* 4790 (1964).

12. Ivanov, S.K. and Shopov, D. *Dokl. Bolg. Akad. Nauk 18,* 845 (1965).

13. Burn, A.J. *Tetrahedron 22,* 2153 (1966).

14. Ivanov, S.K. *Dokl. Bolg. Akad. Nauk 20,* 1153 (1967).

15. Ivanov, S.K. *Dokl. Bolg. Akad. Nauk 20,* 1283 (1967).

16. Shopov, D., Ivanov, S.K. and Ivanova, V.V. *Rev. Inst. fr. Pét. 22,* 1520 (1967).

17. Shopov, D. and Ivanov, S.K. *Int. Oxid. Symp. San Francisco Preprints 1,* 675 (1967).

18. Ivanov, S.K. and Kateva, I. *Dokl. Bolg. Akad. Nauk 21,* 681 (1968).

19. Burn, A.J. *Advan. Chem. Ser. No. 75*, 323 (1968).

20. Shopov, D. and Ivanov, S.K. *Iz. Otd. Khim Nauki Bulg. Akad. Nauk 2*, 619 (1970); *Chem. Abstr. 73*, 44611m (1970).

21. Liston, T.V., Ingersoll, Jr., H.G. and Adams, J.Q. *Am. Chem. Soc. Div. Petr. Preprints 14(4)*, A83 (1969).

22. Shkhiyants, I.V., Sher, V.V., Nechitarlo, N.A. and Sanin, P.I. *Neftekhimiya 9*, 616 (1969).

23. Vinogradova, V.G. and Maizus, Z.K. *Kinet. Kat. 13*, 298 (1972).

24. Ivanov, S.K. and Youritsin, V.S *Dokl. Bolg. Akad. Nauk 25*, 795 (1972).

25. Howard, J.A., Ohkatsu, Y., Chenier, J.H.B. and Ingold, K.U. *Can. J. Chem. 51*, 1543 (1973).

26. Shkhiyants, I.V., Dzyubina, M.A., Sher, V.V. and Sanin, P.I. *Neftekhimiya 13*, 570 (1973).

27. Zverev, A.N., Vinogradova, V.G., Rukhadze, E.G. and Maizus, Z.K. *Izv. Akad. Nauk SSSR Ser. Khim* 2437 (1973).

28. Vinogradova, V.G. and Zverev, A.N. and Maizus, Z.K. *Kinet. Katal. 15*, 323 (1974).

29. Vinogradova, V.G. and Zverev, A.N. *Izv. Akad. Nauk SSSR Ser. Khim.* 2217 (1975).

30. Zverev, A.N., Vinogradova, V.G. and Maizus, Z.K. *Izv. Akad. Nauk SSSR, Ser, Khim* 2224 (1975).

31. Vinogradova, V.G., Zverev, A.N., Maizus, Z.K. and Emanuel, A.N. *Neftekhimiya 15*, 397 (1975).

32. Gervitis, L.L., Zolotova, N.V. and Denisov, Ye. T. *Neftekhimiya 15*, 135 (1975).

33. Ranaweera, R.P.R. and Scott, G. *Eur. Polym. J. 12*, 825 (1976).

34. Howard, J.A. and Chenier, J.H.B. *Can. J. Chem. 54*, 382 (1976).

35. Emanuel, N.M., Maizus, Z.K. and Vinogradova, V.G.

React. Kinet. Catal. Lett. 6, 119 (1977).

36. Chenier, J.H.B., Howard, J.A. and Tait, J.C. *Can. J. Chem. 56*, 157 (1978).

37. Howard, J.A. and Tong, S.B. Unpublished results.

38. Mahoney, L.R., Korcek, S., Hoffman, S. and Willermet, P. *Ind. Eng. Chem. Prod. Res. Dev. 17*, 250 (1978).

39. Burn, A.J., Cecil, R. and Young, V.O. *J. Inst. Petrol. 57*, 319 (1971).

40. Ohkatsu, Y., Kikkawa, K. and Osa, T. *Bull. Chem. Soc. Jpn. 51*, 3606 (1978).

41. Ivanov, S.K. and Kateva, I. *Neftekhimiya 11*, 290 (1971).

42. Ivanov, S.K. and Kateva, Y.D. *Dokl. Bolg. Akad. Nauk 25*, 645 (1972).

43. Ivanov, S.K., Kateva, I. and Shopov, D. *Neftekhimiya 12*, 606 (1972).

44. Ivanov, S.K., Yuritsin, V.S. and Shopov, D. *Collect. Czech. Chem. Commun. 37*, 3284 (1972).

45. Kateva, J. and Ivanov, S.K. *Erdöl Kohle 26*, 77 (1973).

46. Kateva, Y.D. and Ivanov, S.K. *Dokl. Bolg. Akad. Nauk 28*, 83 (1975).

47. Ivanov, S.K. and Kateva, J.D. *J. Polym. Sci. Symposium 57*, 237 (1976).

48. Kateva, J. and Ivanov, S.K. *Izv. Khim. 10*, 535 (1977).

49. Ivanov, S. and Kateva, Y. *React. Kinet. Catal. Lett. 6*, 243 (1977).

50. Howard, J.A. and Chenier, J.H.B. *Can. J. Chem. 54*, 390 (1976).

51. Chenier, J.H.B., Howard, J.A. and Tait, J.C. *Can. J. Chem. 55*, 1644 (1977).

52. Howard, J.A. and Tait, J.C. *J. Org. Chem. 43*, 4279 (1978).

53. Howard, J.A. and Tait, J.C. *Can. J. Chem. 56*, 164 (1978).

54. Brunton, G., Gilbert, B.C. and Mawby, R.J. *J. Chem. Soc. Perkin Trans. 2*, 650 (1976).

55. Rossi, E. and Imporato, L. *Chem. Ind. 53*, 838 (1971).

56. Burn, A.J. and Smith, G.W. *J. Chem. Soc. Chem. Commun.* 394 (1965).

57. Howard, J.A. and Tait, J.C. Unpublished results.

58. Bacon, W.E. and Lesuer, W.M. *J. Am. Chem. Soc. 76*, 640 (1954).

59. Oae, S., Nukanishi, A. and Tsujimoto, N. *Tetrahedron 28*, 2981 (1972).

60. Mahoney, L.R. and Ferris, F.C. *J. Am. Chem. Soc. 85*, 2345 (1963).

61. Chien, J.C.W. and Boss, C.R. *J. Polym. Sci. Part A-1 10*, 1579 (1972).

62. Scott, G. *Br. Polym. J. 3*, 24 (1971).

Frontiers of Free Radical Chemistry

ALIPHATIC AND AROMATIC FREE-RADICAL HALOGENATIONS

JAMES G. TRAYNHAM

Department of Chemistry
Louisiana State University
Baton Rouge, Louisiana

The effect of substituents on the rate or course of a reaction has often been an illuminating line of inquiry about details of reaction mechanism. Halogen substituents sometimes provide a mixture of illumination and irritation because of the mixture of (opposing) inductive and resonance effects they may exhibit. The well-known, fence-straddling position of halogen substituent effects in electrophilic aromatic substitution reactions is illustrative. I want to review the effects of a halo substituent on two free-radical halogenation reactions: substitution in chloroalkanes and substitution in bromobenzenes. In both of these studies, the effect of the halo substituent on the course of the reaction has provided the illumination about details of reaction intermediates or transition states that is the goal of any reaction mechanism study.

FREE-RADICAL SUBSTITUTION IN CHLOROALKANES

Free-radical substitutions in substituted alkanes proceed by separate steps of hydrogen abstraction and subsequent conversion of the alkyl radical intermediate to product. A chloro substituent retards hydrogen abstraction at vicinal secondary positions relative to more remote ones; for example, free-radical chlorination of 1-chlorobutane produces a mixture containing about twice as much 1,3-dichlorobutane as 1,2-dichlorobutane (1). But such selected data may easily be misused (too generalized) or misleading, in much

Relative Rates of Free-Radical Halogenations

$$CH_3\text{-}\underset{\uparrow \atop 1.0}{CH_2}\text{-}\underset{\nwarrow \atop 0.48}{CH_2}\text{-}CH_2\text{-}Cl$$

ISBN 0-12-566550-4

the same way as bond dissociation data have been on the question of relative stabilities of alkyl free radicals (2). Rüchardt has pointed out that the significant decrease in C-H bond dissociation energy along the series CH_3-H to $(CH_3)_3C$-H has been taken to imply stabilization of the radical by alkyl groups, while the near sameness of the C-O bond dissociation energy along the series CH_3-OH to $(CH_3)_3C$-OH and CH_3-OCH_3 to $(CH_3)_3C$-OCH_3 implies that alkyl groups do not affect radical stability significantly (2).

Some years ago, we obtained experimental results which were inconsistent with the prevailing idea that neighboring chloro retards free-radical attack at a vicinal position. As well as I remember, our experiment was, like many others in the history of science, run for the wrong reason, but that did not deter us from using the interesting data. Photoinitiated bromination of trans-1-chloromethyl-4-methylcyclohexane (**1**) produced, not the bromo product resulting from attack at the remote tertiary site, but the one (**2**) resulting from attack at the vicinal one (3).

Cl $\xrightarrow[\text{light}]{Br_2}$ Br, Cl

1 **2**

Was this apparent, favorable effect of neighboring chloro associated with the cyclohexane ring in some way, or was it the first example of a more general effect? We examined a group of acyclic systems in which internal competition between two tertiary sites, remote and vicinal to a chloro substituent, could be assessed. Our major attention focused on 1-chloro-2,3-dimethylbutane (**3**)(4). Photoinitiated bromination of that compound produced a mixture of products (**4**,**5**) in which bromination vicinal to chloro exceeded more remote bromination. Because these initial products were further brominated, and their ratio changed with time of reaction, considerable careful work by my collaborators was necessary before the initial implication of the product ratio could be established in a convincing way.

Vicinal bromination (**4**) occurs about seven times as rapidly as does more remote bromination (**5**). This result is clearly opposite to that reported for bromination of 1-chlorobutane (and other chloroalkanes with secondary vicinal positions) and is confirmed by other intermolecular competitive brominations between selected chloroalkanes and alkanes of corresponding structure (see Table I). Although chloro substituent retards free-radical attack (hydrogen abstraction) at a secondary vicinal position, it enhances free-radical attack at a tertiary vicinal position by a factor of about 4.

Table 1. Relative Rates of Hydrogen Abstraction by Bromine Atoms[a]

Reactants A	Reactants B	Temp. °C	(k_A/k_B) per H
Cl		26	0.29[b]
Cl		26	0.43[c]
Cl		32	3.91[d]
Cl		32	4.40
Cl		19	3.61

[a] *Data from Reference 4.* [b] *Average for all secondary positions.* [c] *Relative reactivity of vicinal CH_2 in A to CH_2 in B.* [d] *The relative reactivity of the 2 position in A.*

Skell has reported that the 2-H in 1-chlorobutane is about 0.11 as reactive as the 2-H in propane (30°)(5). If this retarding effect is taken as a guide to expectations in the tertiary systems we studied, the accelerating effect of neighboring chloro on bromine atom attack at a tertiary position is about 35-40. By comparison, the accelerating effect of a vicinal bromo substituent on attack at a tertiary site is 2-7 times larger.

The larger accelerating effect of bromo on vicinal hydrogen abstraction causes the effect to be discerned in the observed rate data for both intra- and intermolecular competitions at both tertiary and secondary sites (but not at primary sites). The smaller accelerating effect of chloro is discerned in the observed rate data for competitions only at tertiary sites, although calculations of "corrected" acceleration factors from the observed rate data (4) and inductive attenuation factors (5) imply weak acceleration effects (about 2 times) at secondary sites also.

The anchimeric effects of both chloro and bromo substituents on these hydrogen abstraction reactions increase as the incipient radical changes from primary to secondary to tertiary alkyl. A large volume of data in the literature on rates of carbocation reactions leads us to expect the opposite order; *i.e.*, that neighboring group participation will be less significant as we go from primary to secondary to tertiary alkyl reactions sites. How can we account for the neighboring halo participation and for the observed order of effectiveness?

C-H bond breaking in the transition state for hydrogen abstraction by bromine atoms (a slightly endothermic process) is not far advanced; *i.e.*, radical character is developed but little. Polar contributions -- electron depletion from the alkyl center to the bromine -- will stabilize the transition state and will become more significant as the alkyl site is better able to suffer electron depletion (primary<secondary<tertiary). Bromo is larger and more polarizable than chloro and is therefore better able to interact with a weakly deficient, vicinal carbon center. Only when the incipient alkyl radical is tertiary are the polar contributions sufficient for a neighboring chloro to have a net accelerating effect.

The small amount of C-H bond breaking in the bromination reactions is important, even critical, for the polar contributions and therefore for the neighboring halo assistance to show up. For example, the radical character has substantially fully developed at the transition state for thermolysis of symmetrical azoalkanes (C-N bond nearly fully broken), and structural features expected to stabilize polar contributions in the transition state (and thereby increase

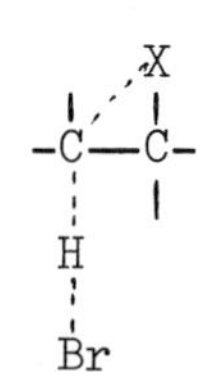

Kinetic assistance appears

X = Br -- secondary

X = Cl -- tertiary

the rate of reaction) have negligible effects (2). Too much bond breaking at the transition state eliminates the significance of polar contributions to the stability of the radical transition state (2).

In summary, the factors which affect neighboring substituent participation (anchimeric assistance) in free-radical hydrogen abstraction reactions are:

(1) *The degree of C-H bond breaking in the transition state.* The transition state must not be too early [too little C-H bond breaking; *e.g.*, attack by Cl· (1)] nor too late [nearly complete radical character developed; *e.g.*, thermolysis of azoalkanes (2)].

(2) *The polarizability of the vicinal substituent.* Bromo bridges better than chloro, and the effect shows up "sooner".

(3) *Structural features favorable to the carbocation condition.*

No one factor alone is enough. Attack by Br· on a bromoalkane meets the first two conditions (C-H bond breaking and polarizability of substituent), but participation by vicinal bromo does not show up in attack at a primary site (6). Benzyl cation is more stable than cyclohexyl cation, but cyclohexane undergoes free-radical chlorination (little C-H bond breaking) faster than does toluene (the reverse is true for bromination) (7).

Thinking about neighboring group participations and polar contributions to the transition states in free-radical hydrogen abstractions as a continuum, then, may usefully guide our expectations about substituent effects on these reactions. For example, the combination of factors might allow vicinal bromo participation to be discerned in attack at a tertiary site by Cl·. The absence of data on this point illustrates the need for continuing balance between availability of chemical coworkers and administrative duties for neighboring group effects to be discerned in a timely fashion.

FREE-RADICAL SUBSTITUTION IN BROMOBENZENES

Now, I want to turn attention to a different area, free-radical aromatic substitution, related to the first only by its illumination, also, of an unexpected effect of a halo substituent. Substituent effects in aromatic substitution reactions have been a focus of one of the major areas of reaction mechanism studies and a basis for a substantial part of organic chemical theory. In spite of the large volume of data and literature on aromatic substitution reactions, and well-known examples of replacements of substituents during those reactions, nearly all discussions, until recently, were concerned only with hydrogen replacements at positions ortho, meta, or para to a substituent. During the past decade, the importance of attack at the ipso (substituted) position during electrophilic reactions has become appreciated (8). Cationic ipso intermediates have been trapped, and their rearrangements have been shown to account for as much as half of the ortho-substitution products obtained in some nitrations (2). Replacement of electronegative substituents rather than hydrogen characterizes nucleophilic aromatic substitutions. Although replacements of a substituent during free-radical aromatic substitutions have been reported by various investigators during the past 76 years, mechanistic speculations have generally been meager or have treated ipso attack and intermediates as unlikely. The current recognition of the significance of ipso intermediates in electrophilic reactions makes them seem more likely in free-radical reactions than they were earlier thought to be. We have obtained evidence for the formation and rearrangement of an ipso intermediate during free-radical chlorination of *p*-bromonitrobenzene (9).

Formation of chlorobenzene from bromobenzene appears to be the first report of a free-radical ipso substitution reaction (10), and halogen exchange during free-radical reactions has been rediscovered from time to time (11,12).

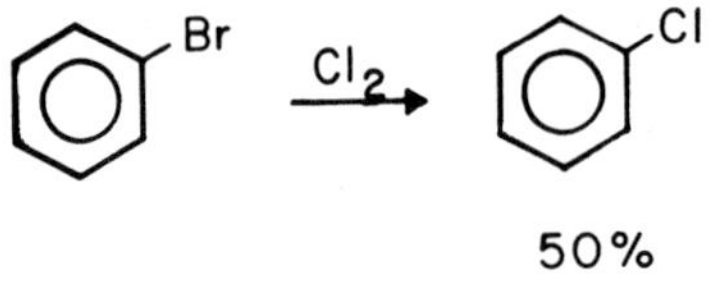

I became interested in this area of research by the chance reading of a *Tetrahedron Letters* report (12) of one of those rediscoveries while preparing for a lecture to a class of graduate students on ipso-nitration. The brief paper reported a substantial amount of data but little speculation about mechanism. The extensive amount of exchange reported for photobromination of dihalobenzenes together with my mistaken perception that the authors also implied some rearrangement of the original halo substitution seemed to fit the mechanistic picture of ipso-nitration, on which I was focusing that afternoon. I included the free-radical bromination data, and the ipso interpretation, in the lecture that night. This stimulus became, for me, a superb example of the contribution of teaching to research (we usually emphasize and hear about the converse).

Subsequently, a review of the literature, aided by friends who heard me talk about ipso substitution and called potentially-interesting references to my attention, revealed a variety of free-radical ipso-substitutions in addition to halogen exchanges (13). Representative of that variety are reports of replacement of halo substituent by aryl (14), alkyl (15), and phenylthiyl (16) radicals; of methoxy by an aryl radical (17); and of a variety of 2-substituents in 1-thia-3-azaindenes (**8**)(benzothiazoles) by adamantyl and acetyl radicals (18). Lead tetraacetate oxidation of several 2'-substituted 2-biphenylcarboxylic acids (**9**) leads to loss of the 2'-substituent as a lactone (**10**) is formed (19), and similar results have been found in mass spectral studies of some 2'-substituted 2-biphenylcarboxylic acids (20).

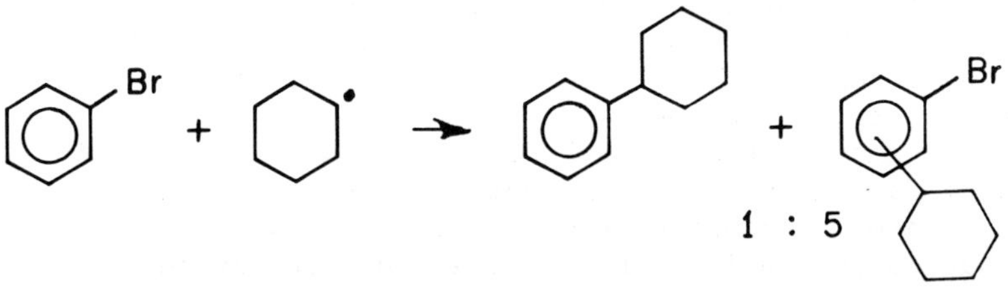

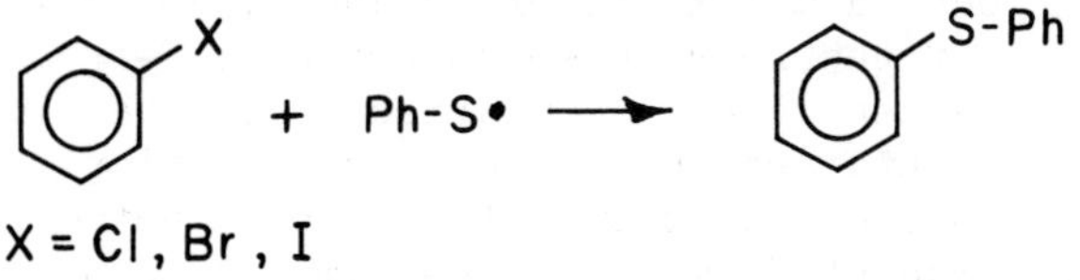

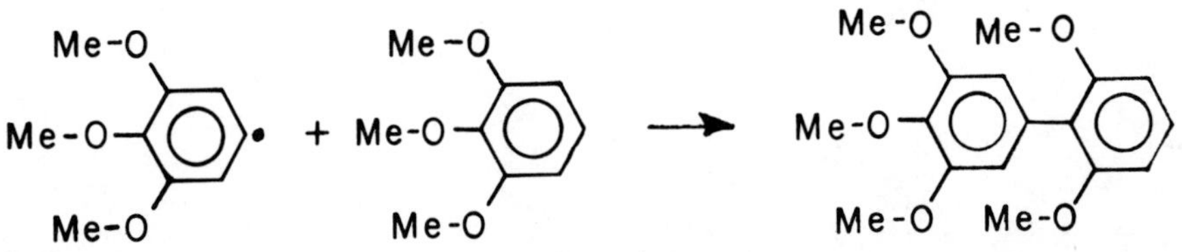

1 : 5

X = Cl, Br, I

In some but not all of these reports the concept of an ipso intermediate (or transition state) was considered, but the term "ipso" was seldom used. Mechanisms other than ipso attack by a free radical to form an ipso intermediate are conceivable, and have been advocated, in many of these cases. Replacement of a substituent is the only result reported for apparent attack at the substituted site, and that behavior alone does not require an ipso intermediate. Rearrangement of the original substituent, paralleling the behavior established for the ipso intermediate in electrophilic nitration, would, we believed, firmly establish the occurrance of an ipso intermediate in a free-radical aromatic substitution reaction.

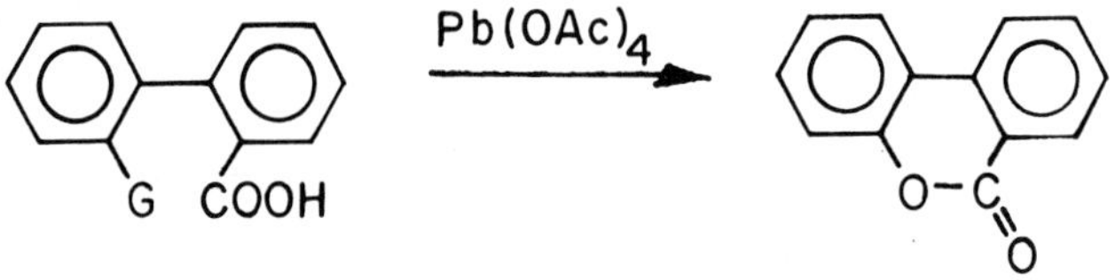

We chose *p*-bromonitrobenzene (11) as a reactant likely to yield rearrangement product. The choice may, again, have been based on doubtful reasoning, but it was, nonetheless, a successful choice. Photoinitiated chlorination of *p*-bromonitrobenzene in carbon tetrachloride solution at room temperature produced a mixture of products, among which was a small amount of the rearrangement product, 2-bromo-1-chloro-4-nitrobenzene (14). Comparison with the simpler product mixture obtained under the same conditions from *p*-chloronitrobenzene (15) showed that the amount of rearrangement product formed from *p*-bromonitrobenzene was most likely about 35 times more than the amount detected in the gc analysis (9).

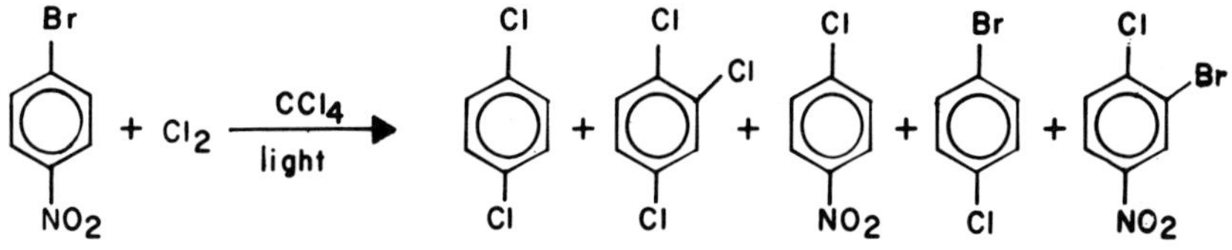

Photoinitiated chlorination of *p*-chloronitrobenzene produces mainly *p*-dichlorobenzene (12) and 1,2,4-trichlorobenzene (13) and detectable amounts of isomeric dichloronitrobenzenes. The ratio of dichlorobenzene:trichlorobenzene was only 1/4 to 1/3 of what it was from *p*-bromonitrobenzene. This difference implies a more favorable route to 1,2,4-trichlorobenzene (13) from *p*-bromonitrobenzene than from *p*-chloronitrobenzene; that is, the trichlorobenzene is produced by a route other than chlorination of *p*-dichlorobenzene (12). We believe that further, facile ipso substitution in the rearrangement product, 2-bromo-1-chloro-4-nitrobenzene (14), provides that route.

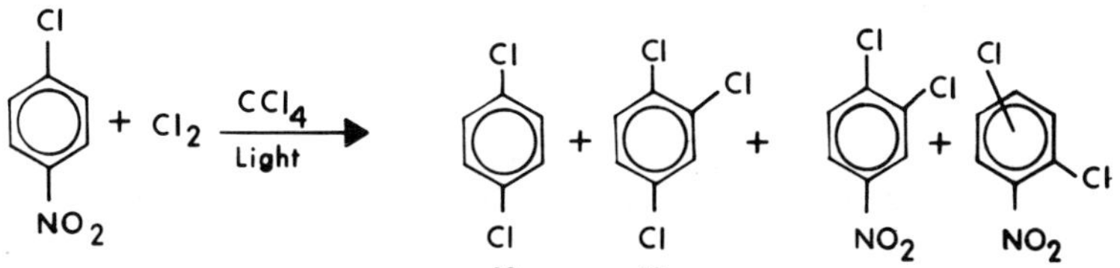

Competitive chlorinations of pairs of substituted halobenzenes (for example, 2-bromo-1,4-dichlorobenzene and *p*-chloronitrobenzene) showed that bromo is much more readily replaced than chloro and that the trisubstituted benzenes are much more reactive than the disubstituted ones. These data support the inference above about the extensive conversion of 2-bromo-1-chloro-4-nitrobenzene (**14**) to 1,2,4-trichlorobenzene (**13**) (9).

Both *p*-bromonitrobenzene and *p*-chloronitrobenzene undergo ipso substitution during photoinitiated chlorination more readily and extensively than any other substitution. The detection of rearrangement product and its implication in the more extensive formation of 1,2,4-trichlorobenzene from *p*-bromonitrobenzene makes a mechanism involving an ipso intermediate (**16**) seem the most likely one. If any of the rearranged intermediate (**17**) loses Br· (to form *p*-chloronitrobenzene) rather than H· (to form rearrangement product, 2-bromo-1-chloro-4-nitrobenzene), the extent to which the formation of that intermediate is perceived will be reduced by that amount (9).

Concurrent with our experimental work, and stimulated by my proposal of the ipso intermediate mechanism, my colleague, R. D. Gandour, calculated the relative stabilities of isomeric intermediates in some prototype aromatic substitution reactions (cationic, free radical, and anionic) (21). By MINDO/3 calculations, ipso intermediates are relatively more favored (compared to ortho, meta, and para ones) in free radical reactions than in cationic ones (21), for which they were first established.

We believe that these data establish the formation and rearrangement of an ipso intermediate in these aromatic free radical chlorinations and strongly imply that similar intermediates play significant roles in other aromatic free radical substitution reactions.

At least one problem remained with the product data in the publication (12) that prompted my interest in this area of research, and puzzling over those data led to insight about an unexpected detail of reaction mechanism. Some of the photobrominations of chlorobenzenes led to both bromo-(exchange) and chloro-substitution products in about equal yields (12). Further, while 1,2-dichlorobenzene (**18**) gave 1-bromo-2-chlorobenzene and trichlorobenzene(s), 1-bromo-2-chlorobenzene (**19**) gave 1,2-dibromobenzene and 1,2-dichlorobenzene (not bromodichlorobenzene) (12). Chlorine can hardly be expected to compete with a much larger amount of bromine in attack on an aromatic substrate, because "equilibrium favors bromine atoms over chlorine atoms by a factor of 10^4 or more," (11f) but a scheme involving transfer of a geminal halogen from an ipso intermediate (**20** or **21**) to a molecule of aromatic substrate, without the generation of chlorine atoms (or molecules) in the reaction mixture, accounts attractively for these results. When Cl is abstracted by the substrate, the exchange product plus a chlorinated product are formed in equal amounts. A substantial body of data in

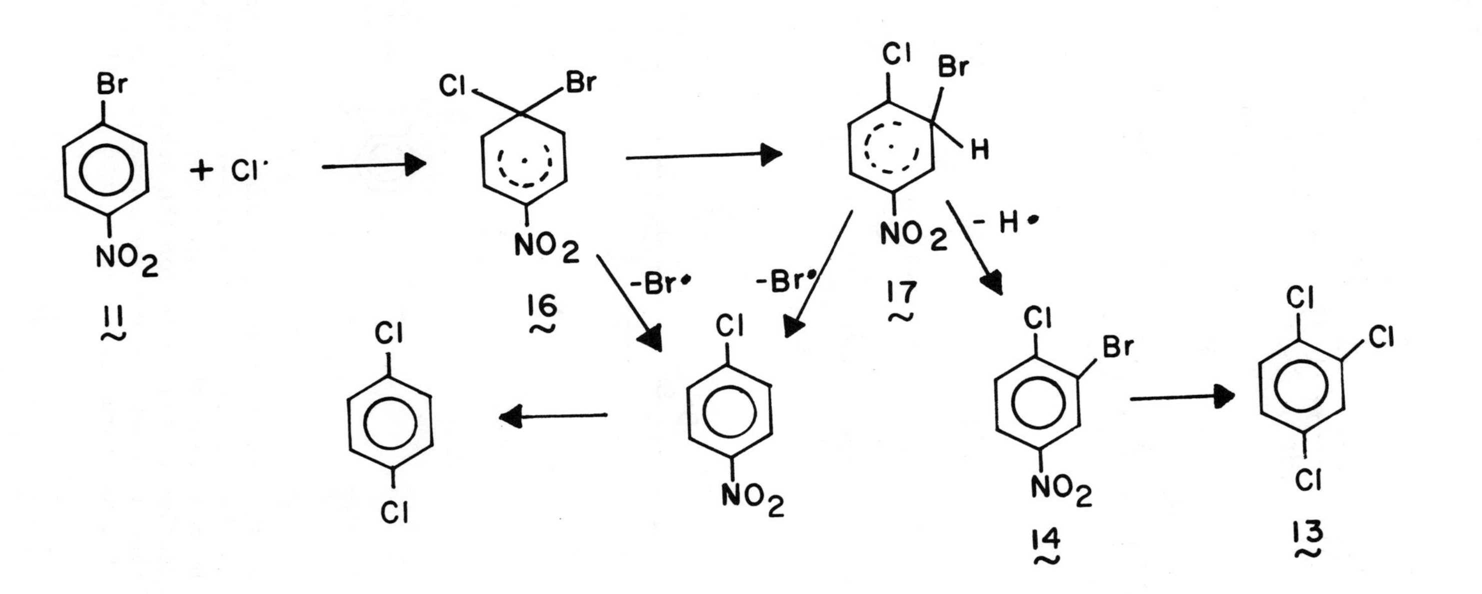

Br
NO2
11
+ Cl·
Cl
Br
NO2
16
-Br•
Cl
Br
H
NO2
17
-Br•
- H•
Cl
Cl
Cl
NO2
Cl
Br
NO2
14
Cl
Cl
Cl
13

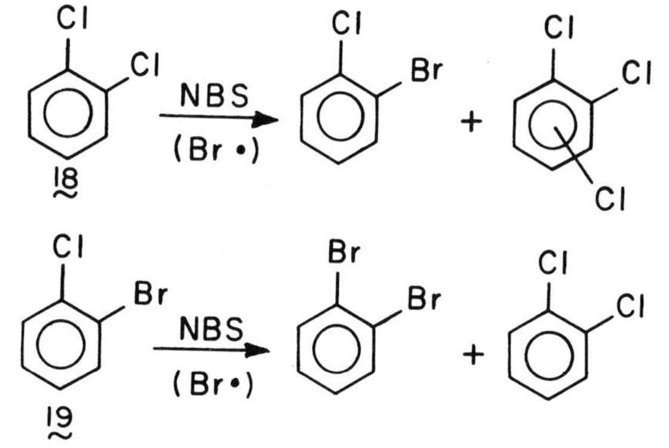

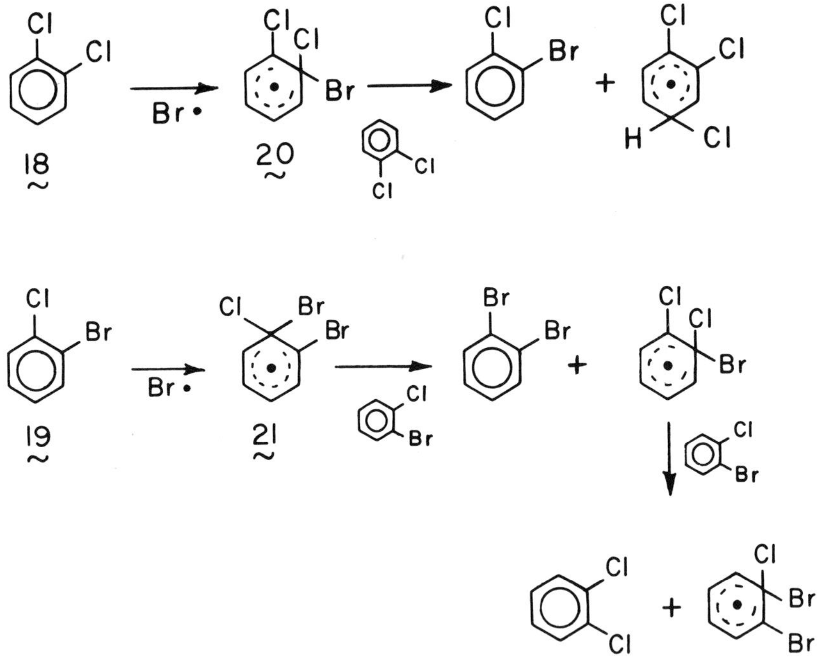

the literature (13) indicates that, in some cases, a geminal substituent must be abstracted from a free-radical ipso intermediate by some reagent, but this scheme is, I believe, the first suggestion of this role for the substrate itself.

To paraphrase a popular beer advertisement, when you've said ortho, meta, and para, you haven't said it all.

I happily and appreciatively acknowledge the significant contributions of my coworkers, Drs. Fritz Schweinsberg and Charles R. Everly, who carried out the experimental work discussed here.

REFERENCES

(1) Thaler, W. A., in "Methods of Free Radical Chemistry," Vol. 2, Huyser, E. S., Ed.; Marcel Dekker: New York, N.Y., 1969, p. 167.

(2) Rüchardt, C. *Angew. Chem. internat. Edit. 9,* 830-43 (1970).

(3) Traynham, J. G., and Hines, W. G. *J. Am. Chem. Soc. 90,* 5208-10 (1968).

(4) Everly, C. R., Schweinsberg, F., and Traynham, J. G. *J. Am. Chem. Soc. 100,* 1200-5 (1978).

(5) Shea, K. J., Lewis, D. C., and Skell, P. S. *J. Am. Chem. Soc. 95,* 7768-76 (1973).

(6) Thaler, W. *J. Am. Chem. Soc. 85,* 2607-13 (1963).

(7) Russell, G. A., and Brown, H. C. *J. Am. Chem. Soc. 77,* 4578-82 (1955).

(8) For a review, with references, see Moodie, R. B., and Schofield, K. *Acc. Chem. Res. 9,* 287-92 (1976).

(9) Everly, C. R., and Traynham, J. G. *J. Am. Chem. Soc. 100,* 4316-7 (1978); *J. Org. Chem. 41,* 1784-7 (1979).

(10) Eibner, A. *Ber. 36,* 1229-31 (1903).

(11) (a) Goerner, G. L., and Nametz, R. C. *J. Am. Chem. Soc. 73,* 2940-1 (1951); (b) Voegtli, W., Muhr, H., and Lauger, P. *Helv. Chim. Acta 37,* 1627-33 (1954); (c) Miller, B., and Walling, C. *J. Am. Chem. Soc. 79,* 4187-91 (1957); (d) Milligan, B., Bradow, R. L., Rose, J. E., Hubbert, H. E., and Roe, A. *ibid. 84,* 158-62 (1962); (e) Milligan, B., and Bradow, R. L. *J. Phys. Chem. 66,* 2118-20 (1962); (f) Echols, J. T., Chuang, V. T-C., Parrish, C. S., Rose, J. E., and Milligan, B. *J. Am. Chem. Soc. 89,* 4081-8 (1967).

(12) Gouveneur, P., and Soumillion, J. P. *Tetrahedron Lett.* 133-6 (1976).

(13) For a review, see Traynham, J. G. *Chem. Rev. 79,* 323-330 (1979).

(14) (a) Claret, P. A., Williams, G. H., and Coulson, J. *J. Chem. Soc. C,* 341-4 (1968); (b) Nonhebel, D., and Walton, J. C. "Free Radical Chemistry; Structure and Mechanism"; Cambridge University: Cambridge, England, 1974, p. 464.

(15) (a) Leffler, J. E. "The Reactive Intermediates of Organic Chemistry"; Interscience: New York, N.Y., 1956, p. 19; (b) Shelton, J. R., and Uzelmeier, C. W. *Recl. Trav. Chim. Pays-Bas. 87,* 1211-6 (1968); (c) Shelton, J. R., and Lipman, Jr., A. L. *J. Org. Chem. 39,* 2386-93 (1974).

(16) Benati, L., Camaggi, C. M., and Zanardi, G. *J. Chem. Soc., Perkins Trans. 1,* 2817-9 (1972).

(17) Paulson, P. L., and Smith, B. C. *J. Org. Chem. 18,* 1403-5 (1953).

(18) Fiorentino, M., Tetaferri, L., Tiecco, M., and Troisi, L. *J. Chem. Soc., Chem. Commun.* 316-7, 317-8 (1977).

(19) Davies, D. I., and Waring, C. *J. Chem. Soc. C,* 1639-42 (1967).

(20) (a) Thomas, C. B., and Willson, J. S. *J. Chem. Soc., Perkins Trans. 2,* 778-82 (1972); (b) Sheley, C. F., and Patterson, R. T. *Org. Mass Spectrom. 9,* 731-4 (1974).

(21) (a) Gandour, R. D., and Traynham, J. G. 173rd National Meeting of the American Chemical Society, New Orleans, LA, March 1977; American Chemical Society: Washington, D.C.; Abstr. ORGN 137. (b) Gandour, R. D. *Tetrahedron,* accepted for publication (1980).

STRUCTURAL AND MECHANISTIC ASPECTS OF METAL COMPLEXES IN FREE RADICAL CHEMISTRY

Jay K. Kochi

Department of Chemistry, Indiana University
Bloomington, Indiana

Summary

The role of metal complexes in free radical chemistry is considered in four perspectives. Part I: Structural Types of Metallo-Radicals includes metal-centered radicals, their structures by esr spectroscopy and energetics of formation from diamagnetic organometals by photoelectron spectroscopy, and carbon-centered radicals, especially with regard to bridging. Part II: Interactions of Free Radicals with Metal Complexes considers mechanisms by which radical reactions occur either at the metal center or at the ligand site. Part III: Electron Transfer and Charge Transfer Processes involves organometals as electron donors in their reactions with outer-sphere and inner-sphere oxidants, like iron(III) and iridium(IV), respectively, as well as in charge transfer processes with tetracyanoethylene, focussing on the properties of organometallic radical-ions. Part IV: Redox Chain Reactions in Metal Catalyzed Processes considers the various catalytic processes involving radical intermediates in oxidation-reduction reactions with metal complexes.

Introduction

The importance of metal complexes to free radical processes lies in their ability to undergo facile reactions resulting largely from the availability of multiple oxidation states, particularly of transition metals. Thus metal complexes such as those of copper(II), cobalt(II) and iron(III) are efficient scavengers of alkyl radicals since these are themselves paramagnetic species with d^9, d^7 and d^5 electron configurations, respectively. For example, the association of a methyl radical with hexaaquocopper(II) to form the metastable methylcopper(III) species:

$$CH_3\cdot + Cu(II)_{aq} \xrightarrow{k} CH_3Cu(III)_{aq} \quad (1)$$

ISBN 0-12-566550-4

is rapid, occurring with a second-order rate constant k of $7.4 \pm 0.6 \times 10^5\ M^{-1}\ sec^{-1}$.(1) The formation of the carbon-metal bond in the equation above is formally considered to be an oxidative addition, since an alternative route to the same species should be available from a methyl cation and copper-(I) or a methyl anion and copper(III), i.e.,

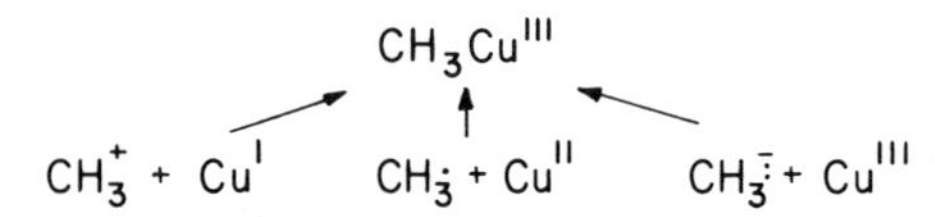

The formalism above emphasizes how the chemistry of free radicals can be related to the more conventional ionic processes. Furthermore, the concept of free radicals as reactive intermediates in chemical reactions is expanded considerably if paramagnetic species centered on the usual main group elements--carbon, oxygen, nitrogen, halogen, etc.--are extended so as to include the metallic elements. In order to pursue this point, we wish to describe some possible, new dirrections for the expansion of the frontiers of traditional free radical chemistry by the use of metals and metal complexes. For convenience in the presentation, the focus here will be on four general areas encompassing: (1) structural types of metallo-radicals, (2) reactions of free radicals with metal complexes, (3) organometal complexes as electron donors in the production of radicals and (4) metal-catalyzed reactions involving free radical intermediates.

I. Structural Types of Metallo-Radicals

If a metal center is included in a conventional organic radical, two principal classes of structures, designated as (1) metal-centered radicals or (2) carbon-centered organometal radicals can be differentiated by the principal site of odd electron density.

A. Metal-Centered Radicals

When there is more than one unpaired electron, as in transition metal complexes, the possibility of zero-field splitting arises and a variety of high spin paramagnetic metal complexes such as those containing nickel(II) (triplet), iron(II) (quintet), etc. are known. However, in considering metal-centered radicals we wish to focus primarily on only those radicals which are either unstable or are transient. Among metal-centered radicals, a variety of neutral uncharged organometal species are known, such as:

$$Et_3Sn,\ CH_3Hg,\ t\text{-}BuOP(CH_3)_2,$$
$$PhSAu(CH_3)PPh_3 \text{ and } ArNiBr_2(PEt_3)_2$$

which are formally related to the parent paramagnetic metal ions: Sn^{3+}, Hg^{+}, P^{4+}, Au^{2+} and Ni^{3+}, respectively. Such species are usually formed by either homolytic fission of a bond to ligand or by oxidative addition of a radical to a diamagnetic metal precursor. Thus trialkylstannyl radicals are intermediates in the radical chain reduction of organic halides RX by trialkylstannanes, in which the propagation cycle includes atom transfer as in:(2)

$$Et_3Sn\cdot + RX \longrightarrow Et_3SnX + R\cdot \quad (2)$$

$$R\cdot + Et_3SnH \longrightarrow RH + Et_3Sn\cdot \text{, etc.} \quad (3)$$

In an alternative way, phosphoranyl(IV) radicals are readily generated by the oxidative addition of various paramagnetic species to phosphines, e.g.,(3)

$$t\text{-}BuO\cdot + P(CH_3)_3 \longrightarrow t\text{-}BuO\dot{P}(CH_3)_3 \quad (4)$$

The esr spectrum shown in Figure 1 consists of a large splitting of 618.5 gauss due to phosphorus which is split

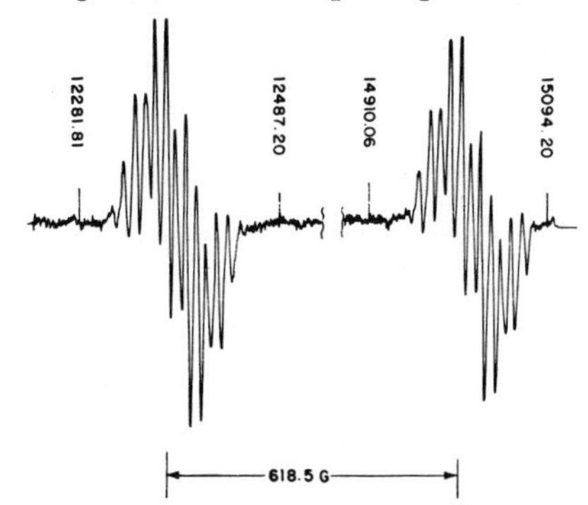

Figure 1. Esr spectrum of $t\text{-}BuOPMe_3$ by the oxidative addition of t-butoxy radical to trimethylphosphine.

further into a complex multiplet consisting of a quartet (4.6 gauss) of septets (2.8 gauss) due to two types of inequivalent methyl groups shown in the trigonal bipyramidal structure below:

(618.5) OBu^t — P — CH_3, CH_3 (2.8 G); CH_3 (4.6 G)

The same metals may also be present in a series of metal-centered radical-ions, cations as well as anions, such as:

$$(CH_3)_4Sn^{+\cdot},\ CH_3HgBr^{-\cdot},\ Ph_3P^{+\cdot},\ CH_3AuPPh_3^{+\cdot}\ \text{and}\ Ar_2Ni(PPh_3)_2^{+\cdot}$$

which are formally related to the metal ions: Sn^{5+}, Hg^{+}, P^{4+}, Au^{2+} and Ni^{3+}. Such radical-ions are usually generated in solution from their neutral diamagnetic counterparts by either electrochemical or chemical oxidation and reduction. Radical-cations are also generated in the gas phase by photoionization, e.g.,(4)

$$CH_3HgCH_3 \xrightarrow{h\nu} CH_3HgCH_3^{+\cdot} + \epsilon \qquad (5)$$

Importantly, the use of monochromatic photons, such as those from the 584 Å resonance line of a helium discharge lamp, is the basis for the photoelectron spectroscopic (pes) determination of the ionization potentials of many organometals. The latter is particularly useful for probing the effects of ligand structure. For example, the He(I) photoelectron spectra of dialkylmercury compounds show two principal bands of interest. The first vertical ionization potential, lying in a rather wide range between 7.57 eV (t-Bu_2Hg) and 9.33 eV (Me_2Hg), is included in a broad unsymmetrical band. A second, weaker band occurring between 14.4 and 15.0 eV is due to ionization from the mercury $5d^{10}$ shell. The effect of alkyl substitution on the first and second ionization potentials of dialkylmercury(II) compounds is indicated in Table I. The plot in Figure 2 shows the effect of alkyl structure on the ionization potentials of a series of homologous $RHgCH_3$, in which R represents increasing α-branching in the order: CH_3, CH_3CH_2, $(CH_3)_2CH$ and $(CH_3)_3C$.

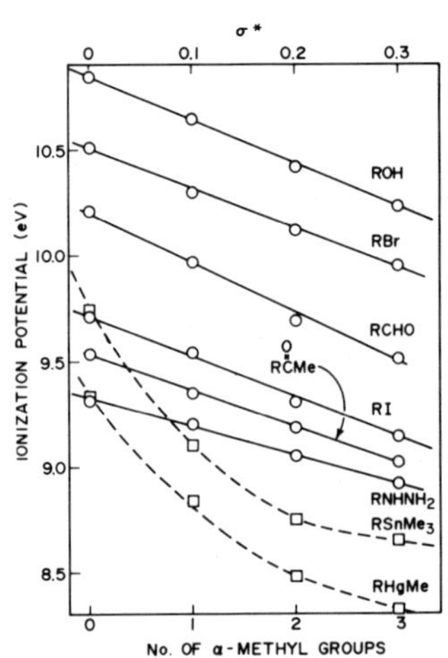

Figure 2. Effect of alkyl structure on the ionization potentials of organometals, and their comparison with other organic derivatives.

Table I. Photoelectron Spectra of Tetraalkyltin Compounds.

	SnR_4	$\Sigma\sigma^*$	IP_1 (eV)			$\overline{IP}$ (eV)
1	Me	0	9.7			9.7
2	Et	.40	8.93			8.93
3	n-Pr	.46	8.82			8.82
4	i-Pr	.76	8.46			8.46
5	n-Bu	.52	8.76			8.76
6	i-Bu	.50	8.68			8.68
7	s-Bu	.84	8.45			8.45
8	neo-Pentyl	.66	8.67			8.67
	Me_2SnR_2	$\Sigma\sigma^*$	IP_1	IP_2	IP_3	
9	Et	.20	9.01	9.28	9.64	9.31
10	n-Pr	.23	8.8	9.2	9.5	9.17
11	i-Pr	.38	8.56	8.99	9.55	9.03
12	n-Bu	.26	8.8	9.2	9.5	9.17
13	t-Bu	.6	8.22	8.74	9.47	8.81
	Me_3SnR	$\Sigma\sigma^*$	IP_1	IP_2		
14	Et	.10	9.1	9.5		9.37
15	i-Pr	.19	8.9	9.45, 9.76		9.37
16	n-Bu	.13	9.0	9.49		9.33
17	i-Bu	.125	9.05	9.50		9.35
18	t-Bu	.30	8.50	9.62		9.24
19	Et_3MeSn	.30	8.95	9.3		9.07

It is particularly noteworthy that the ionization potential decreases by more than 20 kcal mol^{-1} simply by replacing one methyl group in Me_2Hg with a t-butyl group. Such a large electronic effect occurs even without directly altering the nature of the bonding to the mercury center, and it emphasizes the importance of the donor property of alkyl ligands on the redox properties of metal complexes. The ionization process in organometals such as Me_2Hg proceeds from a molecular orbital (HOMO) that has substantial metal-carbon bonding characteristics as qualitatively portrayed in the simple linear combination of group orbital diagram:

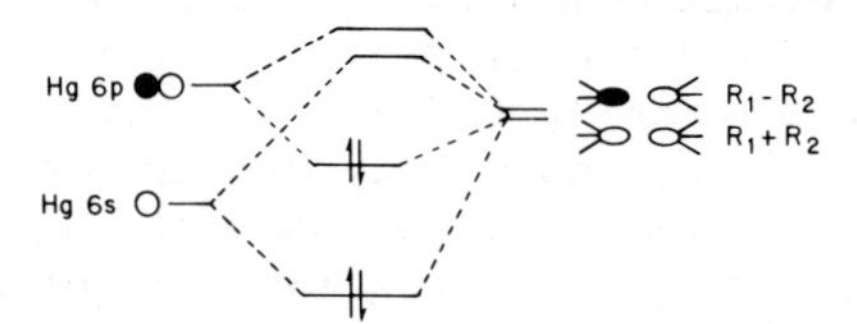

Indeed, the ionization of alkyl radicals [i.e., $R\cdot \rightarrow R^+ + \epsilon$] serves as a reasonable model for the ionization from the HOMO of organomercurials as shown in Figure 3.

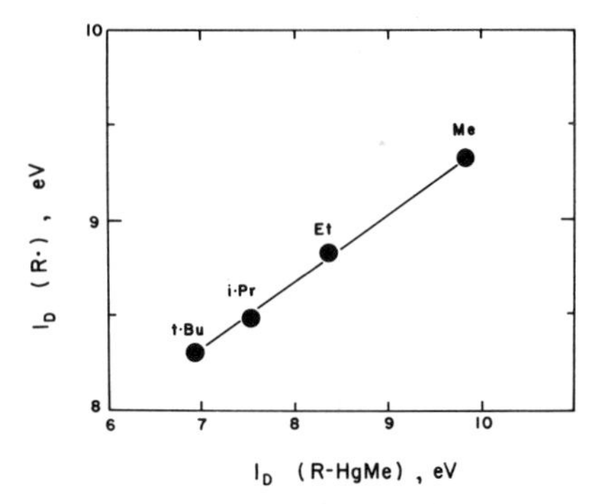

Figure 3. Structural relationships in the oxidation potentials of alkyl radicals and alkylmercury compounds (see Table I for numbering of compounds).

Essentially the same conclusion derives from the examination of the pes of the tetraalkylmetals of the Group IVA elements. Thus the vertical ionization potentials of the tetramethyl derivatives, CMe_4, $SiMe_4$, $GeMe_4$, $SnMe_4$ and $PbMe_4$ decrease monotonically in the order: 10.96, 10.57, 10.23, 9.70 and 8.81 eV, respectively,(5) indicating that ionization is associated with electrons localized relatively close to the central (metal) atom. Indeed, the lowest pes bands have been assigned to the metal-carbon σ-bonding orbitals since semiempirical calculations for tetramethyltin with tetrahedral (T_d) symmetry are in agreement with a highest occupied molecular orbital which is triply degenerate ($3t_2$).

The lowest vertical ionization potentials of three series of homologous tetraalkyltin compounds, viz., R_4Sn, $RSnMe_3$ and R_2SnMe_2, are listed in Table I.(6) For the symmetrical tetraalkyltins, R_4Sn, the values of the first vertical ionization potentials are more or less linearly related to the sums of the Taft polar substituent parameters (σ^*) of the alkyl groups, as shown by the straight line which may be drawn through these points. However, considerable scatter is encountered when the same plots of the data are attempted for the two series of the methyl-substituted analogs, viz., $RSnMe_3$ and R_2SnMe_2. In these unsymmetrical tetraalkyltins, symmetry considerations predict the band **A** of R_4Sn to be split into additional bands. In particular, for the monosubstituted derivatives $RSnMe_3$ with C_{3v} symmetry, band **A** would be split into an a_1 and doubly degenerate e set, whereas for the disubstituted analogs R_2SnMe_2 with C_{2v} symmetry, it would be split into an a_1, b_1 and b_2 set, as the correlation diagram in Figure 4 illustrates for the complete series of five methylethyltin compounds. Indeed, the experimental spectrum for Me_3SnBu^t shown in Figure 5a shows a doublet splitting with the expected 1 : 2 intensity ratio for this low

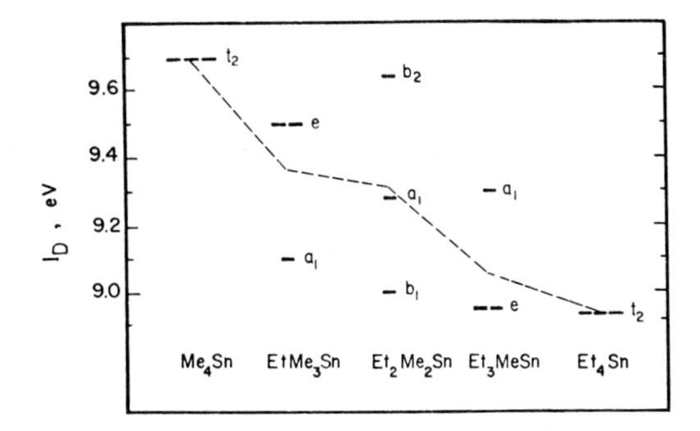

Figure 4. Correlation diagrams for the triply degenerate t_2 molecular orbital of tetramethyltin as a result of successive substitution of ethyl for methyl ligands.

energy band. It is noteworthy that a similar splitting pattern is observed with Et_3SnMe but in a reversed, 2:1 intensity ratio. Furthermore, the pes spectrum of $Me_2SnBu_2^t$ in Figure 5b shows two distinct splittings associated with the three energy levels predicted by this simple formulation.

If we take cognizance of these splittings of the HOMO of tetramethyltin, as they are induced by methyl substitutions, it would appear that the Taft σ^* parameter should correlate better with the weighted (center of gravity) average of all the

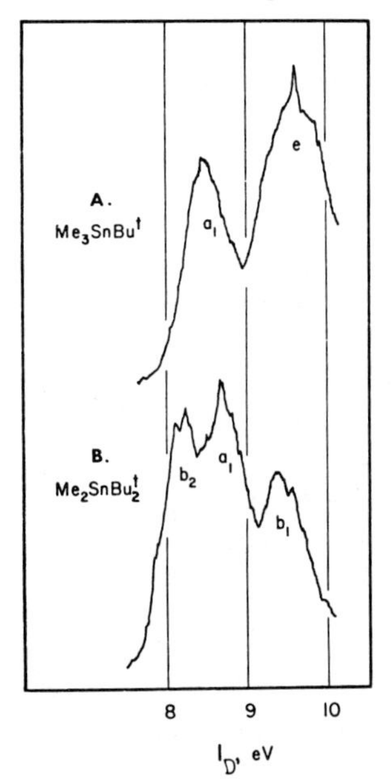

Figure 5. Typical splitting patterns of the lowest energy bands in the photoelectron spectra of unsymmetrical tetraalkyltins: Me_3SnR (upper) and Me_2SnR_2 (lower).

vertical ionization potentials included in the first band A. Such an averaging procedure is tantamount to choosing a single (imaginary) ionization potential, $\overline{IP}$, to represent each tetraalkyltin, irrespective of its substitution pattern. [The dashed line in Figure 4 is drawn through $\overline{IP}$ for each $Me_{4-n}Et_nSn$.] Indeed, Figure 6 shows that the averaged ionization potentials for all the various tetraalkyltins are now well correlated with the Taft σ^* values by a single line.

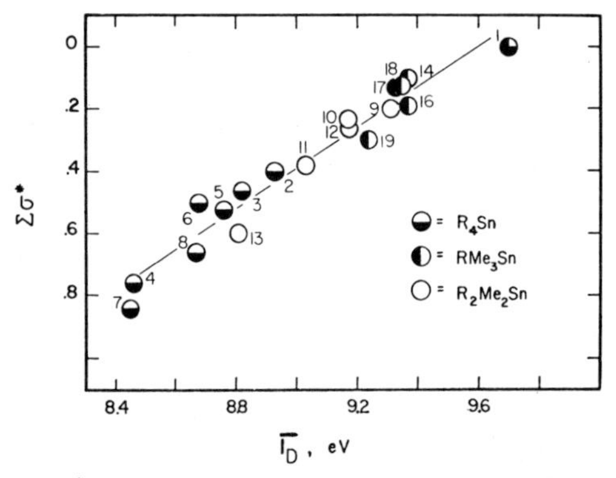

Figure 6. Linear correlation of the weighted average ionization potentials $\overline{IP}$ and Taft $\Sigma\sigma^*$ parameters for the tetraalkyltin compounds listed in Table I.

In the series of monosubstituted tetraalkyltins, $RSnMe_3$, the energy difference Δ between the e and a_1 molecular orbitals (see Figure 4) reflects the perturbation of the triply degenerate t_2 levels in tetramethyltin as a result of successive methyl substitutions at a single methyl ligand [i.e., R = CH_3, CH_3CH_2, $CH_3CH_2CH_2$, $(CH_3)_2CH$, $(CH_3)_3C$, etc.]. As such it is reasonable to expect the magnitude of this splitting to be reflected in the Taft σ^* value for R, as shown in Figure 7. It is noteworthy that the linear correlation passes through the origin, i.e.,

$$\Delta = 3.7\,\sigma^* \text{ eV}$$

in accord with this simple formulation. Thus for a series of unsymmetrical alkyltrimethyltins $RSnMe_3$, the first vertical ionization potential from $HOMO_1$ can be simply related to the ionization potential (IP_1 = 9.70 eV) of tetramethyltin from its Taft σ^* value, i.e.,

$$IP_1\,(RSnMe_3) = 9.70 - 4.3\,\sigma^*$$

Photoelectron spectroscopy of organometals is also a useful technique with which to compare the donor properties of various types of ligands. Thus, hydrogen can be evaluated as a donor ligand by comparing the ionization potentials of the metal hydrides, SiH_4 (12.36 eV) and GeH_4 (11.98 eV), with

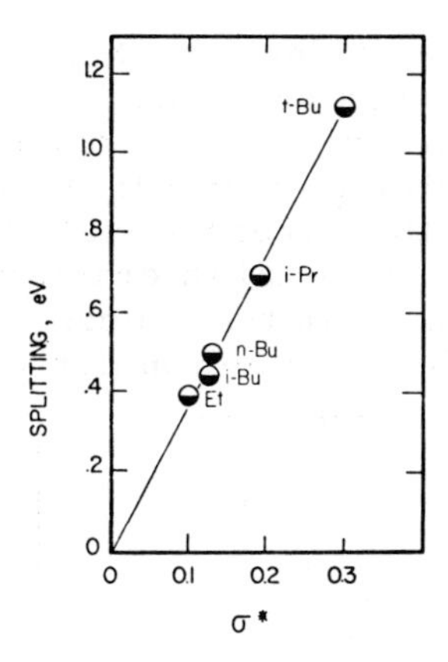

Figure 7. Electronic effects of alkyl ligands (R) measured by Taft σ* values on the splitting of the e and a_1 levels in a series of alkyltrimethyltin $RSnMe_3$ compounds. Note the extrapolation through the origin.

those of the methyl analogs, $SiMe_4$ (10.57 eV) and $GeMe_4$ (10.23 eV), or the ethyl analogs, $SiEt_4$ (9.8 eV) and $GeEt_4$ (9.4 eV).(7) It is clear from both series that hydride is a less effective donor than methyl or ethyl ligands. The difference is also clearly shown by an intramolecular comparison in the series of mixed trialkylmetal hydrides R_3MH shown in Figure 8. Here, the two bands $I_D(1)$ and $I_D(2)$ corresponding to ionization from the carbon-metal and hydrogen-metal

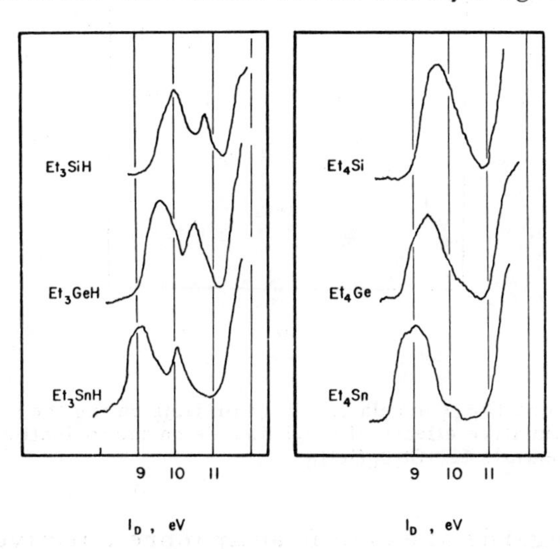

Figure 8. Photoelectron spectra of lowest energy bands for trialkylmetal hydrides (right) and tetra-alkylmetals (left).

σ-bonding orbitals, respectively, for M = Si, Ge and Sn are all clearly resolved with the expected 2:1 intensity ratios [compare the splitting patterns of the orbitals in Figure 4]. In each case the ionization potential $I_D(1)$ of the molecular orbital associated with the carbon-metal interaction is lower than $I_D(2)$ of the hydrogen-metal interaction, and both follow parallel, increasing trends in the order: Sn < Ge < Si as shown in Figure 9. The same notion derives from the cumu-

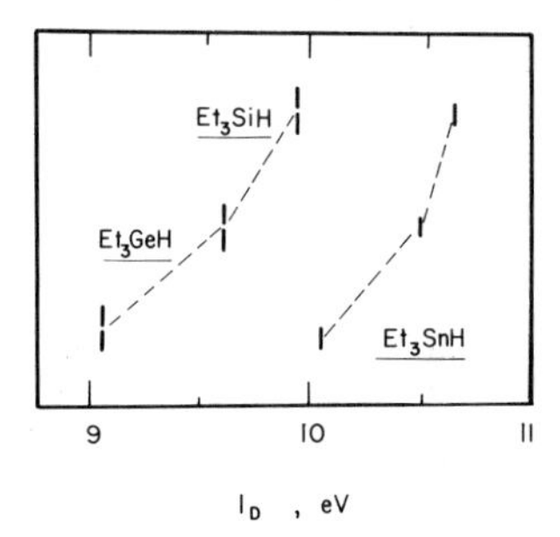

Figure 9. Effect of the metal on $I_D(1)$ [upper energy levels] and $I_D(2)$ [lower energy levels] for trialkylmetal hydrides.

lative effects observed in the magnitudes of $I_D(1)$ as well as $I_D(2)$ in proceeding progressively from $GeEt_4$ to GeH_4 through a series of mixed $Et_{4-n}GeH_n$, where n = 1, 2 and 3, as shown by the correlation diagram in Figure 10.

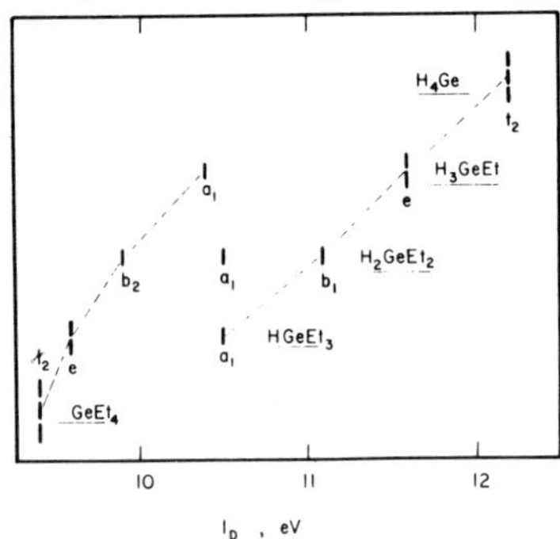

Figure 10. Correlation diagram illustrating the cumulative effects of ethyl ligands on the ionization potentials for $Et_{4-n}GeH_n$.

Alkyl ligands are significantly more effective as donor ligands than halides. For example, the ionization potentials of a series of methylmercury(II) halides(8) are compared in Table II with the I_D of dimethylmercury, the difference between methyl and chloride being more than 35 kcal mol^{-1}. There is also a marked difference among halides as donor

Table II. First Vertical Ionization Potentials (eV) of Binary Mercury(II) Derivatives.

ClHgCl (11.37)	MeHgCl (10.88)	MeHgMe (9.33)
BrHgBr (10.62)	MeHgBr (10.16)	
IHgI (9.50)	MeHgI (9.25)	

ligands, their effectiveness decreasing in the order: I > Br > Cl in Table II, in accord with the trend in electron affinities of the halogen atoms [I (3.17 eV), Br (3.36 eV), Cl (3.61 eV)].(9)

The foregoing discussion emphasizes the major role played by ligands in determining the ionization potentials of metal complexes. Indeed, the effect of ligands can overwhelm even the formal oxidation state of the metal. In Table III are listed a series of transition metal complexes of molybdenum, manganese, iron and nickel, in which the metal center is present in several formal oxidation states. It is clear that the ionization potentials recorded for each complex bears no direct relationship to the formal oxidation state of the metal. For example, the formal oxidation state of the metal in $Mo(CO)_6$ is Mo(0), yet it has the highest oxidation potential, whereas the Mo(IV) complex $Cp_2Mo(CH_3)_2$ has one of the lowest values, reflecting the donor properties of the cyclopentadienyl and methyl ligands.

Table III. Ionization Potentials of Organometals.

Organometal	I_D (eV)	Organometal	I_D (eV)
$Mo(CO)_6$	8.4	$Mn(CO)_5CH_3$	8.65
$Mo(CO)_2(dmpe)_2$	6.00	$Mn(CO)_5CF_3$	9.17
$Mo(CO)Cp_2$	5.9	$Mn(CO)_5H$	8.85
MoH_2Cp_2	6.4	$Mn(CO)_3Cp$	8.05
$Mo(CH_3)_2Cp_2$	6.1	$Mn(CO)_5Br$	8.83
		$Mn(CO)_5SnMe_3$	8.63
$Fe(CO)_5$	8.60		
$FeCp_2$	6.88	$Ni(CO)_4$	8.93
$Fe(CO)_2(CH_3)Cp$	7.7	$Ni(PF_3)_4$	8.82
$Fe(CO)_2(Br)Cp$	7.95	$Ni(bipy)Et_2$	6.4
$[Fe(CO)Cp]_4$	6.45	$Ni(allyl)_2$	7.76

B. Carbon-Centered Organometal Radicals

Organometallic radicals centered on carbon are distinguished from other organic (alkyl) radicals, especially if the metal is situated on either the α- or β-carbon due to the potentially strong proximal interaction of the carbon and metal centers. For example, an α-metalloalkyl radical such as **I** can be considered in two canonical structures as:

$$\cdot C_{\alpha}-M \longleftrightarrow C{=}M\cdot$$
$$\mathbf{I}a \qquad \mathbf{I}b$$

Likewise, a β-metalloalkyl radical can be represented by the limiting valence bond structures below:

$$\cdot C-C_{\beta}-M \longleftrightarrow C{=}C \; M\cdot$$
$$\mathbf{II}a \qquad \mathbf{II}b$$

As the metal is moved progressively further down the alkyl chain from the carbon center, the interaction is expected to diminish, and two separate structures may actually exist, e.g.,

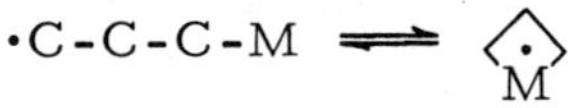

$$\cdot C-C-C-M \rightleftharpoons \text{(cyclic)}\; M$$

Heretofore, α- and β-metalloalkyl radicals are mostly known only for the main group elements such as lithium, magnesium or silicon for **I**, and the Group IVA elements silicon, germanium and tin for **II**. In these examples, the esr parameters and molecular orbital calculations suggest that the species are only perturbed alkyl radicals (i.e., mostly in the form, **I**a and **II**a). For example, the esr spectrum of the α-lithiomethyl radical derived from tetrameric methyllithium shows relatively large α-proton splittings (18.23 G) and small lithium splittings (1.67 G) to 3 equivalent lithium atoms, suggesting that most of the spin density in this species is on the α-carbon.(10)

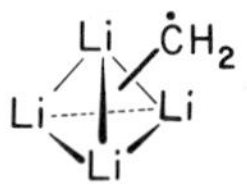

Li $\dot{C}H_2$
Li Li
Li

Furthermore, ^{7}Li and ^{13}C nmr studies of the parent tetrameric methyllithium indicate a near-zero bond order between lithium nuclei indicating that each methyl group interacts with only three lithium atoms. If so, the spin density will not be efficiently transferred by direct spin polarization to the fourth lithium, but rather by a less effective indirect process involving only C-Li bonds. The observation of three equivalent lithium splittings and the absence of either a temperature-dependent proton hfs or selective line broadening in the spectrum cannot be explained by one fixed conformation.

Rather, a rapid rotation ($>10^7$ sec^{-1}) of the methylene group about a sixfold barrier is indicated.

The esr spectra of other (main group) α-metalloalkyl radicals examined heretofore suggest that the spin density in these species also resides mainly on the α-carbon atom, spin delocalization onto the metal being relatively minor.

Organometallic alkyl radicals of type **II** with silicon, germanium, or tin substituents in the β position all exhibit characteristic and unique esr spectra compared to their hydrocarbon analogs,

II

where M = SiR_3, GeR_3, SnR_3. Furthermore, the presence of an oxygen or nitrogen atom in the β position of alkyl radicals does not induce an effect of comparable magnitude.(11) Any interpretation of the esr spectra relative to the structure of these Group IV-substituted β-alkyl radicals must take into account the following observations: (a) an unusually small value for the coupling constants associated with β-hydrogens attached to the carbon atom bearing the heteroatom (Si, Ge, Sn); (b) the relatively large temperature coefficients which are uniquely associated with these coupling constants; (c) the otherwise expected values of the coupling constants for all other β-hydrogens; (d) the large value of the γ coupling constant in β-silyl, -germyl and -stannylethyl radicals. The behavior of the β splittings for these radicals is indicative of the symmetric equilibrium conformations, **III**.

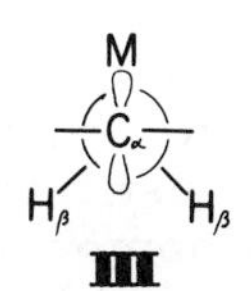

III

In this conformation the heteroatom M is eclipsed by the p orbital of the radical center ($\theta_0 = 2\pi/3$). The proximity of the heteroatom to the p orbital in conformation **III** is also supported by the larger β-silicon coupling constant [a_β(Si) = 37.4 G] in $(CH_3CH_2)_3SiCH_2CH_2\cdot$ compared to the coupling of the silicon atom in the α position in the isomeric radical $(CH_3CH_2)_3Si\dot{C}HCH_3$ [a_α(Si) = 15.2 G].

The results also indicate the presence of substantial barriers which hinder rotation about the C_α-C_β bond. The barrier to internal rotation in a radical such as $(C_2H_5)_3SiCH_2CH_2\cdot$ was determined by fitting the experimental temperature dependence of $a_\beta(CH_2)$. The barrier heights are rather large, being 1.2, 1.6 and 2.0 kcal mol^{-1} for silicon-, germanium- and tin-substituted radicals, respectively, compared to heights of less than 0.5 kcal mol^{-1} for the carbon, oxygen and nitrogen analogs.

The allylic adduct resulting from the addition of trialkylstannyl radical to butadiene also exists in a preferred conformation in which the tin nucleus lies over the π

$$\text{CH}_2\text{=CH-CH=CH}_2 + R_3\dot{Sn} \longrightarrow \text{CH}_2\text{CHCHCH}_2SnR_3 \qquad (6)$$

orbitals.(12) However, unlike the β protons in substituted ethyl radicals, the two protons on carbon-4 in this allylic radical are magnetically inequivalent at the lowest temperatures. The preferred conformations of radicals shown below:

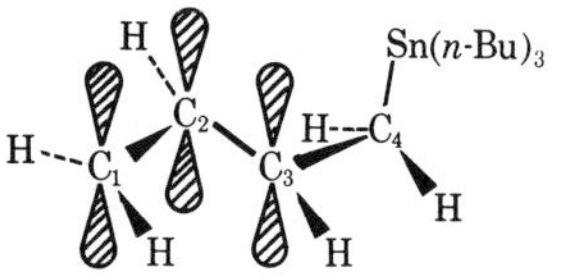

H, C_1, C_2, C_3, C_4, H, H, H, H, $Sn(n\text{-}Bu)_3$

can be deduced more precisely from the values of the hfs of the two protons attached to carbon-4 at lowest temperatures. The angles θ and θ' which describe the dihedral angles between the pair of C_4-H bonds and the odd electron orbital at carbon-3 can be obtained from the general expression invoking a $\cos^2 \theta$ dependence of the β-proton coupling. Hindered rotation about the C_3-C_4 bond leads to selective line broadening in the esr spectra taken at higher temperatures. The rate of this process can be fitted to the experimental line shape with that calculated using the density matrix equation of motion. The barrier to rotation is obtained from temperature-dependence studies and compared with those derived for analogous radicals containing Ge, Si, C and O.

The source of the rather high barriers to rotations in alkyl and allyl radicals β-substituted with Group IVB could lie in hyperconjugative and homoconjugative interactions between the metal and the radical center. For example, two features of the esr parameters of β-substituted ethyl radicals are noteworthy. First, the isotropic g factors of the β-metal substituted ethyl radicals decrease with increasing atomic number of M, and the drop is especially pronounced with Sn. In contrast, the g factors of the metal-centered radicals, in which the odd electron occupies mainly a valence-shell p orbital, increases as the atomic number of the metal increases: $g[CH_3\cdot] = 2.00255$, $g[(CH_3)_3Si\cdot] = 2.0031$, $g[H_3Si\cdot] = 2.0032$, $g[(CH_3)_3Ge\cdot] = 2.01003$, $g[H_3Ge\cdot] = 2.012$, $g[H_3Sn\cdot] = 2.018$.

Moreover, the g factors of α-metal substituted methyl radicals, in which the odd-electron orbital around the metal is mainly its valence-shell d orbital, decrease as the atomic number of the metal increases: $g[(CH_3)_3CCH_2\cdot] = 2.0026$, $g[(CH_3)_3SiCH_2\cdot] = 2.0025$, $g[(CH_3)_3GeCH_2\cdot] = 2.0023$,

$g[(CH_3)_3SnCH_2\cdot] = 2.0008$. The observed trend of the g factors for β-substituted ethyl radicals is thus consistent with the delocalization of the odd electron partly onto vacant d orbitals of the metal. The second feature of these β-substituted ethyl radicals is the large hyperfine splittings by the metal, since the spectrum of $(CH_3)_3SnCH_2CH_2\cdot$ shows the large splittings by ^{117}Sn and ^{119}Sn in natural abundance. The metal coupling constants increase with the atomic number of M and are significantly larger than those observed with the corresponding α-substituted analogs: $a[^{29}Si, CH_3\dot{C}HSiEt_3] = 15.2$ G and $a[^{119}Sn, \cdot CH_2Sn(CH_3)_3] = 137$ G. The large metal coupling constants of β-metal substituted ethyl radicals therefore cannot be explained by a long-range spin polarization or spin polarization of the σ electrons around the metal due to the odd electron density on d orbitals, and suggest that the odd electron is also hyperconjugatively delocalized onto the sp^3 hybrid orbital of the metal.

Whatever the origin to the barrier may be, it is clear from the esr studies that the presence of Group IVA elements in the β-position significantly alters the equilibrium properties of alkyl and allyl radicals. The interaction, however, has not progressed sufficiently far to render these radicals as symmetrically bridged species, even in rather optimum situations such as the adducts to 1,3-cyclopentadiene,(13) e.g.,

Nonetheless, the changes in conformational properties of alkyl radicals effected by these β-metal substituents allude to an electronic origin. The latter is supported by the enhanced reactivity of the methyl hydrogens in the tetraethyl derivatives of silicon, germanium and tin to hydrogen atom abstraction,(14)

$$(CH_3CH_2)_4M + (CH_3)_3CO\cdot \longrightarrow (CH_3CH_2)_3MCH_2CH_2\cdot \quad (7)$$

$$(CH_3CH_2)_4M + (CH_3)_3CO\cdot \longrightarrow (CH_3CH_2)_3M\dot{C}HCH_3 \quad (8)$$

where M = Si, Ge, Sn. The electronic effect of β-metal substituents also pertains to the extreme instability of related radicals, such as the β-mercury analog, to β-elimination:(15)

$$\cdot CH_2CH_2HgCH_3 \xrightarrow{\text{fast}} CH_2{=}CH_2 + \cdot HgCH_3 \quad (9)$$

The esr spectra of these labile radicals have not yet been observed, and what their structure may be remains to be seen (vide infra).

The foregoing examples of metallo-alkyl radicals **I** and **II** refer mainly to those derived from the main group metals. With transition metals, more spin density is likely to be on the metal, and in extreme cases α- and β-metalloalkyl radicals may be considered as carbene (**I**b) and olefin (**II**b) metal complexes, respectively. Recent matrix isolation of the latter as silver and copper olefin complexes indeed confirms the presence of large metal hyperfine splittings.(16)

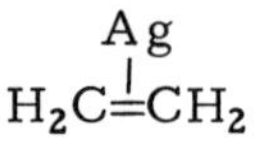

$$\begin{array}{c} Ag \\ H_2C \overset{|}{=} CH_2 \end{array} \qquad \begin{array}{l} \bar{a}_{Ag} \approx 20\,G \\ a_H \leq 1\,G \\ \bar{g} = 2.005 \end{array}$$

Judging from the small, unresolved proton splittings, the ethylene is probably planar and π-complexed in these species.

A spectrum of β-metalloalkyl radicals probably exist, which vary in structure from the rather classical types described above for the β-silicon substituted species at one extreme, to the bridged π-complexed structures, such as the silver-ethylene complex in which the metal atom is more loosely bound, at the other end. Distortion of the classical eclipsed structure [such as that previously proposed for β-chloroalkyl radicals(17)] pertains to β-metalloalkyl radicals with intermediate structures as represented below:

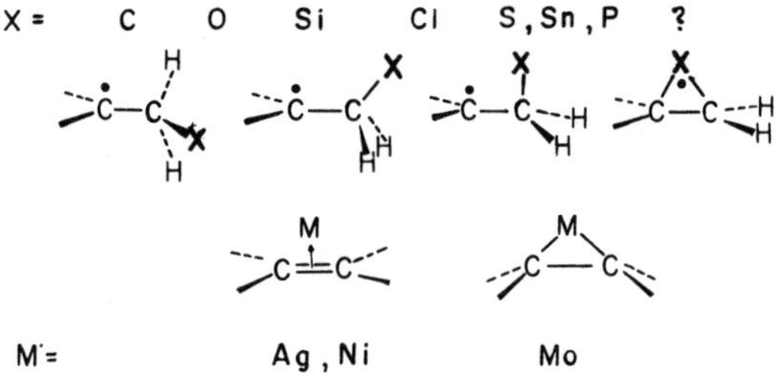

X = C O Si Cl S, Sn, P ?

M = Ag, Ni Mo

The observation of such metastable species, particularly by esr techniques, remains a formidable challenge.

The distinction we have made in the foregoing sections between metal-centered and carbon-centered organometal radicals is useful mainly for the classification of these species, but it is somewhat arbitrary. Thus, the extent to which ligand to metal charge transfer is involved in an organometal radical RM complicates this distinction. It is often unclear whether such an entity should be described as one in which the organic moiety is acting as a radical, carbonium ion or a two-electron ligand with a metal in an unusual oxidation state; for example, in valence bond terminology these structures can be represented respectively as:

$$R\ M^{n} \qquad R^{+}\ M^{n-1} \qquad R^{-}\ M^{n+1}$$

Indeed, a series of related metal complexes has been prepared in unusual oxidation states by successive one-equivalent oxidations and reductions. These compounds are usually characterized by having a ligand in which an electron can be extensively delocalized. The role of alkyl ligands in electron transfer reactions is shown in the homolytic alkyl exchange [k(Et) = 0.1 M^{-1} sec^{-1}],(18)

$$Co^{II}(a) + RCo^{III}(b) \rightleftharpoons RCo^{III}(a) + Co^{II}(b) \qquad (10)$$

where the cobalt nuclei Co(a) and Co(b) are differentiated by the use of similar chelating ligands such as dimethylglyoximato and the cyclohexane analog.

The nature of the carbon-metal bonding, particularly in paramagnetic organometal complexes, is unusual. For example, analysis of the esr parameters for the paramagnetic alkylcobalt cation, $RCo(DMG)_2^+$, derived from the one-equivalent oxidation of alkylbis-dimethylglyoximatocobalt(III) suggests that it be formulated as [R^- Co(IV)] rather than the other limiting valence bond structure [R· Co(III)].(19) In some cases, the valence bond structures represented above may be separated by an energy barrier, as shown by the facile thermal interconversion of two isomeric nickel tetraphenylporphrins,(20)

$$Ni^{II}(TPP^{+\cdot}) \rightleftharpoons Ni^{II}(TPP)^+ \qquad (11)$$

Intramolecular electron transfer has also been observed in cobalt(III) complexes.(21)

$$Co^{III}(O_2CC_6H_4NO_2^{-\cdot})^+ \longrightarrow Co^{II}(O_2CC_6H_4NO_2)^+$$

Intramolecular electron transfer processes involving large energy barriers have been induced photochemically in a variety of metal complexes by excitation of the ligand to metal charge transfer bands, e.g.,

$$Fe^{III}(OH) \xrightarrow{h\nu} [Fe^{II}, OH\cdot] \qquad (12)$$

$$Cu^{II}(Cl) \xrightarrow{h\nu} [Cu^{I}, Cl\cdot] \qquad (13)$$

$$(OC)_5Mn\text{-}Re(CO)_3L_2 \xrightarrow{h\nu} [Mn(CO)_5, Re(CO)_3L_2] \qquad (14)$$

Homolysis induced photochemically is a useful method of generating organometal radicals, and this important subject has been reviewed extensively.(22)

The esr spectrum of a complex between alkylperoxy radicals and cobalt(III) prepared from the reaction of t-butyl hydroperoxide and cobalt(II) acetylacetonate has also been observed.(23) The similarity of the g factors for the complex with that of the uncomplexed alkylperoxy radical indicates minimal spin-orbit coupling associated with the metal nucleus. Analogous species appear to be formed from cobalt(II) acetylacetonate and t-butyl peroxyoxalate, and the g factor suggested an alkoxy radical complex. A phenoxy complex showing splitting by the cobalt nucleus has also been described. Such complexes no doubt play an important role in cobalt-catalyzed autoxidation of hydrocarbons, but their importance has not been fully assessed.

The ability of transition-metal ions to catalyze the dimerization of methyl radicals has been attributed to the formation of a transient alkyl radical complex with the metal, which is stable against abstraction, but long-lived enough to allow radical combination.(24)

$$CH_3\cdot + M^n \rightleftharpoons CH_3M^n \qquad (15)$$

$$2\,CH_3M^n \longrightarrow CH_3CH_3 + 2\,M^n\text{, etc.} \qquad (16)$$

The reversible formation of a σ-bonded complex between benzyl radical and chromium(II) has been described.(25)

$$PhCH_2\cdot + Cr^{II} \rightleftharpoons PhCH_2Cr^{2+} \qquad (17)$$

Similarly, oxidative addition of stable radicals to metal complexes can afford diamagnetic products, presumably in successive steps,(26) e.g.,

$$(CF_3)_2NO\cdot + Ir^{I} \longrightarrow [\text{complex}] \xrightarrow[\text{fast}]{(CF_3)_2NO} Ir^{III}[ON(CF_3)_2]_2 \qquad (18)$$

II. Interactions of Free Radicals with Metal Complexes

Free radicals can react with metal complexes either (a) at the metal center or (b) on the ligand, i.e.,

R· a M b X

Various reactions result from each type of process, as described separately below.

A. Attack at the Metal

Reaction of the free radical at the metal center can lead to either electron transfer, oxidative addition or to displacement. Thus, highly substituted alkyl radicals such as t-butyl appear to undergo direct oxidation with metal complexes, e.g.,

$$(CH_3)_3C\cdot + Pb^{IV} \longrightarrow (CH_3)_3C^+ + Pb^{III} \qquad (19)$$

Electron transfer processes will be dealt with more extensively in the next section. Oxidative addition of free radicals is illustrated in the rapid reaction of alkyl radicals with chromium(II) complexes, e.g.,

$$CH_3\cdot + Cr^{II} \longrightarrow CH_3Cr^{III} \qquad (20)$$

The mechanistic distinction between oxidative addition and displacement as schematically represented by:

$$R\cdot + MX \begin{cases} \nearrow R\dot{M}X \\ \searrow RM + X\cdot \end{cases}$$

depends largely on the lifetime of the adduct. For example, the reaction of t-butoxy radicals with phosphines leads to a paramagnetic phosphoranyl adduct (as described in eq 4) prior to its fragmentation:

$$(CH_3)_3CO\overset{IV}{P}(CH_3)_3 \longrightarrow (CH_3)_3C\cdot + OP(CH_3)_3 \qquad (21)$$

$$(CH_3)_3CO\overset{IV}{P}(CH_3)_3 \longrightarrow (CH_3)_3CO\overset{III}{P}(CH_3)_2 + CH_3\cdot \qquad (22)$$

The liberation of t-butyl radical, observable by its esr spectrum, in eq 21 is an oxygen atom transfer process. The formation of methyl radical in eq 22 corresponds to an overall substitution on the phosphorus center, and in a similar vein, t-butoxy radicals liberate methyl radicals from the boron analog trimethylborane,

$$(CH_3)_3CO\cdot + B(CH_3)_3 \longrightarrow (CH_3)_3COB(CH_3)_2 + CH_3\cdot \qquad (23)$$

However, in this case no evidence for a paramagnetic boron-(IV) adduct showing hyperfine splitting by ^{11}B could be found, even at temperatures as low as -140°C.(27) Production of methyl radicals from phosphine thus clearly results from an oxidative addition–reductive elimination sequence, whereas the absence of an observable intermediate in eq 23 suggests that it be considered an S_H2 displacement. Operationally, the distinction between the two mechanisms must rest heavily on the observation of paramagnetic adducts, and as it becomes possible to shorten the time resolution in these experiments, many if not all processes previously described as S_H2 may be found to involve a transient adduct.

Such a mechanistic distinction may also be approached from a kinetic point of view. For example, alkyl radicals react readily with copper(II) complexes to afford carbonium-type products in high yields together with copper(I), i.e.,

$$R\cdot + Cu^{II} \longrightarrow [R^+] + Cu^{I}$$

which is tantamount to an electron transfer process.(28) However, kinetic studies show that it occurs by a two-step mechanism involving a prior oxidative addition to afford a metastable alkylcopper(III) intermediate, i.e.,

$$R\cdot + Cu^{II} \longrightarrow [RCu^{III}] \longrightarrow [R^+] + Cu^{I} \qquad (24)$$

In particular, the oxidation of β-arylethyl radicals by copper-(II) acetate afforded mixtures of β-arylethyl acetates and styrenes, the molar ratio of which represented the relative rates of oxidative substitution k_s and elimination k_e, respectively. These values of k_s/k_e varied with the ring-substituent and the Cu(II) oxidant. In particular, the amount of ester relative to styrene was optimum with β-anisylethyl radical. The ratio generally correlated with electron release at the

$$ArCH_2CH_2\cdot + Cu^{II}(OAc)_2 \xrightarrow{k_s} ArCH_2CH_2OAc + Cu^{I}OAc \quad (25)$$

$$ArCH_2CH_2\cdot + Cu^{II}(OAc)_2 \xrightarrow{k_e} ArCH{=}CH_2 + HOAc + Cu^{I}OAc \quad (26)$$

para-position of the aryl group. However, the latter conclusion is not necessarily valid since the relative rates of oxidative substitution and elimination really only measure differences in polar effects of these two competing reactions. In order to evaluate polar effects on oxidative substitution and elimination separately, the rate of oxidation of β-arylethyl radicals was measured relative to hydrogen transfer to isobutyraldehyde. The yields of the substituted styrene as well as the β-arylethyl acetate were each determined relative to

$$ArCH_2CH_2\cdot + (CH_3)_2CHCHO \xrightarrow{k_H} ArCH_2CH_3 + (CH_3)_2CH\dot{C}O \quad (27)$$

the corresponding ethylbenzene at various concentrations of isobutyraldehyde and Cu(II). In this manner, it was possible to obtain values of k_e/k_H and k_s/k_H for each β-arylethyl radical. Since the rate of hydrogen transfer from isobutyraldehyde to β-phenethyl radical is the same as its p-methyl, m- and p-methoxy derivatives, the oxidation rates of radicals can be calculated. The absolute value of k_H is not particularly pertinent, since here we are more interested in the relative effect of substituents on the rates of oxidation of phenethyl radicals. The results show that the total rate of oxidation of β-arylethyl radicals given by (k_e+k_s) is remarkably constant for a series of substituents. On the other hand, in varying the substituent from hydrogen to p-methoxy, the rate of oxidative elimination decreases by a factor of approximately 80, whereas the rate of oxidative substitution increases by about a factor of 30. At the same time the relative rates of elimination and substitution change by a factor of 2400. We conclude that the rate-limiting step for oxidation precedes and is separate from the product-forming step. This conclusion militates against a mechanism in which elimination and substitution occur by two independent and parallel processes. It would be highly fortuitous for such concurrent reactions to proceed at widely varying rates, which compensate one another on ring substitution in such a fashion that the total rate remained invariant.

The oxidation of β-arylethyl radicals illustrates the difference between elimination and substitution processes. Deuterium labeling has shown that the α- and β-carbon atoms are extensively scrambled in the β-arylethyl acetate, whereas no such scrambling occurs in the styrene:

$$PhCD_2CH_2\cdot + CuX_2 \rightarrow [PhCD_2CH_2X + PhCH_2CD_2X] + CuX \quad (28)$$

$$PhCD_2CH_2\cdot + CuX_2 \rightarrow PhCD{=}CH_2 + CuX + DX \quad (29)$$

where X = OAc. The loss of identity of the α- and β-carbon atoms during oxidative substitution of β-arylethyl radicals is attributed to the formation of bridged cationic intermediates since the corresponding β-arylethyl radicals do not rearrange under these conditions.

MeO–(+)–CH_2–CD_2 (bridged)   (+)–CH_2–CD_2 (bridged)

Complete equilibration of the α- and β-carbon atoms of the β-anisylethyl moiety during solvolysis of the tosylate derivative has been attributed to the intervention of the delocalized cation. [There is significantly less driving force for the formation of the analogous ethylenephenonium ion in the solvolysis of phenethyl tosylate. Convincing support for this ethylenephenonium ion has only recently been found through solvolytic studies in trifluoroacetic acid.] On the other hand, oxidative elimination to styrene must occur via a different route. The importance of a synchronous elimination of a β-hydrogen is shown by the kinetic isotope effect in the oxidation of β-phenethyl-β,β-d_2 radical. The ratio of the second-order rate constants for oxidative elimination and substitution given by k_e/k_s was readily determined from the relative amounts of styrene and β-phenethyl acetate formed. This value was 33 for the oxidation of β-phenylethyl radical by Cu(II) in 40% v acetonitrile-acetic acid, and 12 for β-phenylethyl-β,β-d_2 radical under equivalent conditions. The isotope effect in oxidative substitution is small and negligible. The observed differences in rates are attributed solely to effects of deuterium substitution in oxidative elimination. On this basis, the kinetic isotope effect for the oxidative elimination of β-phenethyl radical is 2.8. The relatively small isotope effect observed is consistent with a high degree of electron transfer in the transition state for oxidative elimination. This conclusion is also in accord with the rather random loss of β-hydrogens during oxidation of sec-butyl radicals. [Furthermore, the magnitude of the β-deuterium isotope effect for oxidative elimination is intermediate between values observed for the base-promoted elimination and for the unimolecular elimination of a variety of β-arylethyl derivatives. E2 and E1 reactions commonly represent extremes of rupture of the β-hydrogen bond in the transition state for elimination.]

The influence of the Cu(II) complex on the relative rates of oxidative elimination and substitution indicates that it must also be associated with the alkyl moiety in the product-determining step. The mechanism below involving an alkyl-copper(III) intermediate accommodates both of these important limitations.

Scheme I:

$$R\cdot + Cu^{II}(OAc)_2 \rightleftharpoons RCu^{III}(OAc)_2 \qquad (30)$$

$$RCu^{III}(OAc)_2 \xrightarrow{elim.} R(-H) + Cu^{I}OAc + HOAc \qquad (31)$$

$$RCu^{III}(OAc)_2 \xrightarrow{subs.} R^{+}Cu^{I}(OAc)_2\ ,\ etc. \qquad (32)$$

According to this mechanism, oxidative substitution is associated with heterolysis of the alkylcopper intermediate to Cu(I) and a carbonium ion followed by solvation. The importance of this oxidative fragmentation is generally related to the cationic stability of the alkyl fragment and thus constitutes the more important process with cyclobutyl, p-methoxybenzyl, β-anisylethyl and similar radicals. The properties of the resultant carbonium ion (rearrangement, addition, selectivity in solvation, elimination, etc.) are essentially those observed in facile S_N1-E1 solvolyses of the corresponding alkyl tosylates and halides under comparable conditions. Oxidative elimination leading to alkene in eq 31 proceeds from an alkylcopper intermediate by loss of a β-proton synchronously with electron transfer. A possible transition state for such a process is represented below:

$$\left[\begin{matrix} \diagdown\ \ \diagup \\ -C-C- \\ H \quad Cu^{III} \\ AcO \end{matrix} \longleftrightarrow \begin{matrix} \diagdown\ \ \diagup \\ -C-\overset{+}{C}- \\ H \quad Cu^{I} \\ AcO \end{matrix} \longleftrightarrow \begin{matrix} \diagdown\ \ \diagup \\ -C=C- \\ H^{+} \quad Cu^{I} \\ AcO \end{matrix} \right]^{\ddagger}$$

B. Attack at the Ligand

Reaction of the free radical with a metal complex at the ligand site can lead to either ligand transfer, or to fragmentation. Fragmentation is most common with alkyl ligands and results in the cleavage of the carbon-metal bond. For example, abstraction of a β-hydrogen by a radical leads to spontaneous reductive elimination in organomercurials,(29) e.g.,

$$Cl_3C\cdot + CH_3CH_2Hg^{II}CH_3 \longrightarrow Cl_3CH + \cdot CH_2CH_2HgCH_3 \xrightarrow{fast} CH_2{=}CH_2 + Hg^{I}CH_3 \qquad (33)$$

Ligand or atom transfer is a common route by which radicals react with metal complexes, e.g.,

$$CH_3\cdot + IrCl_6^{2-} \longrightarrow CH_3Cl + IrCl_5^{2-} \qquad (34)$$

and it also leads to a concommitant one-equivalent change in the formal oxidation state of the metal. However, atom transfer as described in eq 34 by a concerted process is difficult to distinguish experimentally from a 2-step mechanism involving a prior oxidative addition to the iridium center followed by reductive elimination of methyl chloride.

The mechanistic distinction between these processes has been resolved with copper(II) complexes. Thus, the facile oxidation of alkyl radicals by various copper(II) halides and pseudohalides involves the transfer of a ligand from the copper(II) oxidant to the alkyl radical.(28)

$$R\cdot + Cu^{II}X_2 \longrightarrow RX + Cu^{I}X \qquad (35)$$

where X = Cl, Br, I, SCN, N_3, CN. Free carbonium ions as such are not intermediates since the transfer of chloride, bromide, iodide, thiocyanate, azide and cyanide can be effected in protic media such as acetic acid without direct intervention of the external nucleophile. However, cationoid intermediates are not necessarily precluded, since the subtleties of reactions occurring from ion pairs can exclude direct participation of solvent, as studies of the solvolysis of alkyl derivatives have shown. A much more sensitive probe for the participation of carbonium ions along the reaction coordinate is provided by the study of alkyl moieties prone to cationic rearrangement. Three such systems have been employed in this study: neopentyl, homoallylic and β-arylethyl radicals. A comparison of these alkyl radicals shows that the contribution from a cationic pathway varies with both the alkyl moiety as well as the copper(II) oxidant. Thus, ligand transfer oxidation of neopentyl radical occurs with no evidence of rearrangement, since t-amyl derivatives are absent. On the other hand, cyclobutyl, allylcarbinyl and cyclopropylmethyl cations are more susceptible to rearrangement. The perdominant product obtained from the oxidation of each of these radicals by copper(II) chloride in acetonitrile was that isomeric chloride resulting from the radical with the structure intact. That is, cyclobutyl radical afforded principally cyclobutyl chloride, allylcarbinyl radical gave largely allylcarbinyl chloride. Even cyclopropylmethyl radical produced mainly cyclopropylmethyl chloride, provided the copper(II) chloride concentration was sufficiently high. In every case, the byproduct from the oxidation of each isomeric radical was a mixture of the other two isomeric chlorides:

	$CH_2{=}CHCH_2CH_2$–Cl	◇–Cl	▷–Cl
$CH_2{=}CHCH_2CH_2\cdot + Cu^{II}Cl_2 \longrightarrow$	81%	6%	13%
□· $+ Cu^{II}Cl_2 \longrightarrow$	4%	70%	26%
▷–· $+ Cu^{II}Cl_2 \longrightarrow$	9%	2%	89%

The composition of the products resulting from the oxidation of cyclopropylmethyl radical was highly dependent on the concentration of copper(II) chloride. At very low concentrations of copper(II) the isomeric mixture was the same as that obtained from the oxidation of allylcarbinyl radical. It could be shown from studies carried out at higher concentrations of copper(II) chloride that the oxidation of cyclopropylmethyl radicals also produced a unique mixture of isomeric chlorides. The composition of this mixture was also independent of the concentration of copper(II) chloride. Kinetic analysis of the type described for bromide and thiocyanate led to a second-order rate constant for chloride transfer of $1.1 \times 10^9 M^{-1} sec^{-1}$. C_4H_7-cations are implicated as the precursors for the rearranged products, since the distributions of the isomeric byproducts from the ligand transfer oxidations of all of the radicals are characteristic of the results obtained in solvolytic studies of these systems. Furthermore, any isomerization of the alkyl radical precursors has been taken into account in these results.

The contribution of a polar substituent in promoting the cationic pathway is shown by comparing the ligand transfer oxidation of β-phenethyl and β-anisylethyl radicals. Isotopic labelling allows the examination of carbonium ion intermediates in this system, since complete equilibration of the α- and β-carbons invariably results from a phenonium ion intermediate. [Solvolysis studies show that an even greater driving force for cation formation in the form of the bridged anisonium ion is provided by the presence of a p-methoxy group.] Rearrangement accompanying ligand transfer oxidation of β-phenethyl and β-anisylethyl radicals is summarized in Table IV. No spontaneous rearrangement of either β-phenethyl or β-anisylethyl radical occurs under these conditions.

Table IV. Contribution from the Cationic Path in the Ligand Transfer Oxidation of β-Arylethyl Radicals in Acetonitrile.

Oxidant	Rearrangement	
	$C_6H_5-CD_2CH_2\cdot$	$CH_3O-C_6H_4-CH_2CD_2\cdot$
$Cu^{II}Cl_2$	0%	54%
$Cu^{II}Br_2$	0%	12%

These studies show that two modes of oxidation occur during ligand transfer oxidation of alkyl radicals by copper(II) halides and pseudohalides. One pathway involves direct conversion of the alkyl radical (R•) to the substitution product (RX). The other more indirect route generates a carbonium ion of sufficient integrity somewhere along the reaction coordinate to undergo complete equilibration of the α- and β-carbon atoms of the β-arylethyl moiety and extensive scrambling of the carbon atoms in the cyclobutyl precursors. Stabilization of the alkyl cation in these systems plays an important role in the indirect route, since the neopentyl and β-phenethyl analogs show no tendency to afford carbonium ion intermediates under the same conditions. The results of ligand transfer oxidation of the series of neopentyl, cyclobutyl and β-arylethyl radicals given here are in sharp contrast to the electron transfer oxidation of the same radicals previously examined under comparable conditions. In every example pertaining to the electron transfer oxidation of these alkyl radicals, oxidative solvolysis invariably led to complete cationic rearrangement of the alkyl moiety. Thus, electron transfer oxidation of neopentyl radicals afforded only products derived from the t-amyl cation. Similarly, oxidative substitution of the β-phenethyl radical by copper(II) acetate led to complete equilibration of the α- and β-carbon atoms. Finally, the oxidation of cyclobutyl radicals under the same conditions produced the same extensively rearranged mixture of isomeric C_4H_7-acetates derived by other cationic processes, as described above.

The partial cationic rearrangement observed during the ligand transfer oxidation of certain alkyl radicals and the absence of rearrangement in others indicate that at least two processes are operative. The extent of rearrangement, furthermore, is invariant with the concentration of the copper(II) halide or pseudohalide, but highly dependent on the structure of, and the ligand associated with, the copper(II) nucleus. These results are inconsistent with competition from a kinetically first-order process. It is reasonable that the direct and indirect pathways in ligand transfer oxidation are represented by independent simultaneous second-order processes. The direct process is described as atom transfer and the indirect route is represented as oxidative substitution. These mechanisms will be elaborated further.

Ligand transfer oxidation proceeding via a pathway represented as oxidative displacement is given by eqs 36 and 37 in which copper(II) chloride is used for illustrative purposes. The formation of a carbonium ion intermediate by

$$R\cdot + CuCl_2 \rightleftharpoons RCuCl_2 \quad (36)$$

$$RCuCl_2 \rightleftharpoons RCuCl^+Cl^- \longrightarrow RCl + \overset{I}{Cu}Cl \quad (37)$$

this route is accommodated by a metastable alkylcopper(III) species akin to that presented above in connection with electron transfer processes discussed above.

The alternate route available in ligand transfer oxidation is represented as atom transfer in eq 38. The transition state for atom transfer is free radical in nature, and polar effects are small. Thus, even alkyl radicals with electron-

$$R\cdot + Cu^{II}Cl_2 \rightarrow [R\cdots Cl\cdots CuCl]^{\ddagger} \rightarrow R\text{-}Cl + Cu^{I}Cl \tag{38}$$

withdrawing α-substituents such as cyano, carbonyl and halogen are readily oxidized by copper(II) chloride. The same radicals are, by and large, inert to electron transfer oxidants. [Furthermore, the microscopic reverse process is represented by the direct transfer of a halogen atom from an alkyl halide to chromium(II), and a kinetic study of the reduction of substituted benzyl halides shows no polar effect. The

$$R\text{-}X + Cr^{II} \longrightarrow R\cdot + Cr^{III}X$$

atom transfer mechanism is thus an example of a more general inner-sphere process involving a bridge activated complex. A classic example of the latter is represented below.

$$(NH_3)_5Co^{III}Cl + Cr^{II} \rightarrow [(NH_3)_5Co\cdots Cl\cdots Cr]^{\ddagger} \rightarrow (NH_3)_5Co^{II} + Cr^{III}Cl$$

An atom transfer mechanism, of course, also bears a direct relationship to chain transfer reactions of alkyl radicals with a variety of halogen compounds.] In ligand transfer oxidations, the atom transfer mechanism (eq 38) is generally the energetically more favorable process and usually represents the major course of reaction. The alternative pathway involving oxidative substitution (eq 36 and 37), by virtue of the formation of carbonium ion intermediates, is more applicable to alkyl radicals capable of forming stabilized carbonium ions. Although solvents such as dioxane promote oxidative substitution, the effect is apparent in the ligand transfer oxidation of only radicals such as cyclobutyl and β-anisylethyl, in which the competition between atom transfer and oxidative substitution is already delicately balanced. Thus, oxidative substitution cannot be promoted in neopentyl and phenethyl radicals by solvent changes, and only atom transfer prevails. For a given alkyl radical, the relative importance of atom transfer and oxidative substitution pathways also depends on the ligand involved. For example, the ligand transfer oxidation of cyclobutyl radicals by copper(II) bromide occurs only by transfer. On the other hand, as much as 30% of the oxidation of cyclobutyl radicals by copper(II) chloride takes place by oxidative substitution. Copper(II) thiocyanate occupies an intermediate position. The same general trend of Cu(II)Br > Cu(II)NCS > Cu(II)Cl in atom transfer processes is also established with cyclopropylmethyl and allylcarbinyl radicals. The competition between atom transfer and oxidative substitution is summarized below:

Scheme II:

$$R\cdot + Cu^{II}X_2 \rightarrow [R\cdots X\cdots CuX]^{\ddagger} \rightarrow R\text{-}X + Cu^{I}X$$

and:

$$R\cdot + Cu^{II}X_2 \rightleftharpoons RCu^{III}X_2$$

$$RCu^{III}X_2 \rightleftharpoons RCuX^{+}X^{-} \rightleftharpoons R^{+}Cu^{I}X_2^{-}$$

$$R^{+}Cu^{I}X_2^{-} \rightarrow R\text{-}X + Cu^{I}X$$

Ligand transfer from a metal center to a radical as represented in eqs 34 and 35 has a counterpart in a microscopic reverse process in which a metal-centered radical reacts with an organic halide to transfer a halogen atom,(30) e.g.,

$$RBr + Ag \xrightarrow{k_{Ag}} R\cdot + AgBr$$

$$RBr + Bu_3Sn \xrightarrow{k_{Sn}} R\cdot + Bu_3SnBr$$

$$RBr + Cr^{II} \xrightarrow{k_{Cr}} R\cdot + BrCr^{III}$$

$$R\text{-}I + (NC)_5Co^{II} \xrightarrow{k_{Co}} R\cdot + (NC)_5Co^{III}I$$

The relative second-order rate constants for these halogen atom transfers listed in Table V all show parallel trends with: Et ≈ n-Pr < i-Pr < t-Bu, in accord with the decreasing values of the alkyl-bromine bond strengths. However, the span of rates vary considerably, with the relative reactivities of MeBr/t-BuBr increasing from 7 for BuSn(III) to 150 for $(NC)_5Co(II)$.

Table V. Relative Rates of Bromine Atom Transfer from Alkyl Bromides to Silver(0), Bu_3Sn, Cr(II) and $(NC)_5Co^{3-}$.

Alkyl Bromide	k_{Ag}	k_{Sn}	k_{Cr}	k_{Co}
Et	1.0	1.0		1.0
n-Pr	0.9	1.1	1	0.7
iso-Pr	3.0	3.0	11	20
tert-Bu	19	7.0	58	150

III. Electron Transfer and Charge Transfer Processes

Interest in paramagnetic metal-containing species as described in both of the foregoing sections derives from their behavior as transient intermediates, especially when compared to their diamagnetic counterparts. For example, the diamagnetic, lithium tetramethylaurate(III) and dimethylaurate(I) are thermally quite stable, decomposing at reasonable rates only above 150°C. However, if they are oxidized to the paramagnetic, tetramethylgold(IV) and dimethylgold(II), reductive elimination of ethane is spontaneous, even below 0°C.(31)

$$\underline{(CH_3)_4Au^-} \xrightarrow{-\varepsilon} \underline{(CH_3)_4Au} \xrightarrow{-CH_3CH_3} \underline{(CH_3)_2Au} \xleftarrow{-\varepsilon} \underline{(CH_3)_2Au^-}$$

$$\underline{(CH_3)_2Au} \xrightarrow{-CH_3CH_3} \underline{Au^0}$$

$$\underline{(CH_3)_2Au} \xrightarrow{PPh_3} \underline{CH_3AuPPh_3 + (CH_3)_3AuPPh_3}$$

The difference is also shown in the thermal decomposition of the diethyliron(II) complex, $Et_2Fe(bipy)_2$ at 80°C, which affords disproportionation products, but coupling products on oxidation at 0°C,(32)

$$(bipy)_2Fe(CH_2CH_3)_2 \longrightarrow CH_3CH_3 + CH_2{=}CH_2 + [(bipy)_2Fe]$$

$$\Big\downarrow \underline{E}, -\varepsilon$$

$$(bipy)_2Fe^+(CH_2CH_3)_2 \longrightarrow CH_3CH_2CH_2CH_3 + [(bipy)_2Fe^+]$$

where the oxidant, $\underline{E}$ = $IrCl_6^{2-}$, Ce(IV), Co(III), Cu(II), Br_2, etc.

A number of alkylmetals especially of the main group elements are known as quite stable compounds. For homolytic decomposition,

$$R_nM(g) \longrightarrow M(g) + nR\cdot(g) \qquad \Delta H^\circ/n = \overline{D}(R\text{-}M)$$

the mean bond dissociation energy $\overline{D}$(R-M) for binary alkylmetals such as those given in Table VI can be obtained from

Table VI. Mean Bond Dissociation Energies of Organometals.

$LiCH_3$	59	$C(CH_3)_4$	87	$Ta(CH_3)_5$	62
$Mg(CH_3)_2$	38	$Si(CH_3)_4$	77	$W(CH_3)_6$	38
$Zn(CH_3)_2$	44	$Ge(CH_3)_4$	65	$(CO)_5MnCH_3$	30
$Hg(CH_3)_2$	30	$Sn(CH_3)_4$	54	$Ti(CH_2CMe_3)_4$	50
$B(CH_3)_3$	89	$Pb(CH_3)_4$	40	$Zr(CH_2CMe_3)_4$	60
$Al(CH_3)_3$	62			$Hf(CH_2CMe_3)_4$	64

the enthalpy of formation of the alkylmetal in conjunction with the enthalpies of formation of the alkyl radicals and gaseous metal atoms, i.e.,(33)

$$n\overline{D}(R\text{-}M) = \Delta H^{\circ} = \Delta H_f^{\circ}M(g) + n\,\Delta H_f^{\circ}R\cdot(g) - \Delta H_f^{\circ}MR_n(g)$$

However, these mean bond dissociation energies usually differ significantly from the individual bond energies D_i,

$$R_nM \longrightarrow R\cdot + R_{n-1}M \qquad \Delta H_1 = D_1$$

$$R_{n-1}M \longrightarrow R\cdot + R_{n-2}M\ , \text{ etc.} \qquad \Delta H_2 = D_2$$

where $\Sigma D_i = n\overline{D}$. Indeed, it has been found that D_1 is much larger than D_2 in dichloromercury(II) as shown in Table VII.

Table VII. Bond Energies (kcal mol^{-1}) in Mercurials.

Hg Compound	$D_1 + D_2$	D_1	D_2
Cl-HgCl	106	81	24
CH_3-HgCl		64±2	
CH_3-HgCH_3	58±2	51±2	7±3 (diff.)

Similarly, the first bond dissociation energies for CH_3-HgCl and CH_3-HgCH_3 are significantly larger than the second D_2. In these examples, the differences in bond energies reflect the relative stabilities of diamagnetic mercury(II) complexes compared to the paramagnetic mercury(I) derivatives.

From these illustrations, it is clear that paramagnetic organometal species (whether they be radical-ions derived by electron transfer or neutral radicals formed by homolysis) are much more prone to undergo reaction than their diamagnetic precursors. We now wish to focus on the formation of these paramagnetic species by electron transfer and charge transfer processes, using dialkylmercury, tetraalkyltin and -lead compounds as examples.

A. Electron Transfer

Organometals RM are readily cleaved by iron(III) complexes FeL_3^{3+}, where L = 1,10-phenanthroline and related ligands,(34)

$$RM + 2\,FeL_3^{3+} \longrightarrow [R^+] + M^+ + 2\,FeL_3^{2+} \qquad (39)$$

where $[R^+]$ denotes carbonium ion products. The ease of oxidative cleavage of tetraalkyltin by iron(III) complexes is highly dependent on the donor properties of the alkyl groups as measured by the ionization potentials. Thus, in the homologous series of symmetrical tetraalkyltin compounds R_4Sn, the rates progressively increase with α-methyl substitution from R = methyl < ethyl < isopropyl, roughly in the order of $10^0 : 10^4 : 10^7$. This trend, reflecting an inverse steric effect, is counter to any expectation based on a direct bimolecular scission, and it suggests that the activation process does not involve cleavage of the alkyl-tin bond itself. Instead, electron transfer occurs in a prior, rate-limiting step during oxidative cleavage of organometals. This formulation is in basic accord with the well-established property of tris-phenanthroline and related iron(III) cations to function as oxidants in many inorganic systems. According to the general mechanism presented in Scheme III, the activation process for oxidative cleavage is represented by the electron transfer step 40, which is rapidly followed by homolytic fragmentation of the alkyltin cation-radical [formally an alkyltin(V) species] in eq 41, and further oxidation of the alkyl radical by a second Fe(III) in eq 42.

Scheme III:

$$R_4Sn + Fe^{III} \xrightarrow{k_{et}} R_4Sn^{+} + Fe^{II} \qquad (40)$$

$$R_4Sn^{+} \xrightarrow{fast} R_3Sn^{+} + R\cdot \qquad (41)$$

$$R\cdot + Fe^{III} \xrightarrow{fast} [R^{+}] + Fe^{II} \qquad (42)$$

The mechanism in Scheme III accords with all the observations made in this system, including (1) the stoichiometry, energetics and kinetics of the electron transfer step, (2) the observation of alkyl radicals during oxidative cleavage, and (3) the selectivity observed in the oxidative cleavage of methylethyltin compounds. Each of these will be described more fully in the following discussion.

Electron Transfer as the Rate-Determining Step. The second-order kinetics for cleavage indicate that alkyltin and only one iron(III) are represented in the rate-determining transition state. The other iron(III) required by the stoichiometry must be involved in a fast subsequent step (vide infra). For an electron transfer process to occur between alkyltin and iron(III), the second-order rate constant k_{et} in eq 40 should reflect the ease of electron detachment from alkyltin, as measured independently by the ionization potential in Table I. Indeed, Figure 11 shows the smooth correlation between the vertical ionization potentials of a series of alkyltin compounds and the log k_{et} for oxidative cleavage. The linearity observed for each of the three oxidants, viz., tris-2,2'-bipyridine, 1,10-phenanthroline and 5-chloro-1,10-phenanthroline iron(III), spans a range of more than 10^8 in rates.

The electron transfer between alkyltin and iron(III) in eq 40 is essentially irreversible since the rate of oxidative cleavage is unaffected by the added iron(II) product. The irreversibility derives in part from the metastable nature of the tetraalkyltin cation-radical (vide infra). Indeed, inability to observe the esr spectrum of the alkyltin cation-radical and

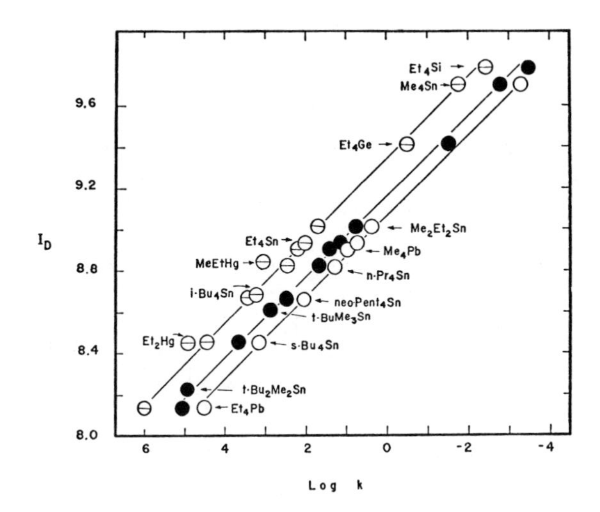

Figure 11. Correlation of the rates (log k) of electron transfer with the ionization potentials I_D for a series of alkylmetals as indicated, using ⊖ tris-5-chloro-1,10-phenanthrolineiron(III), ● tris-1,10-phenanthrolineiron-(III), and O tris-2,2-bipyridineiron(III) as oxidants.

the irreversibility of the cyclic voltammetry indicate that its lifetime is very short. Analogous cation-radicals derived from tetraalkyllead, dialkylmercury and dialkylbis(phosphine)platinum are also unstable.

Alkyl Radicals as Prime Intermediates—Oxidation by Iron(III). The observation of paramagnetic intermediates by spin trapping indicates that alkyl radicals are formed during the oxidative cleavage of alkyltin by iron(III). Furthermore, the scavenging of the alkyl fragments in the presence of molecular oxygen as alkylperoxy products shows that they must depart from tin as the alkyl radicals indicated in eq 41. Accordingly, the isolation of alkyl perchlorates in excellent yields implies that iron(III) is an efficient scavenger of alkyl radicals in eq 42. Indeed, the absence of alkane indicates that hydrogen abstraction from solvent is unable to compete with oxidation,

$$R\cdot \begin{cases} \xrightarrow{Fe^{III}} [R^+] + Fe^{II} \\ \xrightarrow{CH_3CN} RH + \cdot CH_2CN \end{cases}$$

even when oxidative cleavage is carried out with a stoichiometrically limited supply of iron(III), i.e., in the presence of excess alkyltin. Moreover, inability to scavenge all the isopropyl radicals from the oxidative cleavage of i-PrSnMe$_3$ in the presence of excess oxygen,

$$R\cdot \begin{cases} \xrightarrow{Fe^{III}} [R^+] + Fe^{II} \\ \xrightarrow{O_2} ROO\cdot \end{cases}$$

suggests that the oxidation by iron(III) may approach the diffusion-controlled rates. This conclusion is consistent with the second-order rate constant $k \geq 4 \times 10^8 M^{-1} sec^{-1}$ estimated by Walling and Johnson(35) for the oxidation of hydroxy-methyl radical by Fe_{aq}^{3+} which is a significantly weaker oxidant $[E^0 = 0.77 V]$ than $Fe(phen)_3^{3+}$ $[E^0 = 1.22 V]$ in water.(36) Significantly, the facile oxidations associated with these paramagnetic iron(III) complexes are reminiscent of similar high rates of interaction of alkyl radicals with various copper(II), iridium(IV) and chromium(II) complexes.

Selectivity During Fragmentation of Alkyltin Cation-Radicals. Selectivity in the cleavage of alkyl groups from unsymmetrical alkyltin compounds by iron(III) products is represented by the products of cleavage, as illustrated for methylethyltin compounds in eq 43,

$$R_2Sn(Me)(Et) + 2\,FeL_3^{3+} \begin{cases} \xrightarrow{k_{Me}} R_2SnEt^+ + [Me^+] + 2\,FeL_3^{2+} \\ \xrightarrow{k_{Et}} R_2SnMe^+ + [Et^+] + 2\,FeL_3^{2+} \end{cases} \quad (43)$$

where R = Me, Et. The selectivity [S(Et/Me)] represents the ratio of rate constants k_{Et}/k_{Me}.

According to Scheme III, selectivity is established subsequent to the rate-determining electron transfer. During fragmentation of the cation-radical in eq 41, the preference for ethyl cleavage indicated by S(Et/Me) = 27 and 22 for FeL_3^{3+} and $IrCl_6^{2-}$, respectively, in the mixed methylethyltin compounds $Me_{4-n}SnEt_n$, is essentially the same as that observed in the related oxidative cleavage of methylethyllead compounds by $IrCl_6^{2-}$ with S(Et/Me) = 25. Both arise from the fragmentation of the radical-ion:

$$R_2\overset{+}{Sn}(Et)(Me) \begin{cases} \longrightarrow Et\cdot + MeSnR_2^+ \\ \longrightarrow Me\cdot + EtSnR_2^+ \end{cases} \quad (44)$$

Similar selectivities are observed in the mass spectral cracking patterns of methylethyltin compounds although reduced in magnitude. The latter doubtlessly reflects the loss in selectivity of highly energetic species formed by electron impact relative to those cation-radicals formed in solution. The effect of solvation cannot be assessed quantitatively, but the qualitative trends in selectivity, both in solution and in the gas phase, are unmistakable. The prevailing factor which determines the predominance of ethyl over methyl cleavage is the strengths of the relevant alkyl-metal bonds.

These values can be evaluated from the mean bond energies for Et_4Sn and Me_4Sn which are 46 and 54 kcal mol^{-1}, respectively, and for Et_4Pb and Me_4Pb which are 33 and 40 kcal mol^{-1}, respectively.(33)

It is noteworthy that all of these unimolecular selectivities are inverted relative to those observed in other bimolecular processes. For example, the electrophilic cleavage of methylethyllead compounds by acid [S(Et/Me) = 0.11 and 0.021 for HOAc and H_2OAc^+, respectively] and metal ions [S(Et/Me) = 0.018 and 0.022 for CuOAc and $CuCl_2$, respectively] all involve the direct scission of the alkyl-metal bond by the electrophile.(37) As such, the inverted order in selectivity in each of these processes (i.e., methyl cleaved in preference to ethyl) reflects the dominance of steric constraints over electronic effects in bimolecular transition states.

Indeed, differences in selectivity patterns provide one of the best diagnostic methods for distinguishing electrophilic (two-equivalent) from electron transfer (one-equivalent) mechanisms for the cleavage of alkylmetals. More relevant to the issue here, the similar selectivities [clustering around S(Et/Me) = 25] observed for the oxidative cleavage of methylethyltin compounds induced by FeL_3^{3+} and by $IrCl_6^{2-}$ are only consistent with the cation-radical as the common intermediate leading directly to cleavage, as described in eq 44. It is conceivable that the cation-radical $R_4Sn^{+\cdot}$ formed in eq 40 is not free, and the degree to which it is still associated with the reduced iron(II) species would affect its subsequent reactivity. In order to evaluate this problem, let us consider whether the electron transfer step itself conforms to the Marcus criterion for an outer-sphere mechanism. We next compare the oxidation of an alkylmetal RM by iron(III) with that effected by hexachloroiridate(IV),

$$RM \begin{cases} \xrightarrow{FeL_3^{3+}} [RM^{+} \ FeL_3^{2+}] \\ \xrightarrow{IrCl_6^{2-}} [RM^{+} \ IrCl_6^{3-}] \end{cases}$$

Such a comparison also focusses on the ion-pairing energies, since the electrostatic potential in the ion pair derived from iron(III) is repulsive, whereas it clearly changes to an attractive energy in the ion pair derived from iridate(IV).

Outer-Sphere Processes for Electron Transfer from Alkylmetals to Iron(III) Complexes. In the outer-sphere reaction of alkylmetals with iron(III), Marcus theory predicts a slope of 0.5 in the correlation of the rates of electron transfer (log k) with the difference in standard free energy changes of RM and FeL_3^{3+}.

(a) Structural effects of iron(III). For a particular alkylmetal, log k for electron transfer is linear with the standard oxidation potential of the iron(III) complexes. The slope of the correlation in eq 45,

$$\log k \; = \; 8.75 \; E^0 + \text{constant} \tag{45}$$

is equivalent to that of a linear free energy plot with $\Delta G^{\ddagger} = 0.50 \; \Delta G^0 + \text{constant}$, predicted by the Marcus theory for outer-sphere electron transfer. It is noteworthy that the family of lines in Figure 12 for all eleven alkylmetals pass

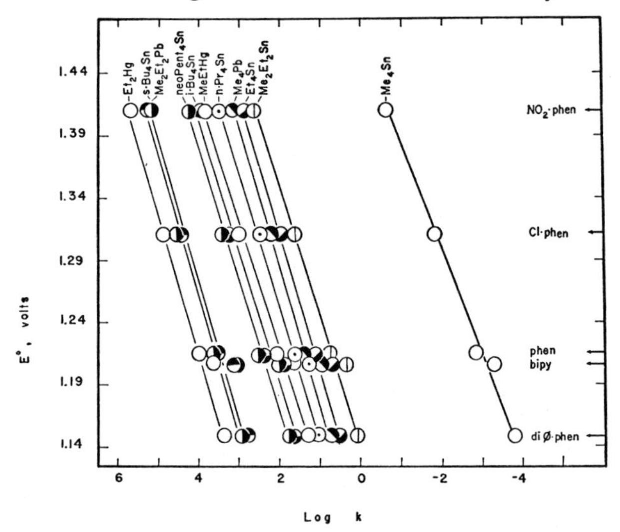

Figure 12. The Marcus plot of standard electrode potentials (E^0 vs. SHE) for various iron(III) complexes with the second-order rate constant (log k) for electron transfer from the alkylmetals listed at the top.

through the experimental points with slopes [8.9±0.4] close to this value. Both two-coordinate dialkylmercury compounds and four-coordinate tetraalkyltin as well as -lead compounds are generally included in this correlation. Furthermore, the points for the iron(III) complex with the most sterically hindered ligand, L = 4,7-diphenyl-1,10-phenanthroline, also fall close to the lines. If the alkylmetal must penetrate the octahedral, tris ligand sphere around iron(III) before electron transfer can take place, the substantial difference in steric effects between 4,7-diphenylphenanthroline and 1,10-phenanthroline should be manifested most either with the linear mercury alkyls or with the highly congested tetraneopentyltin. Thus, the linear relationships generally observed suggest a transition state in which the alkylmetal is located along the periphery of the iron(III) complex. Electron transfer probably could occur via the π-orbitals of the phenanthroline ligand. Indeed, the negative deviations consistently observed with the analogous bipyridine iron(III) complex in Figure 12 accord with the less extensive π-conjugation in this ligand.

(b) Structural effects of the alkylmetal--HOMO and steric effects. For a particular iron(III) complex, log k for electron transfer is also linear with the ionization potential of the alkylmetal. The smooth correlation in Figure 11 includes the tetraalkylmetals of silicon, germanium, tin and lead as well as the two-coordinate dialkylmercury compounds. The linearity spans a range of almost 10^8 in the rates of electron transfer. Furthermore, the correlation,

$$\log k = -4.9\, I_D + \text{constant} \qquad (46)$$

accords with the known relationship between I_D and E^0 and provides additional support for the outer-sphere mechanism.

If only tetraalkyltin and -lead compounds are considered, the linear correlation in Figure 11 is excellent. It suggests that the solvation terms are essentially constant throughout the series of tin and lead compounds. This is not unreasonable since the effective size of the cation-radical from lead(V) is probably not much larger than that of tin(V) due to the lanthanide contraction. However, the most important feature of the correlation in Figure 11 is the striking absence of steric effects with changes in the structures of the alkyl ligands. In particular, increasing the branching of the alkyl ligand at the β-carbon with methyl groups in the homologous series: CH_3CH_2, CH_3CH_2CH, $(CH_3)_2CHCH_2$ and $(CH_3)_3CCH_2$, leads to no deviation from the linear free energy correlation. Even the oxidative cleavage of the sterically hindered tetra-neopentyl is included precisely in the correlations with all three iron(III) complexes. The same applies to α-branching in the series: CH_3, CH_3CH_2, $(CH_3)_2CH$ and $(CH_3)_3C$.

Inner-Sphere Processes for Electron Transfer from Alkylmetals to Hexachloroiridate(IV). Alkylmetals are oxidatively cleaved by hexachloroiridate(IV) by essentially the same mechanism as that described in Scheme III for iron(III). For example, the facile reaction with the homoleptic alkylmetals of mercury and lead has been shown to proceed via a rate-limiting electron transfer.

$$RM + IrCl_6^{2-} \xrightarrow{k} RM^{+\cdot} + IrCl_6^{2-}$$

The products, stoichiometry, and kinetics indicate that the tin derivatives in this study also react by the same mechanism, as shown below:

Scheme IV:

$$R_4Sn + IrCl_6^{2-} \xrightarrow{k_{et}} R_4Sn^{+} + IrCl_6^{3-} \quad (47)$$

$$R_4Sn^{+} \xrightarrow{fast} R_3Sn^{+} + R\cdot \quad (48)$$

$$R\cdot + IrCl_6^{2-} \xrightarrow{fast} RCl + IrCl_5^{2-} \quad (49)$$

The reduction potential of hexachloroiridate(IV) in acetonitrile solution is 0.67 volts, which is less than the E^0 of the iron(III) complexes. However, the second-order rate constants for electron transfer from both tetramethyltin and -lead to hexachloroiridate(IV) are significantly larger than those predicted from an extrapolation of the correlations in Figure 12. Indeed, tetramethyltin reacts about 10^7 times faster than expected. Thus in contrast to iron(III), an inner-sphere contribution to electron transfer is indicated in the case of hexachloroiridate(IV), and it suggests that the alkylmetal can be approached by hexachloroiridate(IV) closer than by iron(III) in the transition state for electron transfer. In other words, steric effects are more important in electron transfer reactions with hexachloroiridate(IV) than those with iron(III). Indeed, the smooth correlation shown in Figure 11 between I_D and log k_{et} for outer-sphere electron transfer with iron(III) is no longer valid. Instead, the rates of oxidative cleavage of the same alkylmetals by hexachloroiridate(IV) are depicted in Figure 13. However, despite the random, "buckshot" appearance of the plot, a closer scrutiny of the data shows a systematic trend among a limited number of related compounds. For purposes of calibration, the dashed line in Figure 13 is the correlation with iron(III), in which the slope is representative of outer-sphere electron transfer from these alkylmetals (vide supra). The correlations of hexachloroiridate(IV) with the methylethyl derivatives of both mercury and lead are fairly linear, with approximately this slope, but not on the same line. Apparently with these less hindered alkylmetals, the rates of electron transfer to hexachloroiridate(IV) are determined more by electronic effects (i.e., the HOMOs described in Figure 6) rather than by steric effects. A greater variety of alkyl structures are included among the tetraalkyltin derivatives and the points in Figure 13 show considerable, but accountable scatter. Thus, the negative deviation from the outer-sphere slope is most pronounced with the α- and β-branched alkyl groups, i.e., the isopropyl, isobutyl and t-butyl derivatives. Clearly, the hindered alkyltin compounds are cleaved by hexachloroiridate(IV) much more slowly than their values of I_D alone would indicate. A similar conclusion may be reached from the varying magnitudes of Δlog k for different alkylmetals. Such a steric effect must reflect the perturbation of the inner

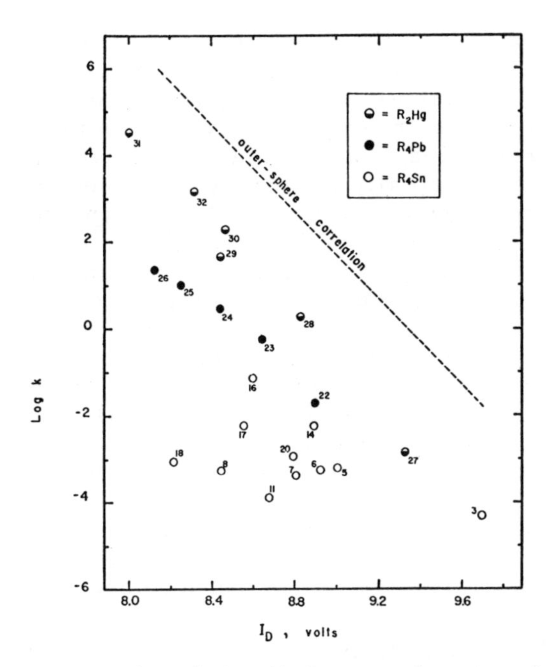

Figure 13. The relationship between the rates of electron transfer (log k) to hexachloroiridate(IV) and the ionization potentials I_D of a series of tetraalkyltin compounds indicated by open circles O. Comparison with ● methylethyllead and ◒ dialkylmercury compounds. The numbers refer to compounds designated in Table VIII. The outer-sphere slope is indicated by the dashed line taken from Figure 11 for tris-5-chloro-1,10-phenanthrolineiron(III).

sphere of the alkylmetal in the transition state for electron transfer. Indeed, this conclusion can be used as an operational criterion for an inner-sphere mechanism of electron transfer from alkylmetals to hexachloroiridate(IV).

A Continuum of Outer- and Inner-Sphere Processes for Electron Transfer from Alkylmetals. The concepts of outer-sphere and inner-sphere electron transfer as we have employed here depend on the availability of various alkyl groups as highly "tunable" probes for steric effects. As such, we might ask how these processes basically differ since the alkylmetal cation-radical is an intermediate which is common to both iron(III) and iridate(IV). Thus, selectivity studies demonstrate that there is no direct, covalent bond formed between the alkylmetal and hexachloroiridate(IV) during inner-sphere electron transfer. Outer- and inner-sphere processes with alkylmetals may be distinguished by the magnitudes of the intermolecular separation between the alkylmetal and the oxidant in the transition states for electron transfer. The driving force as well as electrostatic forces are expected to contribute to the "tightness" of these transition states. In the inner-sphere activated complex, changes in the steric properties of alkyl ligands indicate that the alkylmetal is geometrically perturbed, and we tentatively suggest that a precursor complex is formed in which the tetraalkyltin achieves a quasi five-coordinate configuration reminiscent of a variety of trigonal bipyramidal structures known for tin(IV) derivatives. According to this proposal, substitution-inert organometals can undergo outer-sphere as well as inner-sphere electron transfer. For tetraalkylmetals the inner-sphere process is subject to steric hindrance by the alkyl groups which may be relieved by partial distortion of the configuration at the metal center. This formulation implies that a continuum of outer-sphere and inner-sphere processes is possible for electron transfer which differ principally in geometrical constraints. This problem is discussed further in the next section, in which the same organometals are subjected to charge transfer interactions.

B. Charge Transfer Processes of Organometals with Tetracyanoethylene

Another manifestation of the properties of electron donors is their ability to form charge-transfer complexes. Thus, the addition of dialkylmercury to a solution of tetracyanoethylene (TCNE) results in weak but distinct colors characteristic of the mercurial added. The stability of the band also varies, being the most stable with dimethylmercury. The transient bands from diisopropyl and di-tert-butylmercury were recorded at -77°C, but even at this temperature the solutions bleach rapidly as the charge transfer (CT) complexes undergo further thermal reactions leading to the insertion of TCNE into a single alkyl-mercury bond.(38)

$$\text{R-Hg-R} + (\text{NC})_2\text{C=C}(\text{CN})_2 \longrightarrow \text{RHg-C(CN)}_2\text{-C(CN)}_2\text{-R} \qquad (50)$$

A similar series of observations are also made with tetraalkyltin compounds for which the formation of the charge transfer complex is:

$$R_4Sn + TCNE \overset{K_{CT}}{\rightleftharpoons} [R_4Sn\ TCNE]$$

The bands are broad, as is characteristic of intermolecular charge transfer spectra, and the absorption maxima are highly dependent on the structure of the tetraalkyltin compounds listed in Table VIII.(39)

Table VIII. The Formation and Reaction of Charge Transfer Complexes of Alkyltin and Tetracyanoethylene.[a]

No.	Alkyltin	I_D (eV)	ν_{CT} (10^{-4} cm^{-1})	ϵ_{CT} (M^{-1} cm^{-1})	log K_{CT} (M^{-1})	log k_T (M^{-1} sec^{-1})
1	Me_4Sn	9.69	2.90	500	-0.76	-4.82
2	Et_4Sn	8.90	2.38	167	-0.28	-3.07
3	n-Pr_4Sn	8.82	2.41	29	0.34	-2.59
4	n-Bu_4Sn	8.76	2.40	16	0.89	-2.04
5	$EtMe_3Sn$		2.68	222	-0.61	-4.04
6	n-$PrMe_3Sn$	9.1	2.78	200	-0.61	-4.70
7	n-$BuMe_3Sn$	-	2.84	172	-0.68	-4.54
8	Et_2Me_2Sn	9.01	2.62	143	-0.10	-3.34
9	n-Pr_2Me_2Sn	8.8	2.58	77	0.16	-3.37
10	n-Bu_2Me_2Sn	8.8	2.59	50	0.04	-3.20
11	i-Pr_4Sn	8.46	2.29	95	0.0	-2.85
12	s-Bu_4Sn	8.45	2.33	71	0.39	-2.21
13	i-Bu_4Sn	8.68	2.41	125	-0.52	-3.66
14	i-$PrMe_3Sn$	8.9	2.47	40	0.0	-3.19
15	t-$BuMe_3Sn$	8.6	2.35	-	-	-2.18
16	i-Pr_2Me_2Sn	8.56	2.38	118	-0.02	-2.92
17	t-Bu_2Me_2Sn	8.22	2.38	77	-0.19	-3.43
18	i-Bu_2Et_2Sn		2.38	143	-0.40	-3.48
19	neo-$Pent_3EtSn$			-	-	-4.14
20	neo-$Pent_4Sn$	8.67	-	-		<-6

[a]Charge transfer spectra measured in chloroform solution. Second-order rate constants measured in acetonitrile at 25°C.

According to the valence-bond description, the frequency of the CT band corresponds roughly to the energy required to transfer an electron from R_2Hg to TCNE. [In the following, R_2Hg and R_4Sn are used interchangeably to denote the reactions of organometals RM generally.]

$$[R_2Hg\ TCNE] \xrightarrow{h\nu_{CT}} [R_2Hg^{+\cdot}\ TCNE^{-\cdot}]$$

For weakly associating systems, such as these are, $h\nu_{CT}$ is approximated by eq 51,

$$h\nu_{CT} = I_D - E_A - [G_1 - G_0] \tag{51}$$

where I_D and E_A refer to the vertical ionization potential of R_2Hg and the electron affinity of TCNE, respectively, and G_1, the dominant term in the brackets, involves Coulombic interaction in the excited state. With TCNE as the common acceptor, Figure 14 shows the linear relationship between $h\nu_{CT}$ and the vertical ionization potential for the series of R_2Hg.

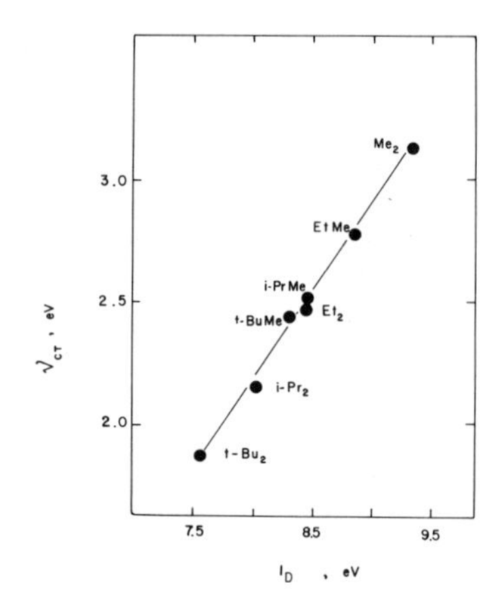

Figure 14. Correlation of the ionization potentials of dialkylmercury compounds with the frequency of the charge transfer band with TCNE.

The interaction of an electron donor with an electron acceptor may lead to a variety of thermal processes including electron transfer, covalent bond formation, etc. For the weak donor-acceptor interactions pertaining to the dialkylmercury-TCNE systems, we use the representation in eqs 52-54, in which insertion occurs subsequent to electron transfer.

Scheme V:

$$R_2Hg + TCNE \rightleftharpoons [R_2Hg\ ,\ TCNE] \quad (52)$$

$$[R_2Hg\ ,\ TCNE] \xrightarrow{k} [R_2Hg^{+\cdot}\ TCNE^{-\cdot}] \quad (53)$$

$$[R_2Hg^{+\cdot},\ TCNE] \xrightarrow{fast} RHg\text{-}TCNE\text{-}R \quad (54)$$

The relationship between the rate (i.e., log k) of the thermal process leading to insertion and the energy of the charge transfer transition in the donor-acceptor complex is illustrated by the potential energy curves below:

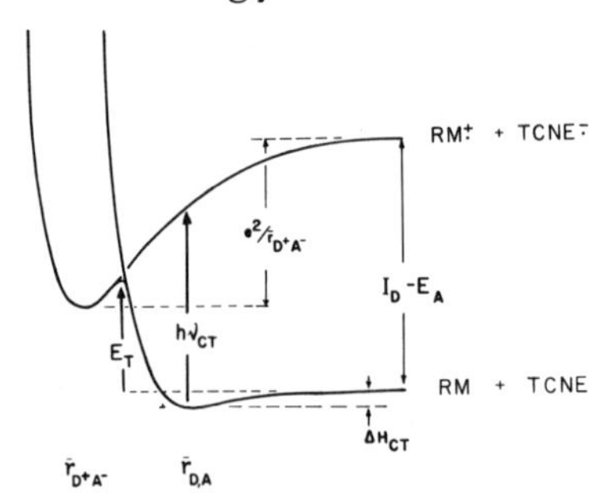

Figure 15. The relationship between thermal (E_T) and photochemical ($h\nu_{CT}$) activation of electron transfer proceeding from the charge transfer complex.

It can be seen from this formulation that the optical transition energy $h\nu_{CT}$ and the thermal activation energy E_T both depend on the potential energy of the ion pair. Lowering the ionization potential of the donor will cause a red shift in the charge transfer band and also lead to an increase in the rate of the thermal reaction. Indeed, there exists a linear correlation between log k for insertion and the frequency of the charge transfer band of a series of dialkylmercury compounds $RHgCH_3$ with TCNE, i.e., $\log k = \alpha \nu_{CT}$, where α is the proportionality constant. According to this mechanism, the actual transfer of an alkyl group from dialkylmercury to TCNE occurs in eq 54, subsequent to the rate-limiting electron transfer step in eq 53. The rapid alkylation of TCNE involves the transfer of an alkyl group from $R_2Hg^{+\cdot}$ which is metastable (vide infra).

The Mechanism of Insertion--Thermal and Photochemical Processes. The observation of the charge transfer complex of alkyltin and TCNE does not, by itself, prove that the complex lies along the reaction pathway to insertion. Complex formation may represent an unrelated side reaction. The difference lies in whether the rate-limiting, second-order rate constant k_T for electron transfer is a product, $K_{CT}k_{et}$, as represented in eqs 55 and 56,

$$R_4Sn + TCNE \xrightleftharpoons{K_{CT}} [R_4Sn\ TCNE] \quad (55)$$

$$[R_4Sn\ TCNE] \xrightarrow{k_{et}} [R_4Sn^{+\cdot}\ TCNE^{-\cdot}] \quad (56)$$

or a simple bimolecular constant representing the direct reaction of alkyltin and TCNE, distinct from the charge transfer complex as in eq 57.

$$[R_4Sn\ TCNE] \xrightleftharpoons{K_{CT}} R_4Sn + TCNE \xrightarrow{k_{et}} R_4Sn^{+\cdot} + TCNE^{-\cdot} \quad (57)$$

The reasons for favoring electron transfer to proceed directly from the alkyltin-TCNE complex derive from (1) the correlation of the formation constant K_{CT} with the phenomenological rate constant k_T, as well as the intimate relationship between (2) the photochemical activation and (3) the thermal activation of electron transfer. Following a discussion of these mechanistic points, we wish to consider (4) the nature and fate of the ion pair as an intermediate common to both thermal and photochemical activation and (5) steric effects involved in the electron transfer within the charge transfer complex.

Correlation of K_{CT} and k_T for the Thermal Insertion Reaction. The measured second-order rate constant k_T for the reaction of alkyltin and TCNE is plotted against the formation constant K_{CT} for the charge transfer complex in

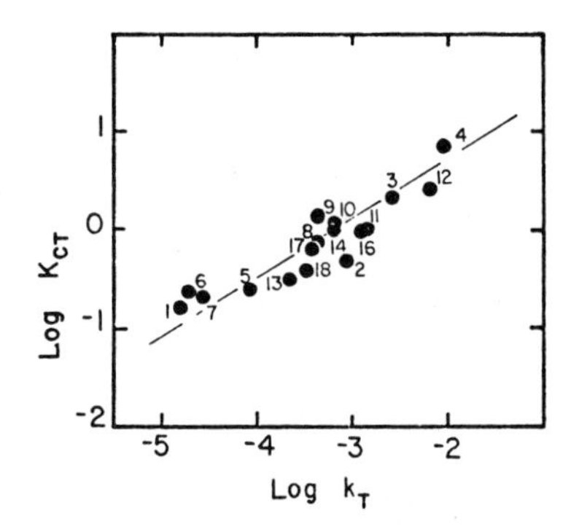

Figure 16. The parallel between formation constants of charge transfer complexes and the thermal rates of insertion. Numbers refer to tetraalkyltin compounds in Table VIII.

Figure 16. Despite the absence of any correlation of K_{CT} with the ionization potential I_D of tetraalkyltin or the frequency ν_{CT} of the charge transfer complex due to steric effects as discussed above, there is a reasonable correlation with the overall second-order rate constant. Such a parallel relationship between K_{CT} and k_T is more in keeping with the charge transfer complex as an intermediate, rather than as an unrelated side product.

Photochemical Activation of Insertion. Photoinsertion, resulting from irradiation directly at 436 nm or at 546 nm, where only the charge transfer absorption occurs, must necessarily proceed via excitation of the charge transfer complex and not that of either alkyltin or TCNE alone.

$$[R_4Sn\ TCNE] \xrightarrow{h\nu_{CT}} [R_4Sn^{+\cdot}\ TCNE^{-\cdot}] \qquad (58)$$

Moreover, esr studies demonstrate that alkyl radicals and TCNE anion-radicals are intermediates formed simultaneously during this photoactivation. They must result from a dark reaction following electron transfer, i.e.,

$$[R_4Sn^{+\cdot}\ TCNE^{-\cdot}] \xrightarrow{fast} [R\cdot\ R_3Sn^{+}\ TCNE^{-\cdot}] \qquad (59)$$

in accord with the known instability of the tetraalkyltin cation-radical. The lifetime of the pair of caged radicals in eq 59 is exceedingly short, and their direct esr observation is only allowed by the physical constraints imposed by the frozen matrix. Radicals produced during charge transfer excitation must be intermediates in photoinsertion since the quantum yield of 0.2, measured as a lower limit for radical production in a frozen matrix at -175°C, is still rather large and approaches the quantum yield of one, measured in solution for the photoinsertion process itself. These observations are readily accounted for by the sequence of steps described in eq 55 and 56. Photoinsertion must then follow directly from the cage collapse of these fragments formed in eq 59, i.e.,

$$[R\cdot\ R_3Sn^{+}\ TCNE^{\overline{\cdot}}] \xrightarrow{fast} [R_3Sn\text{-}TCNE\text{-}R] \qquad (60)$$

The mechanism of photoinsertion is thus represented by the sequence of reactions given by eqs 55, 56, 59 and 60.

Thermal Activation of Insertion. The observed second-order kinetics indicates that both alkylmetal and TCNE are present in the rate-limiting step for insertion. The importance of electron transfer in the transition state is reflected in the parallel relationship between the thermal rates of insertion and the energetics of electron detachment measured independently by the ionization potentials.

The potential energy diagrams in Figure 15 illustrate how photoactivation via the charge transfer transition $h\nu_{CT}$ is related to the thermal activation, designated as E_T. Indeed, among a limited series of methylethyllead compounds with similar steric properties, there is a reasonable linear relationship between $h\nu_{CT}$ and log k_T for photochemical and thermal insertion, respectively. A similar general trend in this correlation also pertains to the alkyltin analogs examined here. This correlation, coupled with the relationship observed between log k_T and the K_{CT}, leads to the conclusion that electron transfer proceeds from the same charge transfer complex,

$$[R_4Sn\ TCNE] \longrightarrow [R_4Sn^{\dot{+}}\ TCNE^{\overline{\cdot}}] \qquad (61)$$

which was proven to be directly involved in photochemical activation. The subsequent reactions leading to insertion are the same as the rapid dark reactions in eqs 59 and 60 presented for the photoinduced insertion. Accordingly, thermal and photoactivation of insertion share common mechanistic pathways. Any difference may lie in the nature of the paramagnetic ion pair resulting from electron transfer within the charge transfer complex as represented in eqs 58 and 61.

Ion Pairs as Common Intermediates in Thermal and Photochemical Insertion. The Franck-Condon limitations placed on the photoinduced electron transfer in eq 58 restrict the intermolecular separation in the excited ion pair to that of the charge transfer complex. In the thermal process, the same or a similar ion pair is also an intermediate derived by electron transfer in eq 61. Since these paramagnetic ion pairs are formed subsequent to the rate-determining step, their properties are best examined either by direct spectroscopic examination or by product selectivity.

Attempts to observe the triplet esr spectrum of the ion pair, produced either thermally or photochemically in solution or in a frozen matrix, were all consistently unsuccessful. The negative results of the CIDNP studies also point out that the $R_4Sn^{+\cdot}$ moiety is very short-lived. The direct comparison between the thermally and the photochemically induced formation of the ion pair, however, can be made by examining the selectivity in the fragmentation patterns of the alkyltin moiety prior to insertion. For example, the insertion into either the Me-Sn or the Et-Sn bond in the series of methylethyltin compounds is governed by the scission of the relevant bond in the paramagnetic alkyltin moiety represented above in eq 44. The extent to which fragmentation of this cation-radical proceeds from an excited state or from a geometrically distorted configuration or is influenced by $TCNE^{-\cdot}$, its counterion within the cage, would be reflected in changes in ethyl/methyl selectivity for insertion. The striking similarities of S(Et/Me) for both the thermal and photochemical processes strongly suggest that insertion proceeds from more or less the same paramagnetic ion pair. It is noteworthy that this selectivity is reasonably close to that (~6) observed in the unimolecular fragmentation of the molecule ion generated upon electron impact in the gas phase.

The observation of stable TCNE radicals in solution, either as $TCNE^{-\cdot}$ and $R_3SnTCNE\cdot$, arises by a side reaction. Integration of the esr signal indicates that these species generally constitute <0.1% of the reaction. However, measurement of the esr linewidth dependence of TCNE radicals as a function of TCNE concentration indicates that they can undergo exchange at rates of $3 \times 10^9\ M^{-1}\ sec^{-1}$. Similar exchanges lead to broadening of the nmr lines in the absence of an acetic acid quench. According to the Scheme presented in eqs 59 and 60, these radicals arise from partial diffusive separation from the cage, i.e.,

$$[R\cdot\ R_3Sn^+\ TCNE^{-\cdot}] \xrightarrow{\text{diffuse}} R_3SnTCNE\cdot + R\cdot \quad , \quad \text{etc.}$$

Unfortunately, our attempts to observe CIDNP effects associated with such a competition have been unsuccessful as yet.

Steric Effects in Electron Transfer from Charge Transfer Complexes. If the ion pairs described in the preceding section resulted from simple electron transfer between alkyltin and TCNE, it is expected that the rates (log k_T) would correlate linearly with the ionization potentials of the alkyltin compounds. Indeed, such a linear correlation can be observed in Figure 17 for the insertion reaction with a limited series of methylethyllead compounds with similar steric properties. If the correlation is extended to the

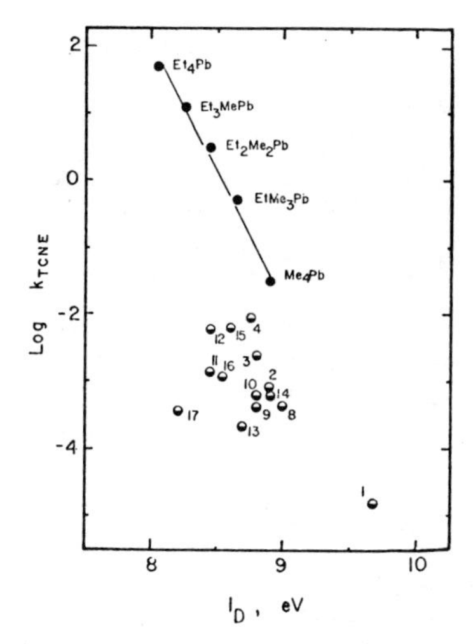

Figure 17. Steric effects in the correlation of the thermal rates of insertion and the ionization potentials of tetraalkyltin designated by numbers in Table VIII. Comparison with methylethyllead compounds.

greater variety of alkyltin structures available in this study, it shows the same general trend, but with considerable scatter. However, a closer examination of the data reveals that deviations are systematic and most marked with sterically hindered compounds, increasing roughly in the order: t-Bu > i-Bu > i-Pr > Et ≫ Me. Tetra-neopentyltin, the most sterically hindered compound, does not react at all.

The same general steric effects are shown in the correlation of log k_T with the charge transfer frequency ν_{CT} and lends further support to the charge transfer complex as an intermediate in both thermal and photochemical insertions. The intermolecular distance between R_4Sn and TCNE in the transition state for electron transfer must be sufficiently small to allow for these variations. Indeed, a comparison with the inner-sphere pathway for electron transfer between the same alkyltin compounds and hexachloroiridate(IV) suggests that TCNE may have penetrated the coordination sphere of alkyltin sufficiently to cause significant distortion of the tetrahedral tin structure.

C. Quantitative Evaluation of Steric Effects in Electron Transfer and Charge Transfer

For TCNE interacting with a series of related alkylmetals, it follows from eq 51 that the difference in the electrostatic terms ΔE between a particular alkylmetal RM relative to a chosen reference RM_0 is:

$$\Delta E \;=\; \Delta I_D - \Delta h\nu_{CT} \tag{62}$$

where ΔI_D is the difference in the ionization potentials between RM and RM_0 and $\Delta h\nu_{CT}$ is the difference in their charge transfer energies. The conversion of this energy difference to a rate factor $\Delta \log k$ is given by eq 63.

$$\Delta E \;=\; 2.3\,RT\;\Delta \log k \tag{63}$$

[At this juncture, it is convenient to consider the rate factor $\Delta \log k$ simply as a contribution of steric effects to the rate of electron transfer.] A corrected value of the electron transfer rate constant $\log k^{corr}_{TCNE}$ may be expressed as:

$$\log k^{corr}_{TCNE} \;=\; \log k_{TCNE} + \Delta \log k \tag{64}$$

where k_{TCNE} is the experimental second-order rate constant. Under these circumstances, $\log k^{corr}_{TCNE}$ represents the electron transfer rate constant under hypothetical conditions of constant steric effects (i.e., relative to that of the chosen standard $RM_0 = Me_4Sn$). Stated alternatively, $\log k^{corr}_{TCNE}$, or its equivalent $\Delta G^{\ddagger}_{corr}$, is the form to be related to the driving force of the electron transfer (ΔG^0), in the absence of steric effects. Figure 18 shows the new correlation of the data previously presented in Figure 17 for the electron transfer rate constants between TCNE and a series of tetraalkyltin compounds. The dashed line in Figure 18 is arbitrarily drawn with a Brönsted slope of $\alpha = 1$ through the point for Me_4Sn.

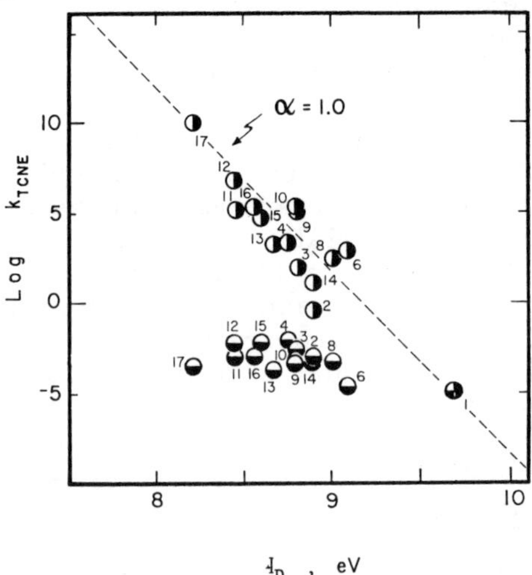

Figure 18. The linear free energy relationship of the electron transfer rate constant for tetraalkyltin and TCNE: ◒ before and ◑ after correction for steric effects.

Let us assume for the moment that steric effects in electron transfer from tetraalkyltin to hexachloroiridate is also the same as those to tetracyanoethylene. Then, the electron transfer rate constant log k_{Ir} can be corrected by an amount Δlog k to afford log k_{Ir}^{corr} in a manner similar to eq 64. The linear free energy plot is presented in Figure 19, with the dashed line showing the correlation of k_{Ir}^{corr} with a Brönsted slope of $\alpha = 1$.

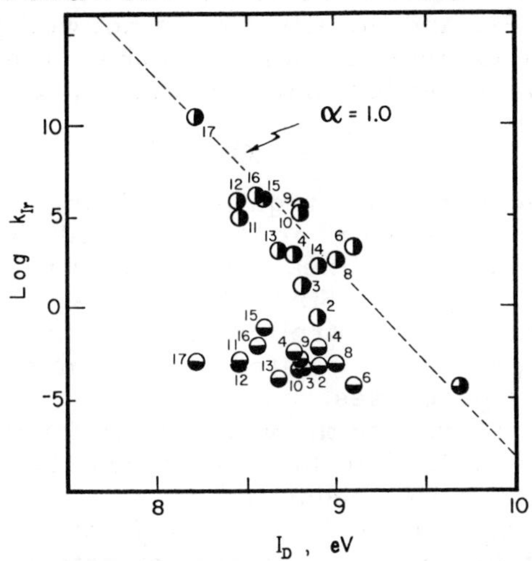

Figure 19. The linear free energy relationship of the electron transfer rate constant for tetraalkyltin and $IrCl_6^{2-}$: ⊖ before and ◑ after correction for steric effects.

It is striking that a single linear free energy relationship, i.e.,

$$\log k \; = \; -9.8\, I_D + \text{constant} \tag{65}$$

can be obtained empirically by using a simple correction, Δlog k from eq 63, to correlate the rates of electron transfer to both TCNE and $IrCl_6^{2-}$ from a wide variety of the organometals R_4Sn, R_4Pb and R_2Hg containing alkyl ligands with diverse steric and polar parameters.

Equation 65 is equivalent to the free energy changes expressed more familiarly as:

$$\Delta G^{\ddagger} \; = \; 1.0\, \Delta G^0 + \text{constant} \tag{66}$$

The coefficient 1.0 in eq 66 is the Brönsted coefficient α, and it differs from $\alpha = 0.5$ for outer-sphere electron transfer found for FeL_3^{3+}, as described above.

It is important to emphasize that the derivation of the linear free energy relationship in eq 65, or its equivalent in eq 66, obtains directly from the experimental data by a purely operational approach with no extensive assumptions.

We now proceed to its possible implications, especially as it may relate to the inner-sphere mechanism for electron transfer.

In the region of weak overlap as in outer-sphere mechanisms, the Marcus eq 45 provides a theoretical basis for electron transfer rates, as shown for alkylmetals and FeL_3^{3+}. When electron transfer involves considerable resonance splitting as in a variety of inner-sphere mechanisms, eq 45 no longer applies, and the situation is not well provided by theory. At one extreme of an inner-sphere mechanism, where $\Delta G^0 \geq \lambda$ (i.e., endergonic processes), Marcus predicted the relationship:

$$\Delta G^{\ddagger} = \Delta G^0 + w^p \qquad (67)$$

where w^p is the work term required to bring the products to within a mean separation $r^{\ddagger}$ in the activated complex. Equation 67 corresponds to a linear free energy relationship with the Brönsted slope $\alpha = 1$. Since the microscopic reverse represents a diffusion-controlled electron transfer, in qualitative terms eq 66 can be related to the Hammond postulate for endothermic processes.

In order to evaluate the work terms for various alkylmetals in eq 67, values of the free energy change ΔG^0 are required. The reversible reduction potentials of TCNE and $IrCl_6^{2-}$ in acetonitrile are 0.46 and 0.67 volts vs. SHE, respectively. Unfortunately the oxidation potentials of tetraalkyltin and -lead are not experimentally measurable due to the irreversibility of the cyclic voltammograms. However, there is an empirical linear correlation relating E^0_{RM} to I_D of these alkylmetals, i.e.,

$$I_D = 1.8\, E^0_{RM} + \text{constant} \qquad (68)$$

which derives from eqs 45 and 46. Thus to evaluate ΔG^0 for various alkylmetals, the absolute measurement of E^0_{RM} is required for only one alkylmetal. We resort again to the comparative method, and arbitrarily set $E^0_{Me_4Sn} = 1.39$ volts to allow:

$$\Delta G^0_{Me_4Sn} = \Delta G^{\ddagger}_{Me_4Sn}\,, \text{ i.e., } w^p_{Me_4Sn} = 0$$

The free energy change ΔG^0 for the other alkylmetals can be derived from eq 68. The difference in work terms designated as Δw^p can be evaluated relative to Me_4Sn, i.e.,

$$\Delta w^p = \Delta\Delta G^{\ddagger} - \Delta\Delta G^0 \qquad (69)$$

where $\Delta\Delta G^{\ddagger} = \Delta G^{\ddagger}_{RM} - \Delta G^{\ddagger}_{Me_4Sn}$ and $\Delta\Delta G^0 = \Delta G^0_{RM} - \Delta G^0_{Me_4Sn}$. According to this formulation, steric effects in an inner-sphere mechanism for electron transfer are embodied in the work term.

The work term becomes largely an electrostatic interaction in the charge transfer model for an inner-sphere mechanism. According to Mulliken, the charge transfer transition in eq 58 corresponds to an electronic excitation from the charge transfer (ground state) complex to an excited ion-pair state. The potential energy surfaces are depicted in Figure 15, where the crossing occurs at the transition state for thermal electron transfer. The activation energy E_T for electron transfer can be represented as:

$$E_T = I_D - E_A - e^2/r^{\ddagger} - \Delta E_{solv} \qquad (70)$$

where $r^{\ddagger}$ is the mean separation of the ion-pair in the transition state and the solvation terms are collected in ΔE_{solv}. The work term in eq 67 may be ascribed totally to an electrostatic potential, i.e., $w^p = -e^2/r^{\ddagger}$ for the ions, $RM^{+\cdot}$ and $TCNE^{-\cdot}$ or $IrCl_6^{3-}$. In solution, this work term must be corrected for solvation, i.e.,

$$w^p = -e^2/r^{\ddagger} + \Delta\Delta H \qquad (71)$$

where $\Delta\Delta H$ is the difference in solvation energy of the products and the transition state. Using the comparative method, we define the work term w^p for alkylmetals relative to that of the reference Me_4Sn, i.e.,

$$\Delta w^p = w^p_{RM} - w^p_{Me_4Sn} = -e^2/r^{\ddagger}_{RM} + \Delta\Delta H_{RM} + e^2/r^{\ddagger}_{Me_4Sn} - \Delta\Delta H_{Me_4Sn}$$

If the changes in solvation energies are constant, which is reasonable for a series of related alkylmetals, then:

$$\Delta w^p = e^2/r^{\ddagger}_{Me_4Sn} - e^2/r^{\ddagger}_{RM}$$

The work term defined by this electrostatic model has the same functional form of ΔE obtained independently in eqs 51 and 62 (where $G_1 = e^2/r_{DA}$) from the charge transfer interaction, i.e.,

$$\Delta E = e^2/r_{Me_4Sn\text{-}TCNE} - e^2/r_{RM\text{-}TCNE}$$

Indeed there is a reasonably good agreement between Δw^p and ΔE for tetraalkyltin and especially for methylethyllead. The general trend of the agreement is unmistakable, and it strongly implies that the mean separation in the charge transfer complex is a factor in the transition state structure, particularly as it relates to steric effects.

The same analysis can be carried out for electron transfer to $IrCl_6^{2-}$, and the agreement between Δw^p and ΔE is also reasonably good. It is noteworthy that the work term for electron transfer between Me_4Sn and TCNE is 4.2 kcal mol^{-1} less than that for Me_4Sn and $IrCl_6^{2-}$. If this difference is wholly attributable to an electrostatic term, a difference of 0.1 Å is required to account for the change in distance [which interestingly is the difference in van der Waal's radii of Cl(1.8 Å) in $IrCl_6^{2-}$ and C(1.7 Å) in TCNE].

D. Facile Cleavage of Organometals in Radical-Ions

Studies with organomercurials and organolead compounds have shown that cleavage of the carbon-metal bond in these compounds occurs readily only after electron transfer. In other words, the radical-ion is much more labile than its diamagnetic precursor, i.e.,

$$\text{R-m} \xrightarrow{-\epsilon} \text{R-m}^{+} \xrightarrow{\text{fast}} \text{R}\cdot + \text{m}^{+}$$

where m = HgR, PbR_3, SnR_3. Similarly, studies with other organotransition metals show the same results,(40) e.g.,

$$(CH_3CH_2)_2Pt^{II}(PPh_3)_2 \xrightarrow{-\epsilon} (CH_3CH_2)_2Pt^{III}(PPh_3)_2^{+}$$

$$(CH_3CH_2)_2Pt^{III}(PPh_3)_2^{+} \xrightarrow{\text{fast}} CH_3CH_2\cdot + CH_3CH_2Pt^{II}(PPh_3)_2^{+}$$

Other modes of cleavage of the alkyl-metal bonds are also available to radical-ions. For example, organocobalt complexes undergo a facile cationic scission of an alkyl group.(41)

$$\text{R-Co}^{III}(DMG)_2 \xrightarrow{-\epsilon} \text{RCo}^{IV}(DMG)_2^{+} \xrightarrow{\text{fast}} \begin{cases} \xrightarrow{H_2O} \text{ROH} + \text{Co}^{II}(DMG)_2 + H^{+} \\ \xrightarrow{Br^{-}} \text{RBr} + \text{Co}^{II}(DMG)_2 \end{cases}$$

On the other hand, stable organo-nickel and gold complexes are known to eliminate dimers rapidly on conversion to the radical-ions:(42)

$$Ph_2Ni^{II}(PEt_3)_2 \xrightarrow{-\epsilon} Ph_2Ni^{III}(PEt_3)_2^{+} \xrightarrow{\text{fast}} \text{Ph-Ph} + Ni^{I}(PEt_3)_2^{+}$$

No doubt other modes of fragmentation of alkyl-metal bonds will be found as additional organometallic systems are examined.

Facile cleavages of the organometals described above depend on the availability of electron acceptors to effect charge transfer. The efficacy of the acceptor depends to a large degree on its electron affinity (compare the illustration of the potential energy curves in Figure 15 and eq 51. Tetracyanoethylene represents a viable example of an olefinic acceptor, whose electron affinity can be modified by replacement with other substituent groups. Carbon tetrachloride, polyhaloalkanes, halogens and alkyl halides, to a lesser degree, can function as electron acceptors. Molecular oxygen, sulfur dioxide, quinones, ketones, nitroalkanes and arenes, imines, esters and carbonium ions are other examples.(43)

For a given electron acceptor, the charge transfer process will also be facilitated by lowering the ionization potential of the organometal donor. Main group organometals such as dialkylmercury and tetraalkyllead are σ-donors (vide supra), whereas the HOMO in many transition organometals are nonbonding d orbitals. In both types of electron donors, it is worth bearing in mind that the ionization potentials are lower in the anionic species, i.e., the metalate complex, compared to the neutral, uncharged counterpart. Thus, organometal anions are generally among the best electron donors (and nucleophiles) available.(44) The judicious use of the myriad combinations of organometal donors with a variety of these electron acceptors will, no doubt, provide for a vast array of potentially interesting reactions to be investigated. In less favorable situations, photochemical promotion could also be employed.(22)

IV. Redox Chain Reactions in Metal-Catalyzed Processes

We now turn to metal catalysis of organic reactions in which the propagation sequences include as their catalytic cycles various combinations of the radical-forming and radical-consuming steps described above. Some examples of metal-catalyzed processes proceeding in this manner are illustrated in Table IX.

Table IX. Redox Chain Reaction: Propagation Steps.

A. Catalyzed Decomposition of Peroxides:(45)

$$RCO_2\text{-}O_2CR + Cu^{I} \longrightarrow RCO_2Cu^{II} + CO_2 + R\cdot$$

$$R\cdot + Cu^{II} \longrightarrow R_{ox} + Cu^{I}, \text{ etc.}$$

B. Addition:(46)

$$Fe^{II} + CCl_4 \longrightarrow Cl_3C\cdot + Fe^{III}Cl$$

$$Cl_3C\cdot + C{=}C \longrightarrow Cl_3C\text{-}C\text{-}C\cdot$$

$$Cl_3C\text{-}C\text{-}C\cdot + Fe^{III}Cl \longrightarrow Cl_3C\text{-}C\text{-}CCl + Fe^{II}, \text{ etc.}$$

C. Sandmeyer Reaction:(47)

$$ArN_2^+ + Cu^{I}Cl \longrightarrow Ar\cdot + N_2 + Cu^{II}Cl$$

$$Ar\cdot + Cu^{II}Cl \longrightarrow ArCl + Cu^{I}, \text{ etc.}$$

D. Oxidative Decarboxylation of Acids:(48)

$$RCO_2Pb^{IV} + R\cdot \longrightarrow R_{ox} + RCO_2Pb^{III}$$

$$RCO_2Pb^{III} \longrightarrow Pb^{II} + CO_2 + R\cdot, \text{ etc.}$$

E. Oxidative Addition:(49)

$$R\cdot + Ir^{I} \longrightarrow R\text{-}Ir^{II}$$

$$RIr^{II} + RBr \longrightarrow RIr^{III}Br + R\cdot, \text{ etc.}$$

F. Reductive Elimination of Alkene:(29)

$$HC\text{-}C\text{-}Hg^{II}R + R\cdot \longrightarrow RH + \dot{C}\text{-}C\text{-}HgR$$

$$\dot{C}\text{-}C\text{-}HgR \longrightarrow C{=}C + Hg^{0} + R\cdot, \text{ etc.}$$

A. Metal Catalysis in Oxidative Decarboxylation

The complex interplay between several paramagnetic intermediates is shown by the copper catalysis of the oxidative decarboxylation of acids by lead tetraacetate (described in entry D in Table IX). Under these conditions the stoichiometry of the reaction is altered to:

$$RCO_2H + Pb^{IV} \longrightarrow R(-H) + CO_2 + H^+ + Pb^{II}$$

whereas in D, R_{OX} is usually acetates (i.e., ROAc). The catalytic mechanism is shown below:

Scheme VI:

$$RCO_2H + Pb^{IV} \xrightarrow{Cu^I} R\text{-}H + CO_2 + Pb^{II} + 2\,H^+$$

Mechanism

$$RCO_2H + Pb^{IV} \longrightarrow RCO_2Pb^{IV} + H^+$$

RCO_2Pb^{IV} → RCO_2Pb^{III} (with Cu^I → Cu^{II}); Cu^{II} → Cu^I (with $R\cdot$ → $R\text{-}H + H^+$); RCO_2Pb^{III} → $R\cdot$ + $CO_2 + Pb^{II}$

B. Metal-Catalysis in Disproportionation and Coupling of Alkyl Groups

The decomposition of organometals described above relates directly to the mechanism of the Kharasch reaction, in which catalytic amounts of transition metal salts are used to effect the reaction of Grignard reagents with alkyl halides.(50) The observation of a complex mixture of alkane, alkene and alkyl dimers earlier led to the belief that alkyl radicals were again prime intermediates. I wish to describe here the mechanistic pathways for two representative catalytic systems: silver for coupling and iron for disproportionation.

$$CH_3CH_2MgBr + CH_3CH_2Br \xrightarrow{Ag} CH_3CH_2CH_2CH_3 + MgBr_2 \quad \text{coupling}$$

$$CH_3CH_2MgBr + CH_3CH_2Br \xrightarrow{Fe} CH_3CH_3 + CH_2{=}CH_2 + MgBr_2 \quad \text{disproportionation}$$

Other catalytic metal systems can be viewed as appropriate combinations of these two processes in a competitive situation.

Silver catalysis in the coupling of a Grignard reagent with an alkyl halide affords a mixture of three alkyl dimers (but no alkenes) when dissimilar alkyl groups are employed. The scrambling of alkyl groups in the coupled products is not due to prior alkyl exchange between the Grignard reagent and the organic halide.(51) Scheme VII summarizes the pathway for the catalytic coupling reaction.

Scheme VII:

$$RMgX + AgX \longrightarrow RAg + MgX_2 \quad (72)$$

$$RAg + R'Ag \longrightarrow [RR, RR', R'R'] + 2Ag \quad (73)$$

$$Ag + R'\text{-}X \xrightarrow{slow} AgX + R'\cdot \quad (74)$$

$$R'\cdot + Ag \longrightarrow R'Ag \quad (75)$$

According to Scheme VII, the scrambling of alkyl groups occurs in eq 73 during the reductive coupling of a pair of alkylsilver intermediates. These alkylsilver species are derived by two independent pathways involving: (a) the metathesis of Grignard reagent with silver(I) in eq 72 and (b) the two-step oxidative addition of alkyl halide to silver(0) in eqs 74 and 75. It can be shown independently that eq 72 is rapid, so the kinetics of the catalytic coupling:

$$\frac{d(R_2)}{dt} = k(Ag^{I})(RBr)(RMgX)^0$$

suggest that bromine transfer in eq 74 is rate-limiting. The effect of the structural variation of R'Br on the second-order rate constant shown in Table V is consistent with that of other halogen atom transfers. Importantly, only that alkyl group R' derived from the organic halide passes through the catalytic cycle as an alkyl radical, but not that R derived from the Grignard reagent. The same conclusion is drawn from the selective trapping of only R'• with 1,3-butadiene and styrene. Furthermore, the coupling of cis-propenyl Grignard reagent with methyl bromide affords cis-butene-2 stereospecifically,

$$\text{cis-CH}_3\text{CH=CHMgBr} + CH_3Br \xrightarrow{Ag^{I}} \text{cis-butene-2} \xleftarrow{Ag^{I}} \text{cis-CH}_3\text{CH=CHBr} + CH_3MgBr$$

whereas that of cis-propenyl bromide and methyl Grignard reagent leads to a mixture of cis- and trans-butene-2 consistent with the fast isomerization of the propenyl radical, known to occur in about 10^{-9} sec.(52)

Iron catalysis in the disproportionation of alkyl groups during the treatment of Grignard reagents with alkyl halides leads principally to a mixture of alkanes and alkenes, with no alkyl dimers.(53) The reactivities of the alkyl bromides [t-butyl > isopropyl > n-propyl] as well as the kinetics of disproportionation are the same as the silver-catalyzed coupling described above and suggest a similar mechanism. Moreover, the CIDNP observed only in the nmr spectrum of the alkene derived from the alkyl halide,(54) together with the known reductive disproportionation of alkyliron support the pathway below:

Scheme VIII:

$$RMgX + Fe^{II}X_2 \longrightarrow RFe^{II}X + MgX_2 \quad (76)$$

$$RFeX + R'FeX \longrightarrow [RH, R'H, R\text{-}H, R'\text{-}H] + 2\,Fe^{I}X \quad (77)$$

$$Fe^{I}X + R'\text{-}X \xrightarrow{slow} Fe^{II}X_2 + R'\cdot \quad (78)$$

$$R'\cdot + Fe^{I}X \longrightarrow R'Fe^{II}X\text{, etc.} \quad (79)$$

It is important here to emphasize the caveat that the qualitative observation of CIDNP alone does not prove the importance of radical processes, since we know from the absence of dimers that bimolecular radical disproportionation cannot represent the principal route to alkenes in this system.

To summarize, the foregoing mechanistic studies of the stoichiometric decomposition of transition alkylmetals to afford disproportionation and coupling products clearly rule out bimolecular reactions of alkyl radicals. [If alkyl radicals are intermediates in these decompositions, alternative pathways must be delineated.] Furthermore, the catalytic disproportionation and coupling of Grignard reagents with alkyl halides (the Kharasch reactions) involve similar organometals as intermediates. But, in addition, free radicals in the catalytic system arise selectively from the organic halide only during oxidative regeneration of the metal species (i.e., in eqs 74 and 78).

V. Conclusions and Prospects

Free radical chemistry can be considerably expanded by including metal centers. In this report I have attempted to describe how the multiple oxidation states of metals allow a variety of new structures and reactions to evolve from the more traditional aspects of free radical chemistry. As reactive intermediates, paramagnetic species play a ubiquitous role in organometallic chemistry and in catalytic processes. No doubt, as further forays are made into this new frontier many unexpected and interesting developments will follow.

Acknowledgment:

I wish to thank all my coworkers mentioned in the references for their intense effort in making this research possible, and in particular C. L. Wong, S. Fukuzumi, W. A. Nugent and H. C. Gardner for their invigorating and creative ideas. I also wish to thank the National Science Foundation and the donors of the Petroleum Research Fund of the American Chemical Society for financial support.

References

1. G. Ferraudi, Inorg. Chem. 17, 2506 (1978).
2. (a) H. G. Kuivila, Acc. Chem. Res. 1, 299 (1968). (b) A. G. Davies, Adv. Chem. Ser. 157, 26 (1976).
3. P. J. Krusic, W. Mahler and J. K. Kochi, J. Am. Chem. Soc. 94, 6033 (1972).
4. T. P. Fehlner, J. Ulman, W. A. Nugent and J. K. Kochi, Inorg. Chem. 15, 2544 (1976).
5. S. Evans, J. C. Green, P. J. Joachim, A. F. Orchard, D. W. Turner and J. P. Maier, J. Chem. Soc., Faraday Trans. II, 905 (1972).
6. C. L. Wong, K. Mochida, A. Gin, M. A. Weiner and J. K. Kochi, J. Org. Chem., submitted for publication.
7. G. Beltram, T. P. Fehlner, K. Mochida and J. K. Kochi, J. Electron Spectrosc., in press.
8. J. H. D. Eland, Int. J. Mass Spectrosc. Ion Phys. 4, 37 (1970).
9. J. L. Franklin and P. W. Harland, Ann. Rev. Phys. Chem. 25, 485 (1974).
10. K. S. Chen, F. Bertini and J. K. Kochi, J. Am. Chem. Soc. 95, 1340 (1973).
11. (a) P. J. Krusic and J. K. Kochi, J. Am. Chem. Soc. 93, 846 (1971). (b) J. K. Kochi, Adv. Free Radical Chem. 5, 189 (1975).
12. T. Kawamura, P. Meakin and J. K. Kochi, J. Am. Chem. Soc. 94, 8065 (1972).
13. T. Kawamura and J. K. Kochi, J. Organometal. Chem. 47, 79 (1973).
14. P. J. Krusic and J. K. Kochi, J. Am. Chem. Soc. 91, 6161 (1969).
15. W. A. Nugent and J. K. Kochi, J. Organometal. Chem. 124, 371 (1977).
16. (a) P. H. Kasai and D. McLeod, Jr., J. Am. Chem. Soc. 97, 6602 (1975). (b) Cf. also H. Huber, G. A. Ozin and W. J. Power, Inorg. Chem. 16, 979 (1977).
17. K. S. Chen, D. Y. H. Tang, L. K. Montgomery and J. K. Kochi, J. Am. Chem. Soc. 96, 2201 (1974).
18. D. D. Dodd, M. D. Johnson and B. L. Lockman, J. Am. Chem. Soc. 99, 3664 (1977).
19. J. Halpern, J. Topich and K. I. Zamaraev, Inorg. Chim. Acta, 20, L21 (1976).
20. D. Dolphin, T. Niem, R. H. Felton and I. Fujita, J. Am. Chem. Soc. 97, 5289 (1975).
21. (a) M. Simic, M. Z. Hoffman and N. V. Brezniak, J. Am. Chem. Soc. 99, 2166 (1977). (b) I. I. Creaser and A. M. Sargeson, J. Chem. Soc., Chem. Commun., 324 (1975).

22. (a) C. R. Bock and E. A. K. von Gustorf, Adv. Photochem. 10, 221 (1977). (b) P. C. Ford, J. D. Peterson and R. E. Hintze, Coord. Chem. Revs. 14, 67 (1974).
23. (a) A. Tkac, K. Vesely and L. Omelka, J. Phys. Chem. 75, 2575 (1971). (b) See, however, K. U. Ingold, J. Phys. Chem. 76, 1385 (1972).
24. J. K. Kochi and F. F. Rust, J. Am. Chem. Soc. 83, 2017 (1961).
25. (a) J. K. Kochi and D. Buchanan, J. Am. Chem. Soc. 87, 853 (1965). (b) R. S. Nohr and J. H. Espenson, J. Am. Chem. Soc. 97, 3392 (1975).
26. B. L. Booth, R. N. Haszeldine and R. G. G. Holmes, J. Chem. Soc., Chem. Commun., 489 (1976).
27. P. J. Krusic and J. K. Kochi, J. Am. Chem. Soc. 91, 3942 (1969).
28. J. K. Kochi, Pure Appl. Chem. Supplement 4, 377 (1971).
29. W. A. Nugent and J. K. Kochi, J. Organometal. Chem. 124, 377 (1977).
30. (a) J. Halpern, Ann. N.Y. Acad. Sci. 239, 2 (1974). (b) H. G. Kuivila, Adv. Organometal. Chem. 1, 47 (1964). (c) J. K. Kochi and J. W. Powers, J. Am. Chem. Soc. 92, 137 (1970). (d) M. Tamura and J. K. Kochi, J. Am. Chem. Soc. 93, 1483 (1971).
31. S. Komiya, T. A. Albright, R. Hoffmann and J. K. Kochi, J. Am. Chem. Soc. 99, 8440 (1977).
32. T. T. Tsou and J. K. Kochi, J. Am. Chem. Soc. 100, 1634 (1978).
33. (a) W. V. Steele, Ann. Rep. Chem. Soc. A71, 103 (1974). (b) J. A. Connor, Topics Curr. Chem. 71, 71 (1977).
34. C. L. Wong and J. K. Kochi, J. Am. Chem. Soc., in press.
35. C. Walling and R. A. Johnson, J. Am. Chem. Soc. 97, 2405 (1975).
36. A. A. Schilt, "Analytical Applications of 1,10-Phenanthroline and Related Compounds," Pergamon Press, 1969.
37. (a) N. A. Clinton, H. C. Gardner and J. K. Kochi, J. Organometal. Chem. 56, 227 (1973). (b) N. A. Clinton and J. K. Kochi, J. Organometal. Chem. 56, 243 (1973).
38. J. Y. Chen, H. C. Gardner and J. K. Kochi, J. Am. Chem. Soc. 98, 2460 (1976).
39. S. Fukuzumi, K. Mochida and J. K. Kochi, J. Am. Chem. Soc., in press.
40. (a) J. Y. Chen and J. K. Kochi, J. Am. Chem. Soc. 99, 1450 (1977). (b) S. Komiya, unpublished results.
41. J. Halpern, M. S. Chan, J. Hanson, T. J. Roche and J. A. Topich, J. Am. Chem. Soc. 97, 1606 (1975).
42. M. Alemark and B. Akermark, J. Chem. Soc., Chem. Commun., 66 (1978). See also reference 32.

43. For listings of electron affinities see G. G. Christophorou, "Atomic and Molecular Radiation Physics," Wiley-Interscience, New York, 1971.
44. (a) Cf. R. E. Dessy and L. A. Bares, Acc. Chem. Res. 5, 415 (1972). (b) W. Tochterman, Angew. Chem. Int. Ed. 5, 351 (1966). (c) G. Wittig, Quart. Revs. 20, 191 (1966). (d) J. F. Normant, Synthesis, 63 (1972). (e) A. Tamaki and J. K. Kochi, J. Chem. Soc., Dalton Trans., 2620 (1973). (f) D. G. Morrell and J. K. Kochi, J. Am. Chem. Soc. 97, 7262 (1975).
45. J. K. Kochi, in "Free Radicals," Vol. I, Wiley-Interscience Pub., New York, 1973, p. 591 ff.
46. M. Asscher and D. Vofsi, J. Chem. Soc., 2261 (1961); 1887, 3921 (1963).
47. (a) S. C. Dickerman, K. Weiss and A. K. Ingberman, J. Org. Chem. 21, 380 (1956). (b) J. K. Kochi, J. Am. Chem. Soc. 79, 2942 (1957). (c) D. C. Nonhebel and W. A. Waters, Proc. Royal Soc. (A) 242, 16 (1957).
48. J. D. Bacha and J. K. Kochi, J. Org. Chem. 33, 83 (1968).
49. J. A. Osborn, "Prospects in Organotransition Metal Chemistry," Plenum Press, New York, 1975.
50. M. S. Kharasch and O. Reinmuth, "Grignard Reactions of Nonmetallic Substances," Prentice-Hall, Inc., New York, 1954.
51. M. Tamura and J. K. Kochi, J. Am. Chem. Soc. 93, 1483 (1971).
52. R. W. Fessenden and R. H. Schuler, J. Chem. Phys. 39, 2147 (1963).
53. M. Tamura and J. K. Kochi, J. Organometal. Chem. 31, 289 (1971).
54. R. B. Allen, R. G. Lawler and H. R. Ward, J. Am. Chem. Soc. 95, 1692 (1973).

THE ROLE OF POLAR EFFECTS AND BOND DISSOCIATION ENERGIES (BDE) IN RADICAL REACTIVITIES: THE IODINATION OF TOLUENES

WILLIAM A. PRYOR
DANIEL F. CHURCH
FELICIA Y. TANG
ROBERT H. TANG

DEPARTMENT OF CHEMISTRY
LOUISIANA STATE UNIVERSITY
BATON ROUGE, LOUISIANA 70803

Summary

This paper reports a new, independent study of the reaction of *tert*-butyl radicals with substituted toluenes; rho is found to be 0.5 at 80°C, in reasonable agreement with the value we reported earlier of 0.99 at 30°C. This confirmation of a positive rho value for the reaction of alkyl radicals with toluenes demonstrates that ring substituents both affect the bond dissociation energy (BDE) of the toluene and stabilize polar resonance structures in the transition state. We also report data on the iodination of substituted toluenes that allow the rho value for the reaction of iodine atoms with toluenes to be determined; rho is found to be −1.6 at 80°C. These iodination data also allow the calculation of the BDE of the benzylic C-H bonds ring-substituted toluenes. We find that electron-donating substituents weaken the BDE of the benzylic bond, and that the change is about 3 kcal/mole over a unit change in sigma constants.

PART I. INTRODUCTION

The rates of radical reactions are sensitive to the presence of polar substituents in the substrate molecule. This observation was first made by Mayo, Price, Bartlett and others in 1945-1947, and it has been supported by extensive findings in more recent years(1). We have felt that one of the most informative methods for probing the reality and quantitative importance of polar effects on radical reactions is by the application of the Hammett

ISBN 0-12-566550-4

sigma-rho equation, and we have published extensively on this subject (2-6). In this article, we review some of our earlier work on Hammett correlations of the reactions of alkyl radicals with toluenes, work that has been questioned recently by Tanner et al. (7), and we report a new study of the reaction of *tert*-butyl radicals with toluenes that is in qualitative agreement with the conclusions of our earlier study. We also describe the reaction of iodine atoms with toluenes and compare the rho value for the iodine atom with that obtained for the other halogens that have been studied. We show that the iodination of toluenes can be used to measure the effect of ring substituents on the BDE of the benzylic C-H bond, data that have not been available hithertofore.

HAMMETT TREAMENTS OF RADICAL REACTIONS: HISTORICAL PERSPECTIVE

The Hammett equation (8) was originally developed as a tool for studying ionic processes, and it was considered surprising that this treatment also can be applied to radical reactions (9). The original explanation of the correlation of radical reactions by the Hammett equation was that the transition states of radical reactions involve dipolar charge-separated forms as resonance structures; i.e., to the extent that radical reactions have dipolar resonance structures they can be treated like ionic reactions in Hammett correlations (9-11).

It was later pointed out, most notably by Russell et al. (12-13), that the Hammett correlation of radical reactions also reflects the reactivity of the radicals involved. For example, the reaction of chlorine atoms with toluenes has a rho value of -0.8 and that of bromine atoms a rho of -1.4 (13). If only the effects of substituents on charge-separated forms in the

$$Cl\cdot + ArCH_3 \longrightarrow [\overset{\delta-}{Cl\cdot} \quad H\text{--}\overset{\delta+}{C}H_2Ar] \qquad (1)$$

$$Br\cdot + ArCH_3 \longrightarrow [\overset{\delta-}{Br\cdot} \quad H\text{---}\overset{\delta+}{C}H_2Ar] \qquad (2)$$

Transition States

transition state were considered, the rho value for abstraction by the more electronegative chlorine atom would be expected to be more negative than that for bromine (12,13). However, Russell (12,13) argued that the reaction of chlorine atoms is much more exothermic than that for bromine atoms and therefore has a transition state that more closely resembles the reactants, as shown in equations 1 and 2 above. By an Evans-Polanyi argument (14), therefore, the substituent effect on the transition state for abstraction by chlorine should be smaller than that for bromine.

The effects of solvents on radical selectivity and reactivity have also been discussed(15-18). Russell and his coworkers proposed (16-18) that complexation of chlorine atoms by aromatic solvents accounted for the greater tertiary-to-primary hydrogen reactivity ratios observed for chlorinations carried out in benzene compared to CCl_4. The influence of aromatic solvents on Hammett correlations of the halogenation of toluenes is more complex, since complexation also alters the electronegativity of the halogen atom, and the problems associated with applying Russell's simple idea of complexation to Hammett correlations have been pointed out (15).

Recently, Zavitsas and Pinto (19) have suggested that Hammett correlations of the reactions of radicals with toluenes reflect only the effect of the substituents on the BDE of the benzylic C-H bonds in the toluenes. They postulated, on the basis of the indirect evidence then available, that electron-donating groups weaken the benzylic bond whereas electron-withdrawing groups strengthen it. Therefore, the reactions of all radicals with toluenes should lead to a negative rho value; all of the rho values reported in the literature at the time Zavitsas and Pinto published their suggestion were negative.

The Zavitsas hypothesis suggests two possible experimental tests. Firstly, since a strict BDE interpretation of rho predicts only negative rho values, a positive rho value for a radical reaction proves other factors also must influence rho. Secondly, the notion that substituents do in fact lead to variations in the benzylic BDE of toluenes was based on indirect evidence, and a direct method for the measurement of the effect of ring substituents on benzylic BDE is of considerable interest.

PART II. HAMMETT STUDIES OF ALKYL RADICALS

As we have indicated above, one crucial test of the Zavitsas hypothesis is to discover reactions that have positive rho values. Since substituent effects on BDE can only lead to negative rho values, a positive rho can be attributed to stabilizing effects of ring substituents on polar structures in the transition state, as shown below for hydrogen abstraction from toluene.

$$\text{Ph-CH}_3 + \cdot\text{X} \rightarrow [\text{Ph-}\dot{\text{C}}\text{H}_2\ \dot{\text{H}}\ \dot{\text{X}} \leftrightarrow \text{Ph-}\overset{-}{\ddot{\text{C}}}\text{H}_2\ \dot{\text{H}}\ \overset{+}{\dot{\text{X}}}] \rightarrow \text{Ph-}\dot{\text{C}}\text{H}_2 + \text{HX} \quad (3)$$

The dipolar structure shown in eq 3 would be important when the attacking radical is nucleophilic (i.e., electron-donating) with respect to the reaction site, a benzylic carbon. Alkyl radicals appeared to be likely candidates for this type of effect since their electron affinities are quite low (20). Thus, the rho values for hydrogen abstraction from toluene by several alkyl radicals were determined, and were all found to be positive (1-4). Recently, however, Tanner et al. (7) have suggested that the positive rho values for hydrogen abstraction by alkyl radicals were experimental artifacts. In view of the importance of this problem, we have re-examined the reaction of *tert*-butyl radicals with toluene (21).

In the original study of abstraction by *tert*-butyl radicals (1,3), the radicals were generated from azoisobutane (AIB) by photolysis in a mixture of a substituted toluene and deuterated thiophenol. Reactivities of the toluenes towards abstraction by the *tert*-butyl radical (R·) relative to abstraction of deuterium from thiophenol were then determined from the ratio of undeuterated to deuterated isobutanes (RH/RD) that were formed (see eqs 4 and 5). The RH/RD ratio was determined by mass spectroscopy.

$$R\cdot + ArCH_3 \longrightarrow RH + ArCH_2\cdot \qquad (4)$$

$$R\cdot + PhSD \longrightarrow RD + RS\cdot \qquad (5)$$

Tanner et al. (7) had the following criticisms of this system: (*i*) The RH/RD ratios obtained by mass spectroscopy were not reliable. (*ii*) The RH/RD ratio could only be used if all of the undeuterated isobutane came from hydrogen abstraction from the benzylic position of toluene (eq 4). Tanner et al., however, found a considerable yield of *tert*-butyltoluene as a product, suggesting that addition of *tert*-butyl radicals to toluene, eq 6, might be followed by aromatization, eq 7, giving RH as a product.

$$R\cdot + PhCH_3 \longrightarrow \text{[cyclohexadienyl radical bearing H, R and } CH_3] \qquad (6)$$

$$\xrightarrow{R\cdot} \text{[R-substituted toluene, } CH_3] + RH \qquad (7)$$

(*iii*) Much of the observed RH could come from proteum impurities in the PhSD solvent. (*iv*) The variation in the yields of RH resulted from variations in the viscosity of the solvent which were due in turn to the variation in viscosities of the substituted toluenes. Tanner et al. showed that we had underestimated the cage yield of RH from the decomposition of AIB, and they suggested that all of the variation in the RH/RD ratio as the substituent in the toluenes is varied was caused by the differing viscosities of the different toluenes. In fact, they felt that no abstraction from the benzylic position of toluene occurs by *tert*-butyl radicals; the *tert*-butyl radicals either add to the toluene ring or react with each other to combine or disproportionate.

We therefore have developed an independent method that is less ambiguous and does not involve thiophenol, cage effects, or mass spectroscopic analysis. Our new method involves photolysis of AIB in a mixture of two toluenes and isolation of all of the major products that contain a benzylic fragment. From the ratio of the yields of the products containing the two different benzylic radicals, the ratio of hydrogen abstraction from the two toluenes by *tert*-butyl radicals can be calculated.

The reactions that occur in this system are shown in eqs 8-19, where QH is toluene, Q'H is a ring-substituted toluene, and R is the *tert*-butyl group. There are 5 major products: the two bibenzyls (Q-Q and Q'-Q'), the cross-bibenzyl (Q-Q'), and two neopentylbenzenes (QR and Q'R) resulting from combination reactions between *tert*-butyl radicals and the two benzyl radicals. These products are formed in eqs 12-16. We find no ring-substituted *tert*-butyl toluene at temperatures from 40°C to 120°C.

$$\text{AIB} \longrightarrow 2\text{R}\cdot + \text{N}_2 \quad (8)$$

$$\text{QH} + \text{R}\cdot \xrightarrow{k_H} \text{Q}\cdot + \text{RH} \quad (9)$$

$$\text{Q'H} + \text{R}\cdot \xrightarrow{k_H'} \text{Q'}\cdot + \text{RH} \quad (10)$$

$$2\text{R}\cdot \longrightarrow \text{H}_2\text{C{=}CMe}_2 + \text{RH} + \text{RR} \quad (11)$$

$$2\text{Q}\cdot \longrightarrow \text{QQ} \quad (12)$$

$$2\text{Q'}\cdot \longrightarrow \text{Q'Q'} \quad (13)$$

$$\text{Q}\cdot + \text{Q'}\cdot \longrightarrow \text{QQ'} \quad (14)$$

$$\text{Q}\cdot + \text{R}\cdot \longrightarrow \text{QR} \quad (15)$$

$$\text{Q'}\cdot + \text{R}\cdot \longrightarrow \text{Q'R} \quad (16)$$

$$\text{R}\cdot + \text{H}_2\text{C{=}CMe}_2 \longrightarrow \text{RCH}_2\text{-}\dot{\text{C}}\text{Me}_2\ (\text{RR}\cdot) \quad (17)$$

$$\text{RR}\cdot + \text{Q}\cdot \longrightarrow \text{Me}_3\text{C-CH}_2\text{-CMe}_2\text{Q}\ (\text{QR}_2) \quad (18)$$

$$\text{RR}\cdot + \text{Q'}\cdot \longrightarrow \text{Me}_3\text{C-CH}_2\text{-CMe}_2\text{Q'}\ (\text{Q'R}_2) \quad (19)$$

(In fact decomposition of AIB in neat benzene leads to only negligible yields of *tert*-butylbenzene.) Two products are formed in somewhat smaller yields; these have the mass and cracking patterns consistent with the addition of an R· radical to isobutylene to give a radical that then terminates with Q· or Q'· as shown in eqs 17-19; products of this type are symbolized as QR_2 and $Q'R_2$ in the equations.[1]

Although we find no *tert*-butyltoluene, Tanner et al. (7) reported this is a major product. Private communications with D.D. Tanner and the exchange of gas chromatographic tracings show that his students had identified neopentylbenzene as *tert*-butyltoluene. (The two elute very close together on many columns.) Since no ring addition occurs, the amount of side-chain abstraction by the *tert*-butyl radical from a given toluene can be obtained from the yield of products containing the benzyl fragment. We find that the amount of side chain abstraction increases (relative to combination and disproportionation of R· radicals, eq 11) as the temperature is increased. At 40°C 16% of the R· abstract benzylic hydrogens; at 60°C this becomes 30%; at 80°C it is 37%; and at 100°C 43% of the *tert*-butyl radicals abstract hydrogen from the side-chain position of toluene. Thus, abstraction of benzylic hydrogens by *tert*-butyl radicals is not a minor process.

The relative rate constants for hydrogen abstraction by R· can be obtained from product composition data. We have expressed these data in terms of a "product ratio", PR, where this is defined as in eq 20.[2]

$$PR = \frac{2[QQ] + [QQ'] + [QR] + [QR_2]}{2[Q'Q'] + [QQ'] + [Q'R] + [Q'R_2]} \qquad (20)$$

[1] We have not yet conclusively eliminated the possibility that the QR_2-type products result from addition of a Q· radical to isobutylene followed by termination of R·, eq 21. However, the MS cracking pattern favors the product shown in eq 18.

$$Q\cdot + CH_2{=}CMe_2 \longrightarrow QCH_2\dot{C}Me_2 \xrightarrow{R\cdot} QCH_2CMe_2{-}CMe_3 \qquad (21)$$

[2] In most of the compounds we studied, we were not able to separate products of the QR_2 type, either because too little was formed or because the glpc columns did not resolve these peaks. However, for three toluenes (the toluene/*tert*-butyltoluene and the toluene/p-chlorotoluene pairs) we were able to do this more complete analysis. In both cases, the ratios of k_H/k_H' were the same, within ± 2%, whether the QR_2 products were included or not

The relative rate constants then are given by eq 22.

$$\frac{k_H}{k'_H} = PR \frac{[Q'H]_o}{[QH]_o} \qquad (22)$$

In control experiments, we showed that the exchange reaction eq 23 can be neglected relative to the combination of the Q• radicals:

$$Q\cdot + Q'H \longrightarrow QH + Q'\cdot \qquad (23)$$

We also showed that the reaction of *tert*-butyl radicals and benzyl radicals occurs virtually exclusively by combination and very little disproportionation (to produce isobutylene and toluene) occurs. The value of k_d/k_c for *tert*-butyl and benzyl radicals is about 0.007 at 80°C.

Using this method, we obtain a rho value of 0.5 at 80°C. This appears to be in reasonable agreement with our earlier report of 0.99 at 30°C, despite the limitations of our earlier method.[3] We feel this new study of hydrogen abstraction by *tert*-butyl radicals from toluenes demonstrates unambiguously that positive rho values do occur. It also is worth pointing out that positive rho values have been reported by several groups of workers for the addition of alkyl radicals either to ring-substituted benzenes or to ring-substituted styrenes[4](24). We have suggested that the transition states for these processes are similar and that they have similar rho values (24,25).

---

[3] The experimental design reported here is completely free of ambiguity due to the effects of viscosity. Tanner et al. pointed out that the viscosity of the solutions of a given toluene in thiophenol vary as the nature of the toluene is varied, and they ascribed all of the variation in the yields of RH/RD to this variation. Dr. Tanner kindly sent us a preprint copy of his manuscript that ultimately was published (7). In our referee report, we pointed out that the runs in which the ratio of QH/PhSH were varied showed that the cage yield of RH does not vary with viscosity. (Viscosity changes as much or more when the ratio of QH/PhSH is varied as when the nature of the QH is changed.) Unfortunately, the data showing the lack of variation in the cage yields with the QH/PhSH ratio were omitted from the published version of the manuscript (7).

[4] The study of chain transfer from by the polystyryl radical with ring-substituted toluenes by Otsu and his coworkers (22) is particularly interesting and warrants further study. They measured the transfer constants at 60° C for toluene, m- and p-xylene, and p-bromo and p-chlorotoluene; they find rho to be +1.2. Despite the acceptable "r" value (0.95), the Hammett plot shows curvature, as might be expected since the abstracting radical and the leaving-group radical are both benzylic. The

PART III. IODINATION OF TOLUENES AND MEASUREMENTS OF THE BDE OF RING-SUBSTITUTED TOLUENES

Zavitsas and Pinto (19) suggested that the BDE of toluenes are weakened by electron-donating groups and strengthened by electron-withdrawing groups. They rationalized this effect of ring substituents in the following way. The methyl group of toluene donates electron density to the phenyl ring by hyperconjugation; this results in the C-H bonds of the methyl group acquiring more "s" character. Since bonds having more "s" character are stronger (26,27), the benzylic bonds in substituted toluenes have larger BDE as the substituents become increasingly better electron acceptors. Zavitsas and Pinto cited two pieces of evidence to support this hypothesis (19). Firstly, the percent "s" character in benzylic C-H bonds determined from C-13 studies increases as the substituents become more electron withdrawing (28). Secondly, the primary tritium isotope effects for hydrogen transfer from thiophenol to a benzylic radical decrease as the ring substituent becomes more electronegative (29), a result that is interpreted (19,29) in terms of a stronger benzylic BDE for electron-withdrawing substituents.

While the evidence cited above is consistent with the hypothesis that the BDE of substituted toluenes vary in the direction suggested by Zavitsas, the direct measurement of these BDE's is desirable. The only data on substituted-toluene BDE's available is that of Szwarc (30,31) obtained from pyrolytic experiments. Szwarc found a slightly lower BDE for the xylenes compared to toluene itself; however, there was a large experimental uncertainty.

THE REACTION OF IODINE WITH TOLUENE AND SUBSTITUTED TOLUENES

Gas Phase Studies

The thermal reaction of iodine with toluene has been studied in the gas phase at 200-300°C by Benson et al. (32). These workers found that a reaction occurs between iodine and toluene to give benzyl iodide and HI (eq 24), and that an equilibrium is established. The position of the equilibrium lies well to the

---

method of measuring transfer constants eliminates many of the kinetic complexities in Hammett studies. It would be interesting to measure the transfer constants for p-methoxystyrene and p-nitrostyrene.

$$I_2 + PhCH_3 = PhCH_2I + HI \quad (24)$$

left, and Benson et al. found that it is possible to follow the appearance of benzyl iodide starting from toluene and iodine. Thus, they were able to investigate the kinetics of the iodination of toluene and to show that the following mechanism obtains.

$$I_2 \rightleftharpoons 2I\cdot \quad (25)$$

$$I\cdot + PhCH_3 \longrightarrow PhCH_2\cdot + HI \quad (26)$$

$$PhCH_2\cdot + I_2 \longrightarrow PhCH_2I + I\cdot \quad (27)$$

At very low conversions, when both the HI and $PhCH_2I$ concentrations are low, the reverse of both eqs 26 and 27 can be neglected.

We have carried out a similar kinetic study of the iodine/toluene reaction in solution over a temperature range from 50° to 90°C (33). At these lower temperatures the thermolysis of iodine (eq 25) does not provide an adequate steady-state concentration of iodine atoms; we therefore generated iodine atoms by photolysis of the iodine molecule. Although we find the overall photo-iodination of toluene in solution to be similar to the gas-phase process, a number of additional complexities are introduced.

Experimental Procedure

Solutions that were 1.5 M toluene(or a substituted toluene) and 10^{-3} M iodine were prepared in an appropriate solvent. (These concentrations are at the upper ends of the ranges of toluene and iodine concentrations used in determining the kinetic orders.) Approximately 2 ml aliquots of these solutions were then sealed in 7 mm (ID) Pyrex tubes after several freeze-degas-thaw cycles. The tubes were placed in a merry-go-round apparatus inside a Rayonet reactor equiped with 16 standard 8-watt fluorescent bulbs, the merry-go-round was immersed in a thermostated oil bath, and a copper chloride solution was used to filter out light with wavelengths shorter than 5000 Å. Individual tubes containing the reaction mixture were removed periodically and analyzed by glpc.

Products of Photo-Iodination of Toluene

The only products observed after photo-iodination of toluene are benzyl iodide and the isomeric ring-substituted iodotoluenes. These products can be accounted for by independent radical and ionic pathways as shown in eqs 28 and 29. (Only one of the possible iodotoluene isomers is shown).

$$C_6H_5CH_3 \xrightarrow{I\cdot} C_6H_5CH_2\cdot + HI \xrightarrow{I_2} C_6H_5CH_2I + I\cdot \quad (28)$$

$$C_6H_5CH_3 \xrightarrow{I_2} [CH_3C_6H_5(H)(I)]^{+} + I^{-} \longrightarrow CH_3C_6H_4I + HI \quad (29)$$

Under our conditions, benzyl iodide accounts for 70-95% of the products. If the reaction is run in the dark, only the ring-iodinated toluenes are produced, and at about the same rate as in the light. Thus, the two processes are independent, one being a radical and one an ionic reaction. No bibenzyl can be detected, even under conditions that favor its formation more than do those used in our kinetic runs. Therefore, the formation of benzyl iodide can be used to monitor the rate of the radical reaction.

Kinetics and Mechanism

We find the same kinetics orders in iodine (1/2) and toluene (1) that were obtained by Benson et al. in the gas phase (32); thus, the mechanism that applies in solution is similar to that which has been demonstrated for the gas phase. However, cage return steps must be included for reactions in solution. The complete mechanism, therefore, is that shown below, where RH is toluene or a substituted toluene, and brackets represent the solvent cage.

$$I_2 \rightleftharpoons [I\cdot, I\cdot] \quad (30)$$

$$[I\cdot, I\cdot] \longrightarrow 2I\cdot \quad (31)$$

$$RH + I\cdot \underset{k_{-H}}{\overset{k_H}{\rightleftharpoons}} [R\cdot, HI] \quad (32)$$

$$[R\cdot, HI] \xrightarrow{k_d} R\cdot + HI \quad (33)$$

$$R\cdot + I_2 \longrightarrow RI + I\cdot \quad (34)$$

$$2I\cdot \xrightarrow{k_t} I_2 \quad (35)$$

Note that we have allowed for the possibility of cage return both in the initial photolytic production of iodine atoms (eqs 30 and 31) and in the hydrogen abstraction step (eq 32). The other steps are sufficiently exothermic in the forward direction (32) so that cage return should not be an important consideration. We represent the net rate of photolytic production of iodine atoms (corrected for cage return) as in eq 36 and the fraction of the

$$k_{\phi} = k_{30}k_{31}/(k_{-30} + k_{31}) \quad (36)$$

hydrogen abstraction reactions that proceed to form product as in eq 37. Thus, the rate expression for the appearance of benzyl iodide can be derived from a steady-state treatment of eqs 30-35 and is given by 38; the observed rate constant, k_{obs},

$$f = k_d/(k_d + k_{-H}) \quad (37)$$

$$d[PhCH_2I]/dt = f(k_{\phi}/k_t)^{0.5}k_H[PhCH_3][I_2]^{0.5} \quad (38)$$

equals $f(k_{\phi}/k_t)^{0.5}k_H$. Values of k_{obs} in several solvents at temperatures are shown in Table I. As can be seen from the raw data, the nature of the solvent has only a small effect on the observed rate constants and essentially no effect on the activation parameters. There are two types of solvent effects we will consider in the sections that follow: the effect of the change in solvent viscosity on the cage return steps and the effect of complexing of both I_2 and I• by benzene solvent.

Table I. Observed Rate Constants for Iodination of Toluene at Various Temperatures[a]

Solvent	Viscosity[b]	Temperature, °C	k_{obs}[c]	E_{obs}	log A_{obs}
Benzene	.33	50	0.41	16.0[d]	2.4
		60	0.94		
		70	1.83		
		80	3.72		
		90	6.84		
Cyclohexane	.38	50	0.47	16.1[d]	2.6
		60	1.58		
		70	2.64		
		80	4.74		
		90	8.89		
Hexadecane	1.45	80	3.18		

*(a)*The error limits associated with the k_{obs} and Arrhenius parameters are $\pm 5\%$. *(b)*In cP at 80°C, data from "Handbook of Chemistry and Physics", R.C. Weast, ed., 55th Ed., CRC Press, Cleveland, Ohio, 1974, pp. F50-F55. *(c)*x 10^{-8} $M^{-0.5}$ s^{-1} *(d)*kcal/mole

Solvent Effects: Viscosity and Cage Return

Changes in the viscosity of the medium can affect the amount of both the cage return steps (eqs 30 and 32) as well as the diffusion-controlled rate of termination (eq 35). Thus, in the equation for k_{obs}, the k_{ϕ},k_t and f terms may all vary with the viscosity of the solvent. However, the data of Rabinowitch and Wood (34), who determined the iodine atom concentration in the photo-stationary state, indicate that the ratio k_{ϕ}/k_t is remarkably independent of viscosity effects. That is to say, while both the k_{ϕ} and k_t terms individually show a solvent effect (35,36), the effects tend to cancel one another. Thus, it appears that most of the observed rate difference between cyclohexane and hexadecane must arise from the viscosity dependence of cage return of the hydrogen abstraction step (eq 32).

The amount of cage return in the hydrogen abstraction step can be estimated in the following manner. From the data of Rabinowitch and Wood, the average value of k_{ϕ}/k_t under our conditions is 2.8×10^{-12} M. By dividing this ratio into our solution phase k_{obs} and comparing the result with the gas-phase rate constant obtained by Benson et al. (32), we can calculate the fraction of hydrogen abstraction reactions that proceed on to products (our f term). Thus, in cyclohexane f has a value of 0.13, while in hexadecane f is about 0.08.

Solvent Effects: Complexation

If the viscosity effect on cage return were the only solvent effect on kobs, then a slightly faster rate would have been expected in benzene as solvent compared to cyclohexane. In fact, just the opposite is observed: the rate in benzene is substantially slower than is the rate in cyclohexane (see Table I). This apparent anomaly can be readily explained by solvent complexation of both molecular and atomic iodine in the aromatic solvent.

Complexation by aromatic solvent has been reported for all of the halogens except fluorine (37-40). Both I_2 (39) and I• (40) undergo rapid equilibrium in aromatic solvents to form 1:1 complexes, as shown in eqs 39 and 40. For both the molecule (39)

$$I_2 + ArH \rightleftharpoons (I_2,ArH) \quad (39)$$

$$I_2 + ArH \rightleftharpoons (I,ArH)\cdot \quad (40)$$

and the atom (40) at 25°C, approximately half of the iodine species is uncomplexed. Such complexation requires us to consider a number of additional possibilities in our photo-iodination mechanism, as shown below. (In this scheme, S is the solvent.) Also note that we have not explicitly shown cage return processes in this scheme, although they are of course

$$I_2 + S \rightleftharpoons (I_2,S) \quad (39)$$

$$I_2 \xrightarrow{k_\phi} 2I\cdot \quad (30,31)$$

$$I\cdot + S \rightleftharpoons (I,S)\cdot \quad (40)$$

$$I\cdot + RH \xrightarrow{fk_H} R\cdot + HI \quad (32a)$$

$$(I,S)\cdot + RH \longrightarrow R\cdot + HI + S \quad (32b)$$

$$R\cdot + (I_2,S) \longrightarrow RI + (I,S)\cdot \quad (34b)$$

$$2I\cdot \xrightarrow{k_t} I_2 \quad (35a)$$

$$2(I,S)\cdot \longrightarrow (I_2,S) + S \quad (35b)$$

$$I\cdot + (I,S)\cdot \longrightarrow (I_2,S) \quad (35c)$$

still involved, as discussed in the previous section.

A modified rate expression that includes a term for solvent complexation is shown in eq 41, where K is the equilibrium constant for complexation of I_2 by the solvent.

$$\frac{d[RI]}{dt} = f\,(k_\phi/k_t)^{0.5}\,(1+K)^{-1}\,k_H\,[RH]\,[I_2]^{0.5} \quad (41)$$

Several assumptions have been used to derive this rate expression. Firstly, only the uncomplexed iodine molecule is photolyzed. (This is a reasonable assumption based on the data of Strong and Rand (40).) We also assume that the complexed iodine atom is too unreactive to compete with uncomplexed iodine atom in the hydrogen abstraction step (eqs 32a,b). On the other hand, iodine atom transfer (eq 34) and termination (eq 35) are both very exothermic processes, and it is unlikely that the energy difference between complexed and uncomplexed iodine will be reflected in a significant rate difference. Thus, k_t in benzene is a composite representing reaction by both complexed and uncomplexed iodine species.

We can compare the rate constant in benzene with that in cyclohexane. Since the viscosities of benzene and cyclohexane are about equal at 80°C, the only effect on the benzene solvent is on K. Using a value of K of 0.17 in benzene (37), and the value of k_{obs} for cyclohexane shown in Table I, we calculate k_{obs} in benzene to be 4.0×10^{-8} $M^{-0.5}$ s^{-1}; this is in good agreement with the observed value of 3.7×10^{-8} $M^{-0.5}$ s^{-1} at 80°C (see Table I).

Table II. Reactivities, Activation Energies and Heats of Reaction in Halogenation of Hydrocarbons in the Gas Phase[a]

							Evans-Polanyi Parameters					
							without $PhCH_3$			with $PhCH_3$		
X	log k_H[b]		C_2H_5-H	$(CH_3)_2CH$-H	$(CH_3)_3C$-H	$PhCH_2$-H	α[c]	C[d]	r[e]	α[c]	C[d]	r[e]
F[f]	9.8	ΔH[g]	-37.8	-40.8	-43.8	-	0.05	2.14	0.866	-	-	-
		E_H	0.3	0.0	0.0							
Cl[f,h]	10.3	ΔH[g]	- 5.0	- 8.1	-11.1	-18.0	0.15	1.79	0.980	0.00	0.67	0.34
		E_H	1.0	0.7	0.1	1.0						
Br[i,j,k]	1.4	ΔH[g]	10.6	7.6	4.6	2.0	1.03	2.88	0.996	0.81	4.86	0.968
		E_H	14.0	10.4	7.8	7.2						
I[l,m]	-8.4	ΔH[g]	26.6	23.6	20.6	13.6	0.92	3.00	0.997	0.94	2.48	0.999
		E_H	27.5	24.4	22.0	15.2						

*(a)*Units of ΔH and E_H are kcal/mole. *(b)*Rate constants for hydrogen abstraction from ethane at 300°K. *(c)*Slope resulting from Evans-Polanyi treatment of E_H and ΔH. *(d)*Intercept resulting from Evans-Polanyi treatment of E_H and ΔH. *(e)*Correlation coefficient. *(f)*Fettis, G.C. and Knox, J.H. Progr. Reaction Kinetics 2, 1 (1964). *(g)*Calculated from BDE values compiled in "Handbook of Chemistry and Physics", R.C. Weast, Ed., CRC Press, Cleveland OH, 1974, p. F220. *(h)*Russell, G.A., Ito, A. and Hendry, D.C. J. A. Chem. Soc. 85, 2976 (1963). *(i)*Amphlett, J.C. and Whittle, E. Trans. Faraday Soc. 64, 2130 (1968). *(j)*King, K.D., Golden, D.M. and Benson, S.W. Trans. Faraday Soc. 66, 2794 (1970). *(k)*Russell, G.A. and Desmond, K.M. J. Am. Chem. Soc. 85, 3139 (1963). *(l)*Knox, J.H. and Musgrove, N.A. Trans. Faraday. Soc. 63, 2201 (1967). *(m)*Walsh, R., Golden, D.M. and Benson, S.W. J. Am. Chem. Soc. 88, 650 (1966).

Comparison of Solution and Gas-Phase Kinetics

In cyclohexane as solvent, where complexation is not important, the observed Arrhenius parameters can be expressed in terms of the parameters for the individual steps in the mechanism as shown in eqs 42 and 43. These equations can be simplified

$$E_{obs} = E_H + 0.5E_\phi - 0.5E_t + E_f \qquad (42)$$

$$\log A_{obs} = \log A_H + 0.5 \log (A_\phi/A_t) + \log A_f \qquad (43)$$

as follows. Noyes et al. (36) concluded that the photostationary concentration of iodine atoms, and therefore the ratio k_ϕ/k_t, is essentially temperature independent. Thus eq 42 can be simplified to eq 44. Then, since both diffusion and cage return are fast

$$E_{obs} = E_H + E_f \qquad (44)$$

and have only small activation energies, the net activation energy for the cage return factor E_f can be assumed to be small relative to E_H. Thus, to a good approximation, E_H equals E_{obs} (eq 45).

$$E_{obs} = E_H \qquad (45)$$

We can calculate log A_{obs} in the following way. We can assume that log A_f is approximately zero since f involves a ratio of unimolecular rate constants for reactions of the cage. The ratio k_ϕ/k_t is approximately temperature independent, and therefore is given by A_ϕ/A_t, the value of which is about 3×10^{-12} M under our conditions. With these assumptions, eq 47 gives log A_{obs} as 8.4 in cyclohexane and 8.2 in benzene; within the assumptions used, these are in satisfactory agreement with the value of 8.8 obtained in the gas phase (32).

BDE's of Substituted Toluenes

The kinetics of iodination of toluenes can be used to derive the BDE of the benzylic C-H bonds (41-47). The Evans-Polanyi relation, eq 46, relates observed activation energies to the heats of reaction(14).

$$E = \alpha\Delta H + C \qquad (46)$$

The Evans-Polanyi equation gives an increasingly better correlation of data as reactions become less exothermic; in extremely exothermic reactions, changes in heats of reaction produce only minor changes in activation energies. In Table II we have compiled data on the halogenation of alkanes (at primary, secondary, and tertiary hydrogens) and of toluene. Note that the reactivity of the halogen atoms decreases markedly on going from fluorine and chlorine to bromine and iodine. Also note how the Evans-Polanyi correlations improve as the reactivity of the halogen decreases. In fact, only in the case of the iodine atom do the toluene data fit the correlation determined from the alkane data, undoubtedly reflecting the fact that iodine abstractions are the most endothermic of the group. Thus, iodination is the most generally useful reaction for determining BDE values using kinetically-derived activation energies.

The observed rate constants for the iodination of toluenes at 80°C in both cyclohexane and benzene are shown in Table III. The values of k'_H/k_H for the substituted toluenes can be derived from the k_{obs} values if the cage fractionation term (f) and k_ϕ/k_t are constant and do not vary significantly as the ring substituent is varied. Clearly k_ϕ/k_t will be nearly constant as substituents are changed; both of these rate constants have very small activation energies and their ratio is nearly independent of viscosity. Although the cage fraction term, f, would be expected to vary with changes in viscosity, we have found that at 80°C, 1.5 M solutions of p-chlorotoluene in benzene or cyclohexane have viscosities that are within 8% of those for comparable solutions of toluene.[5] A viscosity change of this magnitude corresponds to less than a 1% change in the observed rate. Further, since both E_{-H} and E_d are very small (1-3 kcal/mole), it is clear that any effects of substituents on these already very small values must be small. Thus, we can assume that the cage fractionation factor also is constant throughout a series of substituted toluenes in a given solvent.

The lack of solvent effect on the relative values of k_H values was confirmed by studying the simultaneous, competitive iodination of pairs of toluenes in the same solution (an "internal competition" method). These experiments (33) gave the same ratio of reactivities for pairs of toluenes as were obtained from the k_{obs} values for that pair obtained in separate experiments ("external competition"). Clearly, then, solvent effects on complexation of iodine by the toluenes and the effects of varying the toluenes on the viscosities of the benzene and cyclohexane solvents must be small.

The BDE values for substituted toluenes can then be calculated as follows:

(1) Since the pre-exponential for eq 32, A_H, is virtually independent of the substituent, the changes in activation energies for the substituted toluenes may be calculated from k'_H/k_H and the activation energy for toluene itself, eq 47. Activation energies calculated in this way are shown in Table IV.

$$E'_H - E_H = -2.303 \ RT \log (k'_H/k_H) \quad (47)$$

[5] These two QH's represent the extremes in viscosities.

Table III. Reactivities of Substituted Toluenes Toward the Iodine Atom at 80°C.[a]

Substituent	k_{obs} ($\times 10^{-8}$ $M^{-0.5}s^{-1}$)[b]	k'_H/k_H[c]	σ^+[d]	σ[d]
	Solvent: Cyclohexane			
p-Me	11.1[e]	2.4[e]	-0.31	-0.17
p-Bu^t	8.2	1.7	-0.26	-0.20
m-Me	5.3[e]	1.1[e]	-0.07	-0.07
H	4.7	(1.00)	0.00	0.00
p-Cl	4.3	0.91	0.11	0.23
m-Cl	1.0	0.22	0.40	0.37
m-Br	0.95	0.20	0.41	0.39
	Solvent: Benzene			
p-Me	9.5[e]	2.6[e]	-0.31	-0.17
p-Bu^t	7.3	2.0	-0.26	-0.20
M-Me	3.8[e]	1.0[e]	-0.07	-0.07
H	3.7	(1.00)	0.00	0.00
p-Cl	3.6	0.98	0.11	0.23
m-Cl	0.86	0.23	0.40	0.37
m-Br	0.74	0.22	0.41	0.39

(a) $[I_2]$ = 0.0010 M and $[X\text{-}C_6H_4\text{-}Me]$ = 1.5 M. *(b)* The observed rate constant $k_{obs} = f(k_\phi/k_t)^{0.5}k_H$. The uncertainty associated with k_{obs} is less than 5% except for p-chlorotoluene in cyclohexane where the uncertainty is 7%. *(c)* Reactivity relative to unsubstituted toluene. *(d)* Values from J.E. Leffler and E. Grunwald "Rates and Equilibria of Organic Reactions", John Wiley and Sons, New York, 1963. p. 204. *(e)* The statistically greater number of methyl hydrogens has been taken into account by dividing the raw rate by 2.

Table IV. Bond Dissociation Energies for Substituted Toluenes Estimated from their Relative Reactivities Towards Iodine Atom.[a]

Solvent =	Cyclohexane		Benzene	
Substituent	E_H[b]	BDE[c]	E_H[b]	BDE[c]
p-Me	15.5	85.2	15.3	85.0
p-Bu^t	15.7	85.5	15.5	85.5
m-Me	16.1	85.9	16.0	85.8
H	16.1	85.9	16.0	85.8
p-Cl	16.2	86.0	16.0	85.8
m-Cl	17.2	87.1	17.0	86.8
m-Br	17.2	87.1	17.0	86.8

*(a)*All values in kcal/mole. *(b)*Calculated from relative rate data at 80°C using equation 47. *(c)*Calculated from activation energies using equation 49.

Table V. Rho Values from the Hammett Treatment of Relative Reactivities of Iodine Atom Towards Substituted Toluenes at 80°C.

Solvent	Substituent Constant	ρ	r[a]
Cyclohexane	σ	-1.6 ± 0.2	0.93
	σ^+	-1.4 ± 0.1	0.98
Benzene	σ	-1.6 ± 0.2	0.93
	σ^+	-1.4 ± 0.1	0.98

*(a)*Correlation coefficient

(2) From the correlation of gas-phase data for the iodination of alkanes and aralkanes (Table II), the Evans-Polanyi equation relating E_H values with the corresponding heats of reaction is shown in eq 48.

$$E_H = 0.94 \ \Delta H + 2.48 \quad (48)$$

This equation can then be used to calculate the ΔH values for substituted toluenes from the E'_H values determined above.
(3) Finally, the ΔH values obtained from eq 48 can be used to calculate the benzylic C-H BDE values using eq 49. Using a value of 71.4 kcal/mole for the BDE of HI (48), the BDE's calculated in this manner are shown in Table IV.

$$\text{BDE } (PhCH_2\text{-H}) = \text{BDE (H-I)} + \Delta H \quad (49)$$

The data in Table IV clearly show that the BDE's of substituted toluenes do change substantially (ca. 3 kcal/mole per sigma unit) over the range of substituents commonly utilized in Hammett treatments of reactivity.[6] Furthermore, the trend is in the direction predicted by Zavitsas (19), with electron-donating substituents giving weaker C-H bonds than electron- withdrawing substituents.

Rho Values of Hydrogen Abstraction from Toluenes by Halogen Atoms

In Table V we show the rho values obtained from the Hammett treatment of our data for hydrogen abstraction from toluenes by iodine atom. Note that identical results are obtained with either cyclohexane or benzene as solvent. The sigma-plus substituent parameters (49) give a slightly better correlation; this may reflect radical resonance stabilization effects in the product-like transition state, rather than a contribution from charge separated forms (see below).

Table VI compares our rho values for hydrogen abstraction by iodine with those for abstraction by bromine and chlorine. In every case, the rho values shown have been calculated using the sigma substituent parameters and omitting from the correlation

[6] A least squares comparison of our calculated BDE's with the Hammett sigma parameter gives eq 50.

$$\text{BDE} = 2.8\,\sigma + 85.9 \quad (50)$$

The uncertainty in the BDE calculated by this method is approximately 1 kcal/mole.

Table VI. Comparison of Rho Values for Hydrogen Abstraction from Toluenes by Halogen Atoms.

X	t(°C)	Solvent	Rho[a]	EA[b]	ΔH[c]
Cl[d]	60	CCl_4	-0.46	3.61	-18
Cl[d]	60	PhH	-0.98	3.61	-18
Br[e]	80	CCl_4	-1.6	3.36	- 2
Br[d]	60	PhH	-1.6	3.36	- 2
I[f]	80	c-C_6H_{12}	-1.6	3.07	+13
I[f]	80	PhH	-1.6	3.07	+13

*(a)*These rho values were all calculated using σ substituent parameters and omitting p-methoxy toluene data from the correlations. *(b)*Electron Affinity in electron volts. *(c)*Heat of reaction for $PhCH_3 + \cdot X \longrightarrow PhCH_2\cdot + HX$; kcal/mole. *(d)*Hradil, J. and Chvalovsky, V. Coll Czech. Chem. Commun. 33, 2029 (1968). This group is the only one to have studied both CCl_4 and benzene as solvents; other groups have reported rho values not always in agreement with the data of these workers. For a review see: Pryor, W.A., Lin, T.H., Stanley, J.P., and Henderson, R.W. J. Amer. Chem. Soc. 95, 6993 (1973). *(e)*Walling, C., Rieger, A.L. and Tanner, D.D. J. Am. Chem. Soc. 85, 2139 (1963). *(f)*Pryor, W.A., Tang, F., Tang, R. and Church, D.F. to be submitted.

those para substituents for which resonance effects would be important. Table VI shows rho values in different solvents as well as the electron affinities of the three halogen atoms and the heats of reaction of these atoms with toluene. We will discuss the dependence of rho on the electron affinity and heat of reaction data below, but first some remarks about solvent effects on rho are necessary.

Solvent Effects. The most well-known explanation for solvent effects on chlorination reactions is the suggestion that Cl• are complexed by aromatic solvents, as discussed above. However, several arguments suggest that complexation can not satisfactorily explain the observed solvent effects on rho shown in Table VI.

(1) Only a small fraction of the chlorine atoms may be complexed. Atomic iodine is only slightly more complexed than is I_2 (50); since only about 25% of Cl_2 is complexed in benzene solvent at 25°C, it is likely that only a small fraction of Cl· is complexed. Since complexed chlorine atoms would be expected to be less reactive than uncomplexed atoms, the small fraction of chlorine atoms that are complexed may be kinetically negligible.

(2) In complexes with aromatic solvents, chlorine acts as the acceptor species; thus in the complex, chlorine will bear more negative charge relative to the uncomplexed chlorine atom and would be less electronegative. This leads to the prediction of a less negative rho for abstraction by chlorine in an aromatic solvent, rather than the more negative rho that is in fact observed.

(3) In the only systematic study of complexation effects on rho values for bromination and chlorination, Hradil and Chvalovsky (15) found that rho values did become somewhat less negative on going to less complexing solvents for both bromine and chlorine, in agreement with the hypothesis that chlorine and bromine atoms become more selective as the aromatic solvent becomes a better donor and is better able to complex halogen atoms.[7] However, the reactivity data of these workers are not in accord with the hypothesis that complexation makes the halogen atoms less reactive. These workers conclude that "the effect of the medium cannot be accounted for by the simple concept of complex formation between the halogen radical and the organic solvent."

As was reviewed above, Russell (16-18) originally suggested that the effect of aromatic solvents on chlorination selectivities could be explained by complexation of the chlorine atom by the solvent. However, another possible explanation is complexation or solvation of the HCl that is produced in the hydrogen-abstraction step. This latter suggestion also rationalizes the increased preference for tertiary aliphatic hydrogens in benzene solvent, since complexation of the HCl in the transition state would lead to a larger polar contribution and a greater tendency to abstract tertiary hydrogens. Complexation of the HCl predicts a larger polar effect and explains the more negative rho value observed in benzene solvents that are more basic, in accord with the data (15). However, complexation of HCl and the more traditional Russell argument differ in that complexation of HCl would make Cl· more reactive in benzene whereas complexation of Cl· would make it less so. The data of Hradil and Chvalovsky (15) appear to show that the selectivity and reactivity of Cl· both increase in benzene, but the data do not appear to be sufficiently self-consistent to establish this point with any certainty.[7]

[7] See footnote d in Table VI.

Electron Affinities and Heats of Reaction. Table VI shows the electron affinities (EA) of the halogen atoms and their heats of reaction with toluene. Since the iodine reaction is more endothermic than is that for bromine, a BDE argument alone (19) would predict that rho should be more negative for iodine atoms than for bromine atoms; in fact, both are found to have about the same rho value. This enhanced relative rho value for bromine atoms can be rationalized by a larger polar contribution to the reactions of bromine atoms than for iodine. In fact, since the iodine reactions are very product-like, they involve little polar contribution and have rho values determined virtually exclusively by BDE. This, of course, is why the Evans-Polanyi correlation of iodine data is so good and why our calculation of relative BDE values works as well as it does. Thus, we suggest that polar effects on transition states for hydrogen abstraction reactions are most important for the most symmetrical transition states, where bond-breaking and bond-making have proceeded about equally. This hypothesis suggests that polar effects are most important for bromine atoms, but are relatively unimportant for chlorine atoms (which have reactant-like transition states) and iodine atoms (which have product-like transition states).

CONCLUSION

These factors lead us to conclude that observed rho values for radical reactions can be qualitatively interpreted as shown in Table VII. This table shows the contribution from the substituent effects on BDE increasing monotonically as the reaction becomes more endothermic. On the other hand, polar effects are most important for reactions that are approximately thermoneutral where bond-breaking and bond-making are approximately equal. Polar forms will be less important for very exothermic and endothermic reactions where a reactant- or product-like transition state is less polarizable (51).

When polar effects in the transition state do become an important factor, it is likely that several characteristics of the radical must be considered in order to rationalize observed rho values (51). The electron affinity of the radical relative to the reactive site (e.g., a benzylic carbon in abstractions from toluenes) has a major effect upon the stabilization of charge-separated forms; the positive rho values associated with the reactions alkyl radicals reflect their very low electron affinities. However, electron affinity arguments alone are not always sufficient to explain observed polar effects. For example, we have compared the rho values for hydrogen abstraction from toluene by thiyl radicals and bromine atoms (51). Thiyl radicals

Table VII. Effects of BDE and Polar Effects on Rho Values for Radical Reactions.

Heat of Reaction	Effect of Rho	
	Substituent Effect on BDE	Dipolar Resonance Forms
Very exothermic	Not important	Not important
Thermoneutral	Moderate effect (gives negative rho)	Large effect (gives positive or negative rho depending on charge type)
Very endothermic	Large effect (gives negative rho)	Not important

give a rho value that is somewhat more negative than is the rho value for bromine. Although hydrogen abstraction by both of these species is nearly thermoneutral, bromine has a much larger electron affinity than does a thiyl radical (51,52), leading to the prediction of a more negative rho for abstraction by bromine. The similar rho values observed for bromine atoms and thiyl radicals can be rationalized as arising from the similar polarizability of these two species (51).

REFERENCES

1. Davis, W.H. and Pryor, W.A. *J. Amer. Chem. Soc. 99*, 6365 (1977).
2. Pryor, W.A., Fuller, D.L. and Stanley, J.P. *J. Amer. Chem. Soc. 94*, 1632 (1972).
3. Pryor, W.A., Davis, W.H. and Stanley, J.P. *J. Amer. Chem. Soc. 95*, 4754 (1973).

4. Pryor, W.A. and Davis, W.H. *J. Amer. Chem. Soc. 96,* 7557 (1974).
5. Pryor, W.A., Davis, W.H. and Gleaton, J.H. *J. Org. Chem. 40,* 2099 (1975).
6. Davis, W.H., Gleaton, J.H. and Pryor, W.A. *J. Org. Chem. 42,* 7 (1977).
7. Tanner, D.D., Samal, P.W., Ruo, T. C-S. and Herriquez, R. *J. Amer. Chem. Soc. 101,* 1168 (1979).
8. Hammett, L. "Physical Organic Chemistry", McGraw-Hill, New York, 1940, Chap. VII.
9. Walling, C., Briggs, E.R., Wolfstirn, K.B. and Mayo, F.R. *J. Amer. Chem. Soc. 70,* 1537 (1948).
10. Walling, C., Seymour, D. and Wolfstirn, K.B. *J. Amer. Chem. Soc. 70,* 1544 (1948).
11. Walling, C., Seymour, D. and Wolfstirn, K.B. *J. Amer. Chem. Soc. 70,* 255 (1948).
12. Russell, G.A. *J. Org. Chem. 23,* 1407 (1958).
13. Russell, G.A. and Williamson, R.C. *J. Amer. Chem. Soc. 86,* 2357 (1964).
14. Evans, M.G. and Polanyi, M. *Trans. Faraday Soc. 34,* 11 (1938).
15. Hradil, J. and Chvalovsky, V. *Coll. Czech. Chem. Commun. 33,* 2029(1968).
16. Russell, G.A. and Brown, H. C. *J. Amer. Chem. Soc. 77,* 4031 (1955).
17. Russell, G.A. *J. Amer. Chem. Soc. 79,* 2977 (1957).
18. Russell, G.A. *J. Amer. Chem. Soc. 80,* 5002 (1958).
19. Zavitsas, A.A. and Pinto, J.A. *J. Amer. Chem. Soc. 94,* 7390 (1972).
20. Dewar, M.J.S. and Rzepa, H.S. *J. Amer. Chem. Soc. 100,* 7 (1978).
21. Pryor, W.A., Tang, R.H., Tang, F.Y. and Church, D.F. to be submitted.
22. Yamamoto, T., Nakamura, S., Hasegawa, M. and Otsu, T. *Kogyo Kagaku Zasohi 72,* 727 (1969).
23. Giege, B. and Meister, J. *Anger. Chem., Int Ed. Eng. 16,* 178(1977).
24. Pryor, W.A., Davis, W.H. and Gleaton, J.H. *J. Org. Chem. 40,* 2099 975).
25. Pryor, W.A. and Davis, W.H. Abstract ORGN-59, 173rd American Chemical Society National Meeting, New Orleans, Louisiana, March 22, 1977.
26. Skinner, H.A. and Pilcher, G. *Quart. Rev (London) 17,* 264 (1963).
27. Cox, J.D. *Tetrahedron 18,* 1337 (1962).
28. Yoder, C.H., Tuck, R.H. and Hess, R.E. *J. Amer. Chem. Soc. 91,* 539 (1969).
29. Lewis, E.S. and Butler, M.M. *Chem. Commun., 941,*(1971)

30. Szwarc, M. *J. Chem. Phys. 16,* 128 (1948).
31. Szwarc, M. *Chem. Rev. 47,* 75 (1950).
32. Walsh, R., Golden, D.M. and Benson, S.W. *J. Amer. Chem. Soc. 88,* 650 (1966).
33. Pryor, W.A., Tang, F., Tang, R. and Church, D.F. to be submitted.
34. Rabinowitch, F. and Wood, W.C. *Trans. Faraday Soc. 32,* 547 (1936).
35. Booth, D. and Noyes, R.M. *J. Amer. Chem. Soc. 82* 1868 (1960).
36. Roxman, H. and Noyes, R.M. *J. Amer. Chem. Soc. 80,* 2410 (1958).
37. Andrews, L.S. and Keefer, R.M. *J. Amer. Chem. Soc. 73,* 462 (1951).
38. Keefer, R.M. and Andrews, L.J. *J. Amer. Chem. Soc. 72,* 4677 (1950).
39. Benesi, H.A. and Hildebrand, J.H. *J. Amer. Chem. Soc. 71,* 2703 (1949).
40. Rand, S.J. and Strong, R.L. *J. Amer. Chem. Soc. 82,* 5 (1960).
41. Eckstein, B.H., Scheraga, H.A. and van Artsdalen, E.R. *J. Chem. Phys. 28* (1954).
42. Fettis, G.C. and Knox, J.H. *Progr. React. Kin. 2,* 1 (1964).
43. Kerr, J.A. *Chem. Rev. 66,* 465 (1966).
44. Hartley, D.B. and Benson, S.W. *J. Chem. Phys. 39,* 132 (1963).
45. Nangia, P.S. and Benson, S.W. *J. Amer. Chem. Soc. 86,* 2773 (1964).
46. Golden, D.M., Walsh, R. and Benson, S.W. *J. Amer. Chem. Soc. 87,* 4053 (1965).
47. Golden, D.M. and Benson, S.W. *Chem. Rev. 69,* 125 (1969).
48. "Handbook of Chemistry and Physics", R.C. Weist, ed., CRC Press, Cleveland, OH, 1974, p. F204.
49. Brown, H.C. and Okamoto, Y. *J. Amer. Chem. Soc. 80,* 4979 (1958).
50. Strong, R.L. and Perano, J. *J. Amer. Chem. Soc. 83,* 2843 (1961).
51. Pryor, W.A., Gojon, G. and Church, D.F. *J. Org. Chem. 43,* 793 (1978).
52. Janousek, B.K., Reed, K.J. and Brauman, J.I. *J. Amer. Chem. Soc. 102,* 3125 (1980).

INDEX

S

T

V

Z